From Nazis to Nasa

From Nazis to Nasa

THE LIFE OF
WERNHER VON BRAUN

BOB WARD

FOREWORD BY **JOHN GLENN**

SUTTON PUBLISHING

First published as *Dr Space: The Life of Wernher von Braun* in 2005
by Naval Institute Press, Annapolis, Maryland USA.

First published in the UK in 2006 by
Sutton Publishing Limited · Phoenix Mill
Thrupp · Stroud · Gloucestershire · GL5 2BU

British Library Cataloguing in Publication Data
A catalogue record for this book is available from the British Library.

ISBN 0-7509-4303-3

CURRENCY CONVERSIONS
1946 $1 = about 5s
1956 $1 = about 6s 6d
1976 $1 = about 55p
2005 $1 = about 50p
1930 $1 = about 4.20 Reichmarks
1941 $1 = about 2.50 Reichmarks

Typeset in 11.5/13.5pt Photina.
Typesetting and origination by
Sutton Publishing Limited.
Printed and bound in England by
J.H. Haynes & Co. Ltd, Sparkford.

Contents

Dedicated to the memory of my Father,
Robinson Jester Ward, who was,
and perhaps remains, my Muse
in the robust appreciation of Humour
in this tragicomedy we call Life.

Foreword

My most memorable recollection of Wernher von Braun goes back to when we first met. It is my view that if there is one thing that people of large accomplishment have in common it is that they are not limited to just one particular area of expertise. Most such people are not unidimensional. They are curious about everything around them, about how to do or design or make things better, in whatever area that may be.

Back when the first seven of us astronauts were selected in the Project Mercury days, one of the early things we did was start going to Huntsville, Alabama. That was the first time I met Wernher. He was quite well known then, of course, but none of us had ever met him. We were in and out of Huntsville every few weeks for two or three years, and we got to know Wernher and all of his top people.

On one of our earliest visits he invited us to his home on the hill. He had a little library there. We were having drinks, and when I walked into his library I fully expected to see shelves full of books on mathematics, rockets, engineering and subjects like that, because he was known for his expertise in those areas. But I was surprised to find that it contained perhaps even more books on religion, comparative religion, philosophy, geography, geology and politics and a whole realm of other subjects.

That did surprise me. I even mentioned it to him right there, and he just pulled a book from his library and started talking about it. He particularly loved to talk about religion and philosophy. That is something he had in common with other people of great accomplishment, that curiosity about everything around him.

In speeches I have used that example – of what I found the first time we ever went to von Braun's home – to illustrate the point that this is a characteristic of most such people. He was just as curious about matters of religion and politics and philosophy and government as he was interested in how to build a better rocket.

John Glenn

Preface

Some of us fall by the wayside,
Some of us soar to the stars . . .

*Tim Rice, 'The Circle of Life',
sung by Elton ('Rocket Man') John*

Wernher von Braun was a rocket scientist, the most famous – some would say infamous – of them all, the man who led the engineering and scientific teams that were most responsible for giving rise to that still-familiar expression, 'You don't have to be a rocket scientist to know that. . . .'

At times, most of us ponder the unknowable meaning of life, or at least the meaning, if any, of our own lives. From boyhood on, von Braun knew the meaning and purpose of his life: to develop and build the rockets that would enable humanity to explore space, to go to the moon, to Mars and beyond. On two continents, through cycles of peace and war and Cold War, he lived his life's dream and made it a reality. He did so at a sometimes lethal price along the way. He did so even though he needed deathbed reassurance that the space glories achieved had been worth the price – that he and his Old Country and New World colleagues had 'done the right thing' in using the military as the means to their end.

Beyond his burning drive and opportunistic flair, what was Wernher von Braun like as a human being? Some who knew him well view him as a space-age Renaissance man whose intellectual interests and personal pursuits seemed as diverse and as vast as the universe. They say his enviable human qualities – a sense of fairness and of compassion, an unbridled *joie de vivre*, uncommon powers as a speaker and listener and retainer of information, an ability to make the fullest use of his time – were as vital to his success as were his technical brilliance and leadership prowess. And some say, genius or not, he was light-years away from personal perfection.

In speeches and letters during the push towards lunar triumph, von Braun's ringing recruitment cry was: 'Come join us! We're going to the moon!' He believed that if man ever stopped exploring the unknown, he would cease to be man. He believed that in exploring the cosmos, man would find . . . himself. A perhaps surprisingly religious man, he believed every man and woman, of whatever faith, owes a final accounting to the Creator for what he or she does with the precious gift of life.

Edward G. Uhl, a US Army engineering officer in the Second World War, inventor of the anti-tank bazooka rocket and later founder of an aerospace company, came to know and admire von Braun and ultimately to recruit him for the corporate world. At the 1993 Von Braun Exploration Forum in Huntsville, Alabama, Uhl observed:

[Von Braun] was indeed a scientist, a rocket engineer, a teacher, an astronomer – the list goes on. And he was a leader. . . . People wanted to follow him. . . . When we won World War II, we got no territory, we got no ships, no factories, no gold, no war spoils. We got one very important asset. We got a team of 117 professional scientists and engineers, led by a pied piper, Wernher von Braun. And that team helped the United States become the space leader of the world.

I began writing about Wernher von Braun in 1957, shortly after joining the staff of the *Huntsville (Alabama) Times*, and continued to cover him and his team, as a reporter, editorial page writer-editor, and correspondent for national technical and trade publications, until his death in 1977. We became acquainted in a workaday way early on in that span of two decades. Between 1979 and the late 1990s, nearly a dozen books were published about von Braun and his rocket team. Some, written by unabashed admirers, portrayed von Braun in a most favourable light; others, by acknowledged critics, presented dark interpretations of his place in history. But none had dealt fully with the total man: not only the professional figure, but also the personality, the character, the 'human side' of this complex man.

In 1998, I decided to write a biography of von Braun. As a newspaperman who had closely observed, reported on and otherwise written about him in Huntsville and Washington, DC for twenty years, I believed that I could bring new insights and journalistic objectivity to the story.

So, with a 'starter supply' of material gathered over decades, I set out to conduct a series of interviews with the dwindling number of contemporaries of von Braun around the United States and in his twenty-year home town of Huntsville. One revealing interview led to three more, then to a dozen and then to scores more. Tracking down some sources involved lengthy detective work. Archival research of correspondence, speeches, articles and other material took many months. The biography project took on a life of its own. The years 1998 and 1999 came and went, as did 2000, and 2001 and 2002. . . . The number of in-person and telephone interviews, taped and transcribed, exceeded 100. Interviewees were, with the rarest of exceptions, cooperative and forthcoming, and some were amazingly candid about intimate details. There was too much for a single book, and the one you are reading is about half the length of the original manuscript. I have put myself in a few places in the story, mostly instances where I had direct contact with Dr von Braun.

Writing a book may be a solitary pursuit, but researching one is not, nor is turning a manuscript into a published work. From the material-gathering process to an author's need for sounding-boards to the creation of the finished, bound volume, it is a collaborative enterprise.

I am indebted to many people for their help and advice. Foremost are friends and space-writing colleagues Ernst Stuhlinger and Frederick I. Ordway III, both long-time associates of von Braun. Their steadfast assistance in multiple ways during the seven years spent writing this book was boundless. And their selfless help was given, in a sense, to a competing author: their two-volume biographical memoir of 1994, *Wernher von Braun: Crusader for Space*, remains in print. Their work, and the books of other biographers and historians, formed an impressive body of work to build on with this new biography.

Adding their insights to the record were the more than 100 men and women interviewed for this book. I thank every one of these contemporaries – friends, acquaintances, associates, colleagues, critics – of von Braun. They include such well-known figures as Walter Cronkite, John Glenn and Hugh Downs, and scientists James

Van Allen and the late William Pickering and I.M. Levitt. They also include wonderfully helpful von Braun secretaries Dorette Kersten Schlidt, Bonnie Holmes and Julie Kertes; von Braun special assistants Frank Williams, Jay Foster, Jim Daniels and Thomas Shaner; most of the dwindling band of surviving Peenemünde rocket experts who followed von Braun from war-ravaged Germany to America in 1945–6 and later; and many American-born insiders from his US Army and NASA (the National Aeronautics and Space Administration) days of the 1950s, '60s and '70s.

I extend special thanks to John Glenn – former astronaut, senator and then astronaut again – for his foreword, as well as for our interview, and to Michael Collins, the Apollo 11 astronaut, gifted author and former director of the Smithsonian Institution's National Air and Space Museum, for his contribution to the epilogue.

My great gratitude for their encouragement, assistance and incalculable advice goes as well to my gentlemanly friend, the Second World War naval aviator and early space-age catalyst, Frederick C. Durant III; to the late Dr Carsbie C. Adams, health-care administrator, space writer and von Braun family friend; to my late, distinguished lawyer friend, Patrick W. Richardson, long-time personal attorney to Dr von Braun; to the thirty-year Associated Press correspondent at Cape Canaveral, the late Howard Benedict; to authors Ben Bova and Diane McWhorter; and to Ruth von Saurma, von Braun associate and friend and translator *par excellence*.

Particular appreciation for their help must also be expressed to Edward O. Buckbee, founding executive director of the US Space and Rocket Center; to Irene Powell Willhite, the centre's indefatigable archival curator; to the Huntsville-Madison County Public Library's Ranée Pruitt, skilled archivist, and the ever-helpful Thomas Hutchens of its Heritage Room; to David Christensen, veteran space engineer-manager, writer and prodigious saver of space stuff; to NASA retiree Robert Lindstrom for sharing a trove of von Braun correspondence; and to friends and colleagues at the *Huntsville Times* for their valued support over the long haul.

My research assistant in the critical latter phases of this trajectory, Kathy Woody, has my undying thanks for her sunny disposition and inordinate skills as a fact-checker, ferreting out photo images, taming computers and organising material and this writer. My deep thanks go also to my son, Robert Hanly Ward, for his mega-help with my innumerable computer gremlins. For their inestimable help in bringing this book into being, I thank my supportive editors at the Naval Institute Press, especially Mark Gatlin, without whom there would be no book, and Linda O'Doughda, as well as freelance editor John Raymond. My thanks likewise go to Jonathan Falconer, Senior Commissioning Editor, and Clare Jackson, Project Editor, at Sutton Publishing for their work on this edition. My most heartfelt appreciation, however, must go to my sharpest, most honest editor – now and through the years – my beloved wife, Barbara Ann Byrne Ward.

Bob Ward
Huntsville, Alabama

That Accursed Blessing

It was spring 1961 and the Soviet Union continued to trounce the United States in what was called the 'space race'. Since 1957 and the shocking appearance of Sputnik I overhead, the Russians had scored one space first after another. The last straw came on 12 April of US President John F. Kennedy's first year in office when cosmonaut Yuri Gagarin orbited the planet for one full revolution. America was again humiliated – and threatened – by the Soviets' obvious lead in rocket boosting power.

Former naval officer JFK, smarting also from the recent Bay of Pigs fiasco of the US-backed invasion of Cuba by Cuban exiles, would soon speak of space as 'a new ocean' that America 'must sail on . . . in a position second to none'. For now, he knew that the United States must seize the initiative with a dramatic new space goal that could be achieved before the Russians. Among a handful of top American aerospace leaders privately surveyed for their best ideas was the chief rocket man of the young NASA, who laid out a cogent case for the objective that was chosen.

And so, on 25 May 1961, just three weeks after US Navy Cdr Alan Bartlett Shepard Jr's successful non-orbital arc beyond the atmosphere had given America all of 15 minutes of manned space flight, Kennedy stood before a joint session of Congress. The thirty-fifth president challenged the nation to 'commit itself to achieving the goal before this decade is out, of landing a man on the Moon and returning him safely to earth'.

Little more than eight years later, on 16 July 1969, Apollo 11's three astronauts – Neil A. Armstrong, a former Navy combat aviator-turned-civilian, and two West Pointers-turned-US Air Force pilots, Edwin E. 'Buzz' Aldrin Jr and Michael Collins – were powered towards an imminent lunar landing by the Saturn V launcher. NASA's chief rocket man and his team had created the largest, most potent rocket ever built.

The man behind this achievement was an improbable figure, a former enemy scientist and Nazi Party member who had developed missiles for Hitler's Germany and then for the US Army. He was Wernher von Braun, and reaching the moon – and beyond – had been his dream since boyhood.

In his day, von Braun enjoyed popularity equal to that of a film star. Yet despite all the publicity and the flood of fan mail he received, he seemed, at times, scarcely to realise the extent of his fame, preferring to focus on the reason for it. For example, during his NASA days in the 1960s, his friend Edward G. 'Ed' Uhl, the head of the aerospace company Fairchild Industries, invited him to a jaguar hunt in the Yucatán region of Mexico. Sportsman von Braun, who had been raised on the blood sports of wildlife-hunting and bird-shooting in Germany, eagerly accepted. For what Uhl called 'a poor-man's hunt', it would be just the two of them. Uhl's friends were cutting mahogany on some land they leased there and had offered their lumber camp as a base for a hunt in the surrounding jungle.

Uhl recalled that they flew into Merida, where a Second World War jeep with a Mayan driver met them.

Off we went to camp. It was way back in the Yucatán, right where Belize, Guatemala and Mexico all come together. After several dusty hours of driving we came to a little village with grass huts. One, we could tell, was a store, because they had wild turkeys hanging by their necks out front. But they had Cokes, so we stopped. They didn't have any ice, but the Cokes had cooled in the spring.

While we were chatting and drinking the Cokes, one by one these young people came out of their huts and would say, 'Dr von Braun, would you sign this, please?' I couldn't believe it! I said to myself, 'The US government has an advance team out here!' But it couldn't be. But there they were, deep in the jungle, these Indian kids. Some had his picture! Others had notebooks or just a piece of paper – and they all wanted his signature and they all knew of him.

Later, as they drove on, von Braun remarked, 'Isn't it wonderful how young people are turned on by space?'[1]

Von Braun's favourite autograph story concerned a talk he gave to a school in the United States in the late 1950s. Afterwards, he was surrounded by a group of teenage girls. One girl pressed forward and excitedly asked, 'Dr von Braun, could I have two autographs?' 'Well, yes,' he replied, signing the first of the two blank cards she offered. 'But why do you want two?' The girl answered sweetly, 'Because for two of yours I can . . . [get] one of Elvis Presley's.'[2]

During his more than three decades in the United States, from 1945 until his death in 1977, von Braun achieved fame and widespread recognition, in contrast to his largely secret existence in Germany. Hollywood made a film about his life when he was still in his forties. His face graced magazine covers and newspaper front pages. Television programmes featured him in prime time. Books about him spread his renown. And all this was before he spearheaded the triumphant first manned voyages to the moon. From the 1950s onwards, von Braun, 'Dr Space', was the pre-eminent figure in US rocketry and the exploration of space.

He was a hero to millions – rare for a man of science and technology, and rarer still for a former wartime enemy of the United States.

Von Braun saw his fame as both a blessing and a curse. His celebrity gained him a platform from which to call for the exploration and exploitation of space – and aroused resentment among some colleagues. It opened doors and brought a parade of the similarly renowned to his door, along with unceasing demands on his time from a mostly admiring public. Although he called autographs 'this plague', he apparently complied with practically all such requests; it was, in part, good public relations for the space programme. When the demand got too heavy in the early 1970s, photographs for fans were signed by auto-pens, and recipients never knew the difference.

Paradoxically, this extreme celebrity came to a man who spent his career emphasising the 'we' over the 'I' in the engineering and scientific teams, large and small, that he led – while making sure he was the undisputed leader of those teams. The military and civilian programmes he helped advance were unfailingly successful at a time of enormous national and global tension. At the same time, he was a gregarious, charismatic pied piper of space.

There was hardly anywhere on earth he could go without being recognised. For example, in the winter of 1961/2, von Braun once found himself dodging recognition in exotic Kathmandu during a nearly three-week, around-the-world trip for business and sightseeing with his friend Carsbie C. Adams. Adams, a health-care facilities entrepreneur, had co-authored a book with von Braun,[3] and the von Braun family often visited Adams at his plantation in Georgia. Now, on the morning of their departure from Nepal, the two men were eating breakfast in their hotel when they met a woman from New York. Adams recalled: 'Von Braun introduced himself with a rather quick mumble as "Don Brown". I spoke up immediately and quite distinctly introduced myself. I knew that von Braun desired to remain anonymous – due to the CIA's serious concern about his being so close to Red China [25 miles away]. They had not wanted him to go to Nepal in the first place, since they thought he might be kidnapped.'

The woman persisted, asking von Braun again for his name. 'This time,' Adams recalled, 'he pronounced his name correctly and distinctly, and she said, "Oh, von Braun, like the von Braun in rocketry." And von Braun said quietly, "Yes, like the von Braun in rocketry."' Adams almost broke up laughing but somehow kept a straight face. Von Braun did the same, continuing to look down at his plate. He soon finished his breakfast, excused himself and went up to his room to pack. With von Braun gone, the woman said to Adams, 'You know, he does favour [look like] the von Braun, but he's much too young.'[4]

In 1966, en route to the South Pole, von Braun and his chief scientist, Ernst Stuhlinger, checked in without fanfare at the White Heron Lodge in Christchurch, New Zealand, the staging area for US-sponsored scientific activities in Antarctica. After a full day of briefings and trying on polar gear, the two men finally sat down for a quiet dinner at the lodge. A band played onstage in the dining-room. Von Braun whispered to his companion how relaxing it was to be somewhere he wasn't recognised and where no one asked for an autograph. He thought nothing of it when the female vocalist began singing about going 'to the moon with you', as Stuhlinger remembered it. But when she finished she headed straight for their table. 'Dr. von Braun,' she gushed, 'I sang this one just for you, and I mean it! Would you have a seat for me on your moon ship?!' For the rest of the evening a trapped von Braun signed autograph after autograph.[5]

Von Braun had a knack for attracting attention in public and social settings wherever he went, at home as well as abroad. Space historian Frederick Ordway remembered his first exposure to von Braun's 'star quality' even before his fame had soared to its highest. The two men had met in the mid-1950s through the New York City-based American Rocket Society. Ordway and his wife, Maruja, hosted a reception and dinner party at their home in Syosset, Long Island. Guests of honour were von Braun and his young wife and cousin, Maria. Ordway was then working for Republic Aviation and had invited people from the company, other technology firms, banks and brokerages – a mix of business and social friends.

'We were having this cocktail party before dinner, with everybody crowded around him,' recalled Ordway. 'He could talk about any and every subject. There he was, this very handsome guy . . . and all these sophisticated Long Island north shore people . . . were just drooling over him.'[6]

A similar stir occurred among the country music crowd in Nashville, Tennessee. In the late 1960s, von Braun, Ordway, and writer David Christensen were working on a series of educational films on rocketry and space exploration. Von Braun was scheduled to record the narration – in German and English – for the series at RCA studios in Tennessee's 'Music City', long a country-and-western performing and recording centre and home of the renowned *Grand Ole Opry* live radio show. Nashville lies 100 miles north of Huntsville, the self-proclaimed 'Rocket City, USA' ever since the US Army in 1949 made the nearby disused Redstone Arsenal its new centre of missile R&D (research and development) and moved in the von Braun team of rocketeers the following year. When von Braun and Ordway arrived at the studios, they found a bevy of country music performers and crew on hand to tape their own television show.

'They treated him like a hero,' Ordway recalled. 'All the country singers were gaping at him and making over him, including some rather scantily clad young things in long black stockings.' Von Braun was all smiles then, and later, after he had finished the taping, when the same thing happened – black net stockings and all – as he left the studio. Their day's work done, von Braun and Ordway had dinner at their hotel and then repaired to the lounge for a nightcap.

Von Braun loosened up, looked across the table, and said with a grin: 'Well, Fred, it was a great day, wasn't it? But I think you and I entered the wrong business.'[7]

The scientist's superstar fame brought many such lighter moments. In 1958, shortly after he had been on the covers of both *Time* and *Life* magazines, von Braun was on a business visit to San Francisco. His then-assistant Francis 'Frank' Williams recalled that they had some free time before catching the 1 a.m. red-eye flight back to Huntsville. They decided to have a quiet drink at the famous Top of the Mark Lounge crowning San Francisco's Mark Hopkins Hotel.

Two men immediately recognised the scientist, and von Braun invited them to drag up some chairs. Others entered the bar, spotted von Braun, came over and were also invited to join the group. Before long, eight or ten men were sitting at the table, with von Braun the centre of attention. The drinks kept coming until von Braun flashed a look at Williams. The young assistant excused himself, found the head waiter and asked him to page von Braun. A telephone was brought to the table, and Williams took the fake call. He announced that there was an immediate meeting they had to attend elsewhere, and the two escaped.

Later that evening, von Braun admitted he was sometimes pleased at being recognised, but that fame certainly had its downside. If he were ever tempted to go somewhere where he wouldn't want to be seen, he would have to resist, he said, because 'some sonuvabitch would recognise me!'[8]

The November 1957 *Life* cover story on von Braun that had recently appeared quoted him as saying: 'I get about ten letters a day. About half come from youngsters who want advice on how to become rocketeers. We tell them to hit math and physics heavily. One lady wrote that God doesn't want man to leave the Earth and was willing to bet me $10 he wouldn't make it. I answered that as far as I knew, the Bible said nothing about space flight but it was clearly against gambling.'[9]

Von Braun's avalanche of fan mail ran the full spectrum. One woman wrote, 'I only have one question before I sign off: is it possible for one to attain sexual pleasure from sending up rockets?'[10] No record exists of any von Braun reply, and most such correspondence was relegated to a file labelled 'Strange'.

Many letter writers offered technical advice, much of it also strange. A New Jersey man wrote that he had the plans for 'a vibrationless gas engine' which 'should solve your problem of building a rocket ship with 36-cylinder or 72-cylinder engines'. From a man in Massachusetts came a letter stating: 'I know how to make an artificial conscious brain. . . . I could make one in less than three years. You must believe me.' This would allow a person to be 'on Earth and on the moon at the same time, for example'. If only von Braun could help him obtain the 'several portable digital computers' necessary for the task. . . .

One of the strangest letters, an ominous one, came from a correspondent in Germany claiming to represent a 'world-famous rock'n'roll group'. All that the band wanted was for von Braun to secretly bring to America a pretty, 14-year-old German girl whom they had marked for stardom as a singer. They would pay him well for his trouble but, if he didn't cooperate, it would be 'death for you and your wife'.

Beginning on 31 January 1958 with the launching of Explorer 1 – which von Braun's US Army team had developed and launched – the heavy attention paid to the German-born rocket scientist, by then a naturalised US citizen, became less uniformly favourable. The bulging postbag now contained a smattering of hate mail addressed to the father of the V2 missile in the Second World War. Since the late 1940s, it had been no secret that von Braun had once been the technical director of the German Army's rocket development base at Peenemünde (a village on an island in the Baltic Sea at the mouth of the Penne Estuary). But his new-found fame – and near adulation – suddenly drew more attention to his past. It was a past that included acceptance, under pressure, of Nazi Party membership and an SS officer's commission. Some writers accused him of involvement in the Holocaust. Others criticised the seeming ease with which he had shifted his allegiance at the end of the war from Germany to the United States. Would he not go elsewhere if the right opportunity presented itself? In the mid-1960s, when von Braun was absent from one of the unmanned Saturn rocket test flights at Cape Canaveral, a Washington, DC space reporter who was well known as a von Braun critic accosted me, a young missile-and-space writer for the *Huntsville Times*, with: 'Von Braun couldn't make it down, huh? What's the matter – he couldn't miss his Chinese lessons?'[11]

At his speaking appearances around the United States, von Braun occasionally attracted protest groups and hecklers. Some shouted 'Nazi!' and 'Sieg heil!' The title of his 1960 movie biography – *I Aim at the Stars* – drew comedian Mort Sahl's addendum, 'but sometimes I miss and hit London!' Tom Lehrer summed up much of this attitude in a satirical ditty that included the lines:

> Wernher von Braun,
> A man whose allegiance
> Is ruled by expedience . . .
> 'Once the rockets go up, who cares where they come down?
> That's not my department,' says Wernher von Braun.[12]

Despite such criticism, von Braun's celebrity and popularity remained high. Other government officials and colleagues in the private sector were sometimes rankled by his style because he thought big and was aggressive in pushing forward his ideas and his own group. 'That did not endear him to other NASA centre directors,' Thomas Shaner, von Braun's assistant at the Marshall Space Flight Centre in Huntsville in 1969–70, remembered. 'There was a lot of animosity toward Dr von Braun from some at NASA headquarters and some at the other field centres. They were extremely jealous of him.'[13]

On 31 January 1959, the first anniversary of the launching of the Explorer 1 satellite into orbit, a satellite mock-up was presented to what is now the National Air and Space Museum in Washington, DC. Maj-Gen John Bruce Medaris, von Braun's boss at the Army Ballistic Missile Agency at Redstone, spoke briefly, followed by Secretary of the Army Wilber M. Brucker and then von Braun.

'When von Braun stopped speaking,' remembered Ordway, 'there was pandemonium. People all rushed down to get his autograph. Secretary Brucker, who had also been governor of Michigan, looked at me and said, with all seriousness, "Nobody gives a goddamn about the Secretary of the Army!"'[14]

Despite hecklers and envious colleagues, von Braun remained in the spotlight. He often had trouble going about his business, or pleasure, unmolested by an admiring, or at least curious, public. Covert operations were the rule, including intricate planning, anonymous reservations, rear entrances and hidden tables in dark corners.[15]

On at least one occasion, the covert operation came from the other side. In 1959, von Braun was invited to Los Angeles to appear on a television programme with 'a panel of missile experts'. When he reached the studio, however, out walked show host Ralph Edwards. Von Braun was the surprised subject of the top-rated prime-time TV show *This Is Your Life!*

The Hollywood-produced film about his life that comedians poked fun at had its world premiere in Bavaria, where von Braun's parents were living, shortly before it opened in the United States in September 1960. *I Aim at the Stars*, directed by Lee Thompson, raised anew the question of von Braun's role in the development of the V2 missile. He had interceded with the film's British director for script changes to make it clear he knew full well, once war began, that the rocket would be used against London. For example, von Braun protested against a script line in which his character says he 'supposes' the missiles will be fired at London. He had the line replaced with 'of course' they will be fired at London. In the finished film, actor Curt Jurgens, playing von Braun, intones: 'When there's a war, every man wants his country to win. I was no exception. . . . If I had refused to do this work, I would have been labelled "enemy of the state". If you'll forgive me, I chose to stay alive and continue with my job.' The movie was not much liked by critics, filmgoers or von Braun.[16]

The excitement surrounding America's early space efforts brought a steady stream of celebrities to visit the scientist in Huntsville. Walt and Roy Disney were good friends. Singer and space fan John Denver, and actors Robert Young, Bette Davis and Gloria Swanson, as well as Boston Pops conductor Arthur Fiedler, came to see him. He had private get-togethers with Billy Graham and Martin Luther King. The pioneering heart-transplant surgeon from South Africa, Dr Christiaan Barnard, visited him. US media superstars Edward R. Murrow and Walter Cronkite dropped in to do stories on him, and popular US TV host Hugh Downs became a friend in the 1970s. Von Braun was a friend of captains of industry Donald Douglas of McDonnell-Douglas, William Allen of Boeing and William Reynolds of Reynolds Aluminum, among many others. He corresponded in the early 1960s with Dr Albert Schweitzer. Jacques Cousteau was a pal.

The rocket scientist was honoured at White House dinners and knew Presidents Truman, Eisenhower, Kennedy, Johnson, Nixon, Ford and Carter. He was on first-name terms with the astronauts and close to many of the leaders of Congress. Renowned Washington hostess Perle Mesta threw a dinner party in his honour.

Leading scientists also visited von Braun, including Edward Teller, the developer of the hydrogen bomb, and the science fact and fiction writer Arthur C. Clarke. Most came for private sessions with von Braun and to get glimpses of the future, often via personally guided tours of his team's facilities. In the late 1960s, when Gen Mark Clark, US Army (Ret.), visited the Marshall Center, von Braun invited the Second World War field commander and Korean War general to have lunch with him and several senior staff people, scientist-manager William R. Lucas recalled. After lunch, Clark stood up and said, 'Dr von Braun, I enjoyed being with you today, and I appreciate what you are doing for my country, sir.'

Quick as a finger snap, von Braun, a naturalised US citizen, replied, 'Sir, it is my country, too.'[17]

Von Braun was periodically invited to join other VIPs at the Bohemian Club, a private California retreat for the rich and/or famous. There, he strengthened relationships with two US Army heroes of the Second World War, Gen James H. 'Jimmy' Doolittle and Gen James Gavin. Although they had been his enemies in the war, they both became admirers and friends of the German rocket man.

In 1962, von Braun presented the Henry Grier Bryant Gold Medal of the Geographical Society of Philadelphia to his astronomer friend I.M. Levitt, the director of the Franklin Institute's Fels Planetarium in Philadelphia. The medal was given in recognition of Levitt's studies of the moon's geography.

Later, in 1970, when the German Society of Pennsylvania awarded its gold medal to von Braun, Levitt returned the favour. Much of his introduction was an erudite tracing of the twin developmental paths of astronomy and space exploration through key historical figures of various nationalities whom, he said, 'Nature had given to mankind'. And then, observed Levitt, there 'arose the need for a man who could synthesise all the elements to produce a space transportation system to fashion the staircase to the stars. Again, Nature was called upon to produce this synthesising genius, and Nature responded with a remarkably able and astute figure in contemporary science. His name is Wernher von Braun.'

Recalling that occasion some thirty years later, at the age of 90, Levitt noted that von Braun and he had been friends since their first meeting in 1952. He recalled von Braun's rare brand of magnetism. 'Wernher was a commanding figure. He was the type that, when he walked in the room, the conversation would stop and eyes would turn toward his entrance. He had a commanding presence, and he was so very, very bright.' An early advocate of space exploration, Levitt began in 1952 to write an internationally syndicated newspaper column on astronomy and future space exploration. Levitt, who was Jewish, said he was convinced von Braun had not been a Nazi adherent at heart. He said he had reason to believe – from independent sources in Europe – that the rocket leader had accepted Nazi Party membership and an SS officer's commission as mere expediencies. For one thing, he said, they gave von Braun a measure of authority when dealing with the unwelcome SS troops stationed at the Peenemünde V2 centre.[18]

Von Braun received countless medals, keys to cities, more than twenty honorary university degrees, and other domestic and international honours. Included were the highest awards given to US civilians in the fields of defence and space. In the 1950s and '60s, the Associated Press named him 'Newsmaker of the Year in Science' more frequently than anyone else. In January 1977, near the end of von Braun's life, President Gerald Ford awarded him the National Medal of Science.

Earlier, in 1959, Indiana's Notre Dame University gave von Braun its Patriot of the Year award, little more than a decade after the Second World War and just four years after he had gained US citizenship. (It was also before the public learned of his connection with the notorious underground Mittelwerk V2 factory near Nordhausen.) 'It is not ordinary for a man who once worked for the enemies of the United States to receive an award which is established to honour the outstanding American patriot of the year,' the award presenter said. 'But then, Dr von Braun is no ordinary man.'

Forty years later, and in a new century, how would Wernher von Braun be viewed? Would his popularity persist, or would his lingering baggage as a 'former Nazi scientist' affect his standing a quarter-century after his death? For the 2003 centennial celebration of the Wright brothers' first powered airplane flight, *Aviation Week & Space Technology* magazine coordinated a worldwide survey by several professional groups to choose the 'Top 100 Stars of Aerospace'. More than 1 million ballots came in from industry professionals in 180 countries, selecting 'the most important, influential, and intriguing personalities in the history of flight'. The results were revealed in June 2003 at the Paris Air Show and in the magazine.[19]

Where did Wernher von Braun come in? Surprisingly to many, he ranked first among space figures, ahead of such trailblazers as Buzz Aldrin, Neil Armstrong, Yuri Gagarin, John Glenn, Robert Goddard, Alan Shepard and Jules Verne. And in the overall category he ranked second, behind only the Wright brothers themselves.

To the Manor Born

Wernher von Braun was born on 23 March 1912, in the city of Wirsitz, in the province of Posen, part of Prussian Germany. The family had estates in both East Prussia and Silesia, and until the age of 12 Wernher lived in those provinces as well. He was the second of three sons born to Baron Magnus Alexander Maximilian von Braun and his wife, Baroness Emmy von Quistorp von Braun. All the von Braun boys became barons at birth. Although of noble rank, a baron was relatively low in the aristocratic pecking order.[1] Nonetheless, the boys were taught to take their station in life, and its attendant responsibilities, seriously.

'My friend was born in Germany . . . the middle son of a baron, and he was therefore expected to study, to work, to learn, to lead, to be honourable,' said American-born Edward G. Uhl in a talk in 1993, in which he summarised the family milieu that Wernher von Braun had entered.[2]

As *Time* magazine wrote in 1958, taking the long view of the family's history, 'Von Braun's origins had deep earthly roots in Prussian Junkerdom. A von Braun fought the Mongols at Liegnitz in 1245, and the family's aristocracy was certified by the centuries.'[3]

His father, born in 1878, had farming and banking interests. 'Papa' became a provincial and then national government official, principally in agricultural ministries. Wernher's mother, who was of aristocratic Swedish–German lineage and had been brought up in England, was said to have possessed a brilliant mind and an intuitive sense about people.

The family's membership of the aristocracy brought with it wealth and status. The von Brauns lived in substantial homes, where servants waited on them, and they had significant landholdings cared for by estate workers. It maintained a cultured atmosphere of good manners, tradition, appreciation of music, art and literature, and a disciplined devotion to education.

His father had imagined that Wernher, as a future 'landed gentleman', might make farming his life's work. But his parents soon realised that their middle son was developing into an unusual child. 'When he was only four he could read a newspaper – upside down as well as right side up,' his father recalled. 'He was always asking questions that his teachers couldn't answer, and he had a remarkable ability to apply himself completely to whatever interested him.'[4] Although Wernher and his brothers all did well academically, Wernher was the most gifted. 'Sigismund and Magnus are clever,' their father later observed, 'but they are ordinary clever people. Wernher is a genius.'[5]

'I don't know where his talent comes from,' his father wondered years later.[6] His mother certainly had a great deal to do with it. She was a compassionate, well-educated woman, who spoke six languages, was well travelled, loved great music and fine art, and was a serious amateur ornithologist and astronomer. 'She had a large household with many servants during the pre-war years in Germany,' Ernst Stuhlinger later wrote, 'but if one of them fell ill, she personally cared for her like a mother.' She 'was at home in the most distinguished circles of society; but she conversed with equal ease and human interest with her gardener or coachman.'[7]

Before the siren's call of space snared him, Wernher showed an interest in and aptitude for music. 'As a little boy,' his mother later remembered, 'Wernher loved the piano and composed his own music. For a time we even thought he would make music his career. But it was not to be.'[8] Baroness Emmy 'not only opened the gates to the world of science for Wernher, she also taught him to play piano', wrote Stuhlinger, who eventually became von Braun's chief scientist and came to know Wernher's parents later in the United States.[9]

Wernher progressed in his musical studies to the point where, in the 1920s, while living in Berlin with his parents, 'he was accepted for piano lessons by the great composer Paul Hindemith (1895–1963)', Stuhlinger later wrote.[10] By the age of 15, von Braun had written three short original pieces for piano. In 1925, he began taking cello lessons and soon joined his school's orchestra. He continued with both instruments into adulthood, playing cello in a string quartet of German rocket professionals who made music for their own pleasure. He astounded new friends decades later when he sat down at a piano and flawlessly performed a long classical piece from memory.[11]

The von Braun family moved to Berlin in 1920 when the baron was promoted from provincial jobs in a series of German cities to national governmental posts. By 1924, he was the Reichsminister of Agriculture during the Weimar Republic, under President Friedrich Ebert, a position he held until 1932, under the administration of President Paul von Hindenburg.

Wernher's first rocket 'tests' took place with a bang when he was a headstrong 12-year-old. In 1924, he got an irresistible idea, probably inspired by the experimental rocket-powered motor cars then making the headlines. He recruited big brother Sigismund, one year older, for the project; Magnus, just 5, stayed at home. The two older boys bought six large skyrockets and lashed them to Wernher's go-cart, which he had given a special paint job for the occasion. After wheeling the cart onto Tiergarten Strasse, Berlin's most upmarket street, the boys lit the fuses, and Wernher hopped aboard for the ride.

Von Braun later painted a vivid picture of the ensuing chaos. 'I was ecstatic. The cart was wholly out of control and trailing a comet's tail of fire, but my rockets were performing beyond my wildest dreams. Finally they burned themselves out with a magnificent thunderclap and the vehicle rolled to a halt. The police took me into custody very quickly. Fortunately, no one had been injured, so I was released in charge of the Minister of Agriculture – who was my father.'[12]

Sigismund, who later became a career diplomat in Germany's foreign service (before, during and after Hitler's rule), added more specific details: the wild rocket ride had, in fact, caused a casualty or two. The runaway cart 'crashed into the legs of a woman, ruining her stockings', and then ploughed into a fruit stand. At the police station, Wernher was admonished not to repeat such dangerous 'experiments'. The baron paid a fine to cover the damages, angrily lectured the boy, and ordered him and his big brother to be confined to the family home for two days.[13] Wernher's mother, equally concerned, took a sympathetic but practical approach. 'The world needs live scientists, not dead ones,' she told him.[14]

Wernher was not quick to learn his lesson. To celebrate the end of his house arrest, he tied even more skyrockets to his cart and set sail once again down the street – with similar results. (Luckily for Sigismund, he had not been invited along.) Little wonder that, as he later admitted, 'Papa always seemed to be wondering what new damned foolishness I was up to!'[15]

Von Braun's schoolwork faltered around this time. He had become a rather average student and that year even failed his maths and physics exams – in large part because of the hours spent building a home-made motor car with a friend.

He was sent away at 13 to study at one of the progressive Hermann Lietz boarding schools, at ancient Ettersburg Castle near Weimar in central Germany. The Lietz schools were famous for their advanced teaching approach, combining a strong emphasis on academic subjects with hands-on exposure to practical crafts. Six hours of classes began early in the morning, and for the rest of the day the boys learned such crafts as woodworking, metalworking, carpentry, stonecutting and masonry. Wernher thrived under this new regimen.

While enrolled at the school, he received an unusual, life-changing gift from his mother. 'For my confirmation,' von Braun recalled in 1958, 'I didn't get a watch and my first pair of long pants [trousers], like most Lutheran boys. I got a telescope. My mother thought it would make the best gift.'[16] It was through this gift, he said, that 'I became an amateur astronomer, which led to my interest in the universe, which led to my curiosity about the vehicle which will one day carry a man to the Moon.'[17]

When he was 15, he read a science fiction article in an astronomy magazine that ignited his zeal to make rocketry and space exploration his life's work. 'I don't remember the name of the magazine or the author, but the article described an imaginary trip to the Moon. It filled me with a romantic urge. Interplanetary travel! Here was a task worth dedicating one's life to! Not just to stare through a telescope at the Moon and the planets, but to soar through the heavens and actually explore the mysterious universe! I knew how Columbus had felt.'[18]

Later that year, Wernher was reading a pamphlet on astronomy and saw a drawing of a rocket speeding through space towards the moon. It illustrated an article about a man who would soon become his mentor, rocket theoretician Hermann Oberth, then a 30-year-old physics professor in Romania. Intrigued, Wernher sent for a copy of Oberth's classic book, *The Rocket into Interplanetary Space*, which had been published two years earlier. He was soon 'shocked to discover that it contained mostly mathematical equations'. A resolve that would come to characterise him emerged. Wernher, who disliked maths, recalled much later, 'I decided that if I had to know about math to learn about space travel and rocketry, then I'd have to learn math.'[19]

At school he turned his mind to maths and physics with a vengeance. In his mid-teens, Wernher persuaded his father to let him transfer to a branch school in the Lietz system on Spiekeroog Island in the North Sea. At 16 he planned to accelerate his studies so he could graduate a year early and join a group of rocket enthusiasts in Berlin. But he changed his plans when the school headmaster asked him to teach the maths classes of a teacher who had fallen ill. He also privately tutored some of the weaker students in the classes he taught, determined that all would pass their exams. Everyone did, and he made the highest marks in his classes.

It was at the North Sea school that Wernher began to develop his leadership, organisational and communications skills – as well as skills in swimming and sailing in the rough seas. His interest in gripping the helm of a sailboat became a love he returned to often in adulthood. Increasingly interested in astronomy and now caught up in his visions of space travel, he decided he had outgrown the small telescope his mother had given him. He persuaded the headmaster to purchase a 5in refracting telescope and then organised his schoolmates to build an astronomical observatory to house it.

About the same time, he began writing about astronautics. In one of his notebooks he wrote (with no little prophecy): 'As soon as the art of orbital flight is developed, mankind will quickly proceed to utilize this technical ability for practical application.'[20] He also wrote a five-page short story, *Lunetta* (*Little Moon*), which told of life aboard an Earth-orbiting station. The piece was published in the school magazine.[21]

Decades later, while living in the United States for a time before returning to Germany, and as her middle son scored one early space success after another, Emmy von Braun reminisced: 'I used to ask him, "Wernher, what do you want to be?" He was about ten at the time, and he would say, "I want to work on the wheel of progress." It sounded so strange coming from a boy of that age. But he did exactly that, didn't he?'[22]

Pioneering Rocketry

In the autumn of 1929 Wernher von Braun was 17. He had graduated from high school with honours and would soon be bound for higher education in Berlin. He knew what he had to do. He would arrange to meet his idol in rocketry, Hermann Oberth, and offer his services for free. The space-flight theorist, mathematician and teacher was an ethnic German, born in Romania. He had recently come to Berlin to do experimental rocket-testing. His writings had sparked a burning ambition in von Braun.

There were three major space pioneers at that time: Konstantin Tsiolkovskii, the Russian space-research theoretician; Robert H. Goddard, the American physics teacher and rocket experimenter; and Professor Oberth, who was the most well known, at least in Germany. The idea of future space travel was a wildly popular topic in Germany in those days. Oberth had helped to make it so. In addition to having written widely on the subject, after coming to Berlin he had also served as technical adviser for the 1929 science fiction hit film, *Frau im Mond* (*Woman in the Moon*). Before and during the making of the film, he conducted primitive rocket experiments at a site outside the German capital.

Oberth had attracted a band of mostly youthful rocket enthusiasts. Young von Braun needed to get an introduction to the famous professor. He decided to go to the Berlin home of one of Oberth's followers, Willy Ley, who was vice president of the Verein für Raumschiffahrt (Society for Spaceship Travel), or VfR, that sponsored Oberth's work. Von Braun had not met Ley before either. When he went to his house in Berlin, Ley was not home. The young visitor was admitted and allowed to wait in the parlour. Ley returned home to find the blond stranger seated at the piano playing gently and expertly Beethoven's *Moonlight Sonata*. Ley complimented von Braun on his playing. Wernher thanked him and explained his mission. The meeting with Oberth was soon arranged.[1]

By the time he had enrolled at Berlin's Charlottenburg Institute of Technology to study mechanical and aircraft engineering in the spring of 1930, von Braun had signed on as an apprentice to Oberth. At the start, his duties included the non-technical task of raising money for rocket research and development. It was a duty he would never be free of throughout his career. 'My first job with Hermann Oberth,' von Braun recalled long afterwards, 'was helping him with a little display on interplanetary rockets in a Berlin department store. I stood on the stand there for eight hours a day telling shopping housewives how an interplanetary rocket would cost 7,000 marks and take a year to build. Now, forty years later, I realise how little one billion dollars will buy and how little you can build in one year.'[2] Also in his 1930 department store pitch, as von Braun remembered it, 'I said, "I bet you that the first man to walk on the Moon is alive today somewhere on this Earth!"'[3] It so happened that Neil Armstrong was at that time an infant in Ohio.

The Oberth group's sponsor, the VfR, was an organisation for students and other rocket enthusiasts. The Oberth assistants and members of VfR in 1930 included such future notables in the field as Rudolf Nebel, Klaus Riedel and Rolf Engel, in addition to Willy Ley. Newcomer von Braun joined the VfR immediately. A chief purpose of the club was, in fact, to raise money to support Oberth's work on his novel rocket-engine concept, the 'cone jet motor', that he was developing and building to test in a Berlin laboratory and in flight trials. 'In those days we lived from begging,' von Braun recalled. 'We had to ask for every single penny to be able to pursue our dreams.'[4] This 'cash barrier', as he later termed it, would often prove the toughest of all obstacles to putting rockets into space.

But there was far more to being Oberth's assistant than fundraising. There were actual rocket-motor test firings, and danger was a constant presence. The thirty or so primitive 'hot' experiments they conducted in 1928–9 were life-threatening. Von Braun later described a typically 'perilous' ignition system test. 'Riedel lighted a rag soaked in gasoline. We opened valves that let the propellants into our tiny motor. Riedel hurled his torch over the motor's mouth and ducked behind a barricade. The jet ignited with a thunderclap that turned into a roar. After 90 seconds our fuel was exhausted and so were we.'[5]

Even though there was a space craze in Germany at that time, many considered Oberth's visionary notions of space travel delusional. Von Braun later observed that he considered himself very lucky to have been among 'the few who thought Professor Oberth made one hell of a lot of sense'. He added: 'Basically what he was trying to prove were four things far ahead of the comprehension of most people at that time. These were that you could build a machine which could climb beyond our atmosphere, that man could leave the gravity of Earth, that man could survive flight in a ship in space, and that the exploration of space could be profitable.'[6]

Despite its defeat in the First World War and the dismal postwar economic period, Germany at that time excelled in science and technology. While studying at the Charlottenburg Institute, von Braun gained an appreciation of industrial craftsmanship via a profound lesson he never forgot. To teach him the practical aspects of engineering, the institute assigned him to work for a few weeks in the machine shop at a large locomotive factory. The shop was filled with gleaming, high-precision machinery, a sight that thrilled von Braun and reassured him that he would enjoy his work there. He happily presented himself to the shop foreman, a burly, mustachioed man with a dirty apron and a stern look.

'Make this into a perfect cube,' ordered the foreman, handing the youth 'a chunk of iron as large as a child's head'. Every angle had to be a perfect right angle, every side equal, every face smooth. The foreman gave the apprentice a file and pointed to a vice. 'Here are your tools,' he said. An angry von Braun stormed over to a workbench and began filing away. After a few days of work he showed his handiwork to the foreman. It was measured. The angles were off. 'Keep filing,' ordered the supervisor. For two weeks, a seething von Braun filed. Again he submitted his work product to the man; again, imperfection and the order, 'Keep filing'. As von Braun's determination grew, the block of metal shrank. Five raw-fingered weeks went by.

'Finally I handed him my supreme effort,' von Braun later recalled. 'It was slightly larger than a walnut. Peering over his dusty glasses, he measured every side. My heart pounded. My reward was one word. "Gut!" He said, "Ja, gut! This is what we mean."' The student had learned 'the value of self-discipline and perfection in small things.'[7]

Later, in 1930, after Professor Oberth had left Germany to resume teaching in Romania, von Braun and the group of rocketeers formed a small company and continued their mentor's experimental work on their own. They operated at an abandoned, 300-acre ammunition storage depot and proving ground at Reinickendorf on the outskirts of Berlin, having talked the municipal authorities into giving them a lease to the place for next to nothing. The group named the site Raketenflugplatz Berlin (Rocket Flight Field Berlin).

The begging and scrounging for money, materials, machinery and labour continued. The young men fast-talked manufacturers out of materials by touting the bright future of rocketry. They recruited unpaid mechanics and other technical workers from the Depression-era ranks of the unemployed by offering them free meals and housing in the concrete igloos and warehouses at the site. Extra money was raised by staging public shows and charging admission to rocket launchings. Sometimes included, at no extra charge, were the fireworks of lift-off explosions, zigzag flights or sudden plunges to the ground, to the chagrin of the rocketeers.[8]

As von Braun juggled his time between his academic studies and practical propulsion R&D at the old dump, the little liquid-fuelled rockets gradually got bigger. Some flew and some failed, and he and his associates learned valuable lessons, not only in technology but in resourcefulness and persistence as well.

Following the custom of university students spending a term or two at another institution to broaden their perspectives, von Braun registered in the spring of 1931 to attend the summer term at the Eidgenössische Technische Hochschule (Federal Technical University) of Zurich. There he met Constantine D.J. Generales, a young medical student from the United States who was also studying in the Swiss city.

The Greek-American soon discovered that this tallish, blond, 19-year-old budding scientist had a fervent interest in rocketry and space travel. Von Braun showed Generales a letter of reply he had received from a young theoretical physicist named Albert Einstein, complete with formulas for the design and propulsion of rockets. The bold German student had written to the respected physicist about his strong interest in rocketry and space travel and asked Einstein for any helpful thoughts or specific information he might be kind enough to share. With his new friend Generales, von Braun also discussed in detail his own concept of sending a manned spaceship to the moon.

'It was my suggestion,' Dr Generales recalled years later, 'that before he attempted a lunar flight, it might be worthwhile to try it with mice as "passengers" first. Wernher agreed it was a good idea. And so we found ourselves spinning white mice on a specially mounted bicycle in Wernher's rooms.'[9] But disaster struck some of the experiments. As the home-made centrifuge, designed to simulate rocket take-offs, spun faster and faster, the blood of 'a number of these unfortunate beasts' was flung against the ceiling of the room – with unpleasantly messy results, as von Braun later reported. 'Our . . . inquisitions were summarily interrupted by my landlady's violent objections to a ring of mouse-blood upon the walls of my otherwise neat Swiss room.'[10] Medical student Generales dissected the mice and reported to his space-minded friend that the high acceleration had caused cerebral haemorrhages in the subject animals.[11]

Von Braun returned to Berlin in 1931, resumed his studies at the Charlottenburg Institute and rejoined the Raketen group. Through the rest of that year and into the next the experimenters test-launched eighty-five rockets to altitudes of up to 1,200ft in free flight.[12]

The year 1932 proved to be even more eventful for von Braun and the group. Aged 20, he received his bachelor's degree in mechanical engineering, with an aeronautical emphasis, from the Charlottenburg Institute. Word spread of the growing successes of the, as always, cash-hungry rocket club. Special visitors came to call. Writer Daniel Lang of the *New Yorker* described one such visit in his 1951 profile of von Braun:

One day in the spring of 1932, a black sedan drew up at the edge of the Raketenflugplatz and three passengers got out to watch a rocket launching. 'They were in mufti, but mufti or not, it was the Army,' von Braun said to me. 'That was the beginning. The Versailles Treaty hadn't placed any restrictions on rockets, and the Army was desperate to get back on its feet. We didn't care much about that, one way or the other, but we needed money, and the Army seemed willing to help us. In 1932, the idea of war seemed to us an absurdity. The Nazis weren't yet in power. We felt no moral scruples about the possible future abuse of our brainchild. We were interested solely in exploring outer space. It was simply a question with us of how the golden cow could be milked most successfully.'[13]

Von Braun and the small group proceeded to milk the military cow well. The visitors in the black sedan that spring day included a general, a colonel and two captains. The junior officer, Capt Walter Dornberger, aged 35, was a key person, the holder of a master's degree in mechanical engineering. He and his superiors observed demonstrations of the 'liquid shell rocket' as von Braun termed it, and received technical briefings from him and other members of the group. The Army contingent was mildly impressed, requested more scientific data and soon awarded the rocketeers a 1,000-mark contract for improved measuring equipment for thrust, propellant flow rates, atomisation of fuel and other needs. Dornberger remained in contact with the group, and in July a demonstration flight of its Mirak II rocket was arranged at the Kummersdorf Army Proving Ground, secluded in a pine forest an hour's drive from Berlin. There the Army had set up a dazzling array of its state-of-the-art measuring instruments. The test did not go well. The rocket rose about 200ft but then veered crazily and crashed to earth. The Army's interest plummeted along with it.

At this point, von Braun, who was barely out of his teens, showed the persistence, leadership and aggressive opportunism that were to become hallmarks of his operating style. He collected all of the test information from the rocket society's experiments and went to see the senior Army officer. Col Karl Becker, who was a scientist and the Army chief of ballistics and ammunition, heard von Braun's appeal for serious financial assistance so that the rocketeers could become more professional in their work. Impressed, Becker agreed, on condition that the group would work in secrecy within the confines of an Army installation. Apparently, the secrecy was for simple military control and security, and not because of any potential violation of the disarmament provisions of the Treaty of Versailles, which did not address rocketry.

When von Braun took the offer back to his colleagues, not everyone wanted to accept. Senior members Nebel and Riedel, for example, objected to it, preferring a struggling enterprise with private support to a restrictive military arrangement. The pragmatic von Braun disagreed. In time, he sold the plan to most of the other members of the group. Eventually, even Nebel and Riedel agreed that he should accept the Army's offer – but without their participation. (It was five years before the pair would rejoin forces with him.)

Capt Dornberger acted as the Army's go-between with von Braun. Aside from the latter's willingness to cast in his lot with the military and the defection of more experienced, senior members of the group, what made von Braun the Army's choice as leader? Years later Dornberger wrote that, at the time, he was deeply impressed 'by the energy and shrewdness with which this tall, fair young student with the broad massive chin went to work, and by his astonishing theoretical knowledge'.[14]

By the autumn of 1932, the German Army's Ordnance Department had put the rocket society under contract to pursue rocketry R&D. It was the beginning of a long relationship, and thus, at the impossibly tender age of 20, von Braun became the civilian head of what was suddenly a closed and classified operation located at the Kummersdorf site. But, as he later pointed out, 'this was not quite so glorious as it sounds'. In the beginning, his total resources, besides his small group of fellow rocketeers, consisted of one mechanic, the use of one rocket test pit and permission to requisition a limited amount of equipment and materials.[15]

Still, von Braun had made his deal. If advancing rocketry towards the ultimate goal of space flight meant joining forces for the present with the military, with those he knew were primarily interested in weapons development, then so be it. They had the funds and facilities, and he and his group needed them to continue. It was that simple.

About the time young von Braun went to work for the German Army, Capt Dornberger helped him enroll at the Friedrich-Wilhelms University of Berlin. There he pursued a doctorate in physics, studying under professors who included Nobel laureates Erwin Schroedinger, Max von Laue and Walter Nernst. While still a doctoral student, von Braun went on holiday in 1934 to a destination that later figured prominently in his life: London, where both his mother and father had lived for a time.[16] His theoretical and applied research, supported by hands-on experiments with a 1933 version of a liquid-fuelled rocket, served as the basis for his doctoral thesis. Because of military security, that document carried the cryptic title 'About Combustion Tests' – and was classified 'Secret' for many years. Before 1934 ended, Wernher's blandly titled thesis earned him a PhD. At the age of 22, when most university students were receiving their bachelor's degrees, he had become Dr Wernher von Braun. His work status also shifted that year, from private contractor to civil service employee of the German Army's Ordnance Department. He would work only for a government, or governments, for almost four decades.

The year before, in January 1933, the Nazi Party won the election and Adolf Hitler was appointed chancellor. Baron Magnus von Braun, the father of the new young civilian rocket chief of Germany's traditionally non-political Army, 'quit all public offices when Hitler came into power' and retired with Baroness Emmy to private life at their estate in Silesia, their middle son later wrote.[17] The estate was Oberwiesenthal, a property of 500 acres in Loewenberg County that the baron had earlier purchased for retirement.[18] The elder von Braun never joined the Nazi Party.

Although rocketeer Rudolf Nebel, a First World War fighter pilot, had joined the Nazis, von Braun had not, and did not for several years more. In fact, 'most of the other [rocket society] leaders, including Willy Ley, were strongly anti-Hitler', according to astronomer and aerospace historian Patrick Moore.[19] As the persecution of the Jews intensified in 1935, von Braun's friend Ley, who had some Jewish ancestors, left Germany for the United States.[20] Other observers characterised the group as a whole as largely apolitical and focused on their work.

An eventful 1934 had brought cheering successes for the von Braun group. Its work that year climaxed with two successful, pre-Christmas launches of A-2 (A for Aggregat [Aggregate]) liquid-fuelled rockets from the Baltic island of Borkum to altitudes of more than 1½ miles. The team apparently had maintained its sense of humour through rocketry's heartaches leading to the twin successes. They had named the two A-2 rockets 'Max' and 'Moritz', after the popular pair of German cartoon characters who were the models for the American Katzenjammer Kids.

Feats such as the A-2 flights soon caught the attention of the Luftwaffe. It offered the von Braun group a project budgeted at 5 million Reichsmarks to develop a rocket-powered fighter plane at a proposed facility larger than Kummersdorf. The German Army would not allow the Luftwaffe to get the upper hand, so it came up with a 6-million Reichsmark allocation in 1935 to support the group's rocket work along existing lines. It was not the last time competing forces manoeuvred to gain the team's services. For von Braun, now 23 years old and responsible for 11 million Reichsmarks in funding, rocketry was no longer small beer. Rather, it had 'emerged into what the Americans call the "big time"', he later observed. 'Thenceforth million after million flowed in as we needed it.'[21] That was true for a time, at least. Many of the millions were spent during the next two years on planning and building a future joint Air Force and Army Rocket R&D centre on the Baltic. Financial support there waxed and waned in later years, depending on the mercurial moods of Hitler and others.

In the mid-1930s, Army civilian employee von Braun did his military service. He served the two one-year stints in, not the Army, but the Air Force – with time off to tend to his work in rocketry. Bitten early by the flying bug, von Braun began with glider lessons, where a classmate was the future German aviatrix, Hanna Reitsch, with whom he remained friends for the rest of his life. He earned his glider pilot's licence at 19. By 21, he had received his regular pilot's licence and had followed his mechanical engineering bachelor's degree with a master's in aeronautical engineering. Later, in the Air Force reserve, he flew military aircraft.

In May 1936, while running things at Kummersdorf, he began pilot training as a Luftwaffe cadet. By the time he had completed training in June 1938, he was an Air Force reservist qualified to fly military aircraft and was detached to the German Army for service as its technical chief at a place called Peenemünde, where the Luftwaffe also had a base. Flying put von Braun into the third dimension – and a step closer to outer space.

Dornberger, promoted to major during the Kummersdorf period, may have been one military man who shared von Braun's and the Kummersdorf group's ultimately peaceful goals. 'Our aim from the beginning was to reach infinite space,' he declared in 1958.[22] Scientist Ernst Stuhlinger later observed that Dornberger was 'an extremely capable and energetic leader who combined the analytical mind of a top-flight engineer with the military authority and courage' that proved essential for success, especially in the dark times that lay ahead.[23] The well-matched space-minded team of Dornberger and von Braun, even during the initial budgetary penny-pinching at Kummersdorf, continued to attract rocket and space enthusiasts to the operation.

One was Arthur Rudolph, who had worked in Berlin – not with Oberth and the VfR, but with another group of German rocket experimenters that included Max Valier and Walter Riedel. He and von Braun had first met in 1932. That was one year after Rudolph had developed and demonstrated his own rocket under another Army contract.[24] He continued to work with others until early in 1935, when, aged 28, he signed on with the expanding Kummersdorf organisation.

He and von Braun soon became friends as well as colleagues. As Rudolph later recalled: 'At heart, von Braun was an astronomer who wanted to journey into space, not just look at it. We would sit up late at the bachelors' quarters [at Kummersdorf] designing rockets and talking of space travel. And besides rockets, we had two things in common: Neither of us liked to go to bed at night or get up in the morning!'[25]

And still later, writing to congratulate von Braun on his 60th birthday in 1972, he reminded von Braun of 'the continuous work at Kummersdorf-Schiessplatz, where you already developed your first concrete ideas on space travel (in the evenings at your quarters at the Officers Club), which 20 years later were published in *Collier's* magazine; or when you tried to convince General Dr. [Karl] Becker that space travel was possible'.[26]

Although he had to work on weapons while his true preoccupation was with space travel, von Braun kept an active sense of humour. One cold day in the winter of 1935/6, when he opened the door to a construction office at Kummersdorf, von Braun found half a dozen staff shivering inside. All of the windows had been thrown open and frigid air was streaming in.

'What's going on here?' asked von Braun. As one of the men present, Otto Kraehe recalled many years later: 'I told him, "We're just letting fresh air in."' Von Braun quickly sniffed out the situation: one or more of the men had broken wind inside the closed office. 'Remember this,' advised the young rocket boss, 'a lot of folks have frozen to death, but nobody ever suffocated from a foul-smelling odor!'[27]

But humour was not always easy to come by in 1935. As the year wore on, the clandestine German military build-up continued. Then, as von Braun recalled, 'One day . . . Dornberger said to me, "The Ordnance Department expects us to make a field weapon capable of carrying a large warhead over a range much beyond that of artillery. We can't hope to stay in business if we keep on firing only experimental rockets."'[28]

By then, von Braun's group had grown to a staff of almost twenty, and preliminary work had started on their next-generation rocket – the A4, which was later renamed the V2. Dornberger and von Braun began to think of finding a place for bigger and better facilities to pursue their burgeoning work.

The *New Yorker* later described the situation. 'He [von Braun] spent the Christmas of 1935 at his father's estate in Silesia, and while there he mentioned that he was scouting for a coastal site that could be used as an experimental station. "Why don't you look at Peenemünde?" his mother asked. "Your grandfather used to go duck-shooting there." Von Braun did so. "It was love at first sight," he told me. "Marvelous sailing."'[29]

And so Peenemünde, on the island of Usedom along the Baltic seacoast, was selected in 1937 as the location for major R&D operations, pilot production facilities and long-range test flights. The site, providing a 250-mile test-flight range in relative seclusion, was secretly acquired. Construction began on a laboratory and other facilities to provide for tests and to build prototypes of A4 missiles, which were then a high priority of the Hitler regime.

Back at Kummersdorf, von Braun's team, now numbering about eighty men, were at work on a fully inertially guided, or self-steering, rocket called the A3. It was designed to go 15 miles up with a 100lb warhead or other payload.

That same year, the Luftwaffe conducted the first successful test flights with a liquid-fuelled rocket engine installed in a stock propeller-driven fighter plane. It was called the HE-112, and it was the rudimentary first jet aircraft. Air Force successes like that led to plans for the Peenemünde rocket centre to be a dual operation of the Air Force and the Army. Von Braun (in his mid-twenties) would serve as technical director on the Army side at Peenemünde-East, while the Air Force occupied Peenemünde-West. In the German society of the 1930s it was unheard of for one so young to hold such a high post.

In the month following von Braun's 25th birthday celebrations in March 1937, the advance group left Kummersdorf and transferred quietly to the base at Peenemünde. Its Army commander was Walter Dornberger, by now the holder of an honorary doctorate in mechanical engineering from the Friedrich-Wilhelms University of Berlin. With the move to Peenemünde, he gained promotion to the rank of colonel.

Peenemünde Priority

Wernher von Braun had his orders: enough of the demonstration rockets. No more experimental small stuff. If he and his team expected increased support from the German Army, they had to build more powerful, longer-range rockets capable of carrying sizeable warheads. It was now early 1937, and Adolf Hitler had been in control for four years. Von Braun and his crew could dream all they wanted of sending rocket ships to the stars, but the military was paying the bills and it expected weapons.

Most of the corps of scientists and engineers at Kummersdorf made the move to the new R&D base that became known as Peenemünde. Occupying the northern end of the sandy, wooded island of Usedom, about 180 miles north of Berlin, the installation would be two years in construction. It became a secret self-contained centre, complete with housing, offices, laboratories, factories, warehouses, power plants, fire stations, stores and, in wartime, a forced-labour camp with barracks for several thousand POWs and other prisoners. Here the Army team would develop the latest and largest in its series of 'A' rockets: the A5 and A4, in that order, while the Air Force would create such weapons as the winged V1 flying bomb or 'doodlebug'. For the Army team, the move marked the start of what ultimately became – following Hitler's mercurial priorities – a large facility with 10,000 workers.

But in early 1937, although war clouds may have been visible on the horizon, no open hostilities had yet erupted. The next year saw Hitler's troops march uncontested into Austria and then into Czechoslovakia's Sudetenland, dragging both nations into the fold of the Third Reich. Open warfare came the year after, with the German invasion of Poland. Even von Braun's father had warned his son as early as 1933 that war lie ahead. 'But our Führer wants peace,' the young rocket zealot had protested. The baron had stiffened, von Braun recalled years later, and responded: 'I've heard that the Berlin Art Gallery now has a huge picture of Hitler in shining armor. Men don't wear armor unless they mean to fight.'[1]

Rocket engineer Ernst Klauss, who joined the Peenemünde Army group during the early build-up in 1937, later recalled an incident that was not uncommon, perhaps, with so many new people arriving at the base. A newcomer himself, Klauss was working overtime on a new valve design one evening shortly after his arrival. He was alone in the office, and the only light came from the lamp at his table.

'Suddenly the door opened and as I looked up from my drawing board, a young man was standing next to me,' Klauss recalled. 'He introduced himself, but I did not understand his name; my mind was still on the job.' The stranger, saying he had wondered who was working late, asked where Klauss had come from, where he had worked before, if he were married and had children, and how he liked the work and living at Peenemünde. The friendly conversation lasted 15 minutes, and then the young man left.

Later that night, a puzzled Klauss decided that his visitor was probably a low-level manager in the office, as he was so young. But he was deeply impressed by the interest the fellow had shown in him and in his personal life. The next morning he asked his supervisor about the visitor and gave a description. 'I learned that the young man was Dr. Wernher von Braun, the boss,' Klauss related.[2]

The technical director's youth – and even younger looks – sometimes caused him problems. These ranged from difficulty at first in commanding the respect of strangers, to not being able to buy a drink. He and Rudolf Hermann, a young aerodynamicist who joined the team at Peenemünde in April 1937, made frequent trips to German armed forces headquarters in Berlin while the base was being set up. After a busy day of meetings, the pair, both accomplished classical musicians, usually attended a concert. They then sometimes had drinks in one of the bars nearby, where the minimum age was 21.

Hermann later related: 'One evening, the barmaid at the entrance . . . asked von Braun: "Are you twenty-one? That is the law. May I see your I.D. card?" Even with the card under her eyes, she would not believe that von Braun was twenty-five. He looked like eighteen with his blond hair, rosy cheeks, and boyish smile.'[3]

Another reality the youthful von Braun had to deal with was his reputation as a heavy spender for rocket development and prototype production. Decades afterwards, one of the German missile experts reminded him of a March 1938 episode. 'In the evening we had a happy get-together, and someone came up with a skit in which you and Colonel Dornberger appeared as defendants. During the funny one-act skit the "prosecutor" asked how it was possible that a youngster of twenty-six could expend so many millions in public funds!'[4]

Dornberger, who remained von Braun's Army boss at Peenemünde, came to know the young director's spending proclivities only too well. 'Von Braun had stressed to me again and again the importance of having our own wind tunnel for supersonic speeds, as designed by Dr. [Rudolf] Hermann. I agreed, but the cost frightened me; the estimate was 300,000 marks. I had had enough experience with building to know that there wasn't the least chance of the cost remaining at that figure, especially with von Braun about.'[5]

There were many tales of von Braun's alleged extravagances at Peenemünde. The comptroller general there, learning that parts of a certain switch on the experimental A5 rocket were gold-plated, was taken aback. 'Why gold-plated?' he wanted to know. 'Solid gold would have been too expensive,' a straight-faced von Braun replied, not bothering to explain the technical reasons. 'That settled the thing right there,' team member Helmut Horn recalled.[6]

In the pre-war days at Peenemünde, von Braun could have as much fun as anyone. One cold night in February 1938, after they'd spent nine busy months there, Rudolf Hermann and his wife threw a party for about twenty guests. It was a costumed affair, and the theme – the first German party on Mars – 'symbolized our deep involvement and enthusiasm for the future of space travel'. As Hermann later reminisced, the bachelor technical director came as the 'distinguished Professor von Braun, with white hair' and looking fully 70 years old; he had just returned from a visit to an outpost on Mars. Party-goer von Braun joined with everyone in the dancing, drinking, lively conversation and general merriment, Hermann noted. After all had left, the host found himself wondering, 'How could you, Wernher, the embodiment of optimism, be so pessimistic as to assume you would be seventy years of age when we land on Mars?'[7]

Von Braun was able to indulge his love of music – and of making music – during pre-war and wartime periods at the missile centre as well as in the city. 'He loved good music,' one of von Braun's personal secretaries at Peenemünde, Dorette Kersten, recalled. 'He played the cello then, for himself and with a chamber music group – a string quartet of engineers and scientists – at the base.'[8] They usually played in a private room at the Officers' Club, where von Braun also took lunch and dinner. Physicist Ernst Stuhlinger remembered the quartet vividly. 'His cello was accompanied by Rudolf Hermann's and Heinrich Ramm's violins, and by Gerhard Reisig's viola, when the four of them played works by Mozart, Haydn, and Schubert.'[9]

As development of the A4 missile proceeded, security at the Baltic base tightened. One day, von Braun, normally calm and unflappable, took a hurried call about a problem that had arisen elsewhere at the rocket centre. He rushed outside to make his way there – only to be stopped by a guard who would not let him pass.

'You aren't wearing your security badge,' the guard said.

'But you know me!' von Braun protested. 'It doesn't matter that I'm not wearing my badge! Now let me pass!'

The guard refused to budge.

Smiling as he reminisced about the incident years later, an eyewitness recalled: 'I thought von Braun was going to knock the guy down, he was so mad! Even afterward, though, he often didn't – or wouldn't – wear his badge.'[10]

The team developing the A4 (later the V2) in the late 1930s at Peenemünde-East called it 'The Project'. It dominated life and work there, even though its first test launch did not come until June 1942. It was a 46ft-tall, 14-ton liquid-propellant (alcohol and liquid oxygen) rocket weapon designed to carry a 1-tonne warhead over a range of about 210 miles. With an arcing ballistic trajectory that would take it deep into the upper atmosphere, approaching the fringes of outer space, it was originally intended by the German Army for attacking battlefield rear areas beyond the reach of the heaviest conventional artillery. There would be no defence against it, except to burrow deep in the ground.

In 1939, the A4 development and further construction at Peenemünde continued by virtue of remaining initial funding, but without the priority it once enjoyed. Hitler and his coterie had lost interest because of the slow pace of progress. Von Braun wrote years later that even the German invasion of Poland and 'the outbreak of war in the autumn of 1939 brought no acceleration of work at Peenemünde'.[11] And in February 1940, with the war going Hitler's way through the use of conventional armaments only, he ended support for all weapons development efforts that could not be fielded within a year.[12]

In the pre-war months of 1939, however, von Braun still had enough residual start-up funds to continue expanding his force of qualified technical personnel, although he had trouble finding them. Ernst Steinhoff, a former engineering professor who then headed the missile guidance department, promised personally to recruit 1,000 competent engineers and scientists – if von Braun would give him a free hand in hiring.

Von Braun agreed, although he seriously doubted whether Steinhoff could do it. The rocket guidance chief immediately began approaching prospective employees – men he knew in private industry, on university faculties, at research institutes and former students. Not long afterwards, when von Braun and Steinhoff ran into each other in Berlin, the latter reported excellent progress to his boss.

'Well, Steinhoff,' said von Braun, still somewhat dubious, 'let me see one of these characters on your list.'[13] Steinhoff promptly arranged a lunch meeting between von Braun and the man who, because of his talent and promise, was second on his long list: Erich Neubert, a young engineer and former pupil of Steinhoff's. Decades later, Neubert recalled being greatly impressed with von Braun. 'And naturally, my wife, whom I had brought along, was fascinated with him, as all the girls [were]!'[14] And von Braun apparently liked what he saw in Neubert. The latter soon signed on at Peenemünde and remained for decades in key positions with 'the team'.

Possibly at the suggestion of Karl Emil Becker, Army general and chief of ordnance, and certainly with the approval of this original rocket supporter, Peenemünde early in the war extended its workforce numerically and geographically.[15] With the imperatives of war helping to lift certain secrecy restrictions, this expansion was accomplished through association with academic institutions and development contracts with private industries throughout Germany. Selected university professors, working under applied-research grants, attacked specific technical problems in their own laboratories. Von Braun periodically visited them to monitor progress, and these outside researchers also visited Peenemünde for consultations. Among these academics was Professor Oberth, von Braun's early idol and mentor. In the tense late 1930s, Oberth had returned to Germany from Romania and had settled in Felixdorf.

Von Braun saw to it that Oberth, basically a theorist, was nevertheless hired as a consultant to the Peenemünde R&D operations. Because Oberth was not part of management there, it is unclear what purpose he served beyond on-site father figure, inspiration as rocket futurist – and feel-good icon for von Braun. The arrangement allowed the professor-mentor to be a frequent observer of rocket test firings and launchings at the base on the Baltic Sea.

Many years later, Oberth wrote to von Braun, 'I will forever remember in the spring of 1940 when you came to see me in Felixdorf. You took me aside, and sitting near the edge of a half-demolished bridge with our legs dangling in the air, we discussed what should happen from there on.'[16]

Up until the late 1930s, von Braun could freely share his ideas on space travel with his fellow missile developers. Years later, rocketeer Richard Lehnert reminded von Braun of 'how you explained to us what it would take to put a man on the Moon and return him to the Earth. For comparison you pointed out that the launch vehicle for a lunar spacecraft would be several times the size of the huge lighthouse on the small island Oie [near the city of Greifswald] in the Baltic Sea.'[17] The men could hardly imagine a moon rocket that was so much larger than the familiar lighthouse, which was close to where A5 and A4 rockets were readied for test flights. Von Braun could.

Another time, rocket team member Klaus Scheufelen and von Braun were sailing a small boat one summer evening towards Oie, where they planned to have dinner at an inn. The usual breeze did not favour them, and so the going was slow. But with a full moon at least they could easily spot the marker buoys.

Suddenly von Braun said, 'This is where we have to go!'

Scheufelen replied that the wind might not be strong enough to get them there.

'No, no,' said von Braun, looking up, 'not to Oie – to the Moon, and beyond. This is just a matter of cost and time.'[18]

But while von Braun, Dornberger (apparently) and many of the professionals on their rocket team privately nurtured dreams of space exploration and travel, most of the time they were fully occupied with the realities of their work for the Army – developing destructive, lethal weapons of war. Despite its best efforts, the team suffered from leaky lines, malfunctioning pumps and valves, and an equipment failure when 'countless hungry little field mice' ate the rubber insulation from the wiring on 'early and crude electronic equipment' at the Oie test launch site, colleague Fritz Mueller later recalled.[19]

And then there was the problem of the slipping 'corset'. The A4's developers invented an ingenious device that encircled the missile and suspended the main body upright. It featured a joint system that allowed pitch, yaw and roll motions during static firing tests. The inventors called it a corset because, like a woman's girdle, it kept a firm yet flexible grip. Ideally, it would allow the missile body to expand and contract amid changing temperatures.

It didn't always work, however, and one morning rocketeer Bernhard R. Tessmann had to report to von Braun that, the night before, a formerly upraised A4 had somehow slipped through the corset and fallen roughly on its tail. 'Well,' said von Braun to the young engineer, 'just be glad it didn't happen to your girlfriend!'[20]

Tessmann, chief designer of the device, recalled that various reviews and kangaroo-court sessions were held after the mishap. 'Everyone wanted to nail me to the cross,' he remembered. Only von Braun, he said, 'had the right feeling for "tolerances", and that is exactly what it was.' It turned out that a test engineer had mistakenly left super-cold liquid oxygen overnight 'in the belly of our girl, and this caused her to shrink'. The corset design was fine, the damaged tail was soon repaired and, with von Braun's support, Tessmann gained promotion and a pay rise.[21]

As the months rolled on, the A4 development team, including the top staff, continued to work long hours. Sometimes they toiled into the night – the technical director's favourite time of day. Kersten recalled that once, when facing a deadline on a certain task, he asked whether she would like to accompany him to the shops and see if the workers could stay the night. 'Their eyes were shining as he came and talked to them personally,' she recalled. 'He touched their shoulders. He made everyone feel so important. They stayed and worked all through the night.'[22]

Her boss was brimming with charm and thoughtfulness, remembered the wartime secretary, who was just 20, blonde, blue-eyed, beautiful and single in 1941 when she took the job. Late in the war she married one of the rocket experts, Rudolf Schlidt. But of von Braun she recalled: 'All the girls were attracted to him. He was young. He looked to the girls like a Greek god, the way he walked, the way he carried himself. He was a genius, and yet he was playful, sportive.' Occasionally, when she and von Braun faced an especially long workday, he would suggest they take a break and go for a relaxing bicycle ride or a brief sail aboard a borrowed boat.[23]

Otherwise, it was strictly business between the two, his former secretary emphasised. Not that von Braun couldn't switch on the charm at will. After hiring the comely Kersten to supplement the work of a matronly gatekeeper who had been with him a while, as well as another attractive young secretary, he told her with a large grin, 'I always have two secretaries – a pretty one and a prettier one.' Still, Kersten maintained, 'I was committed 100 per cent to the job, as was he.' As a single woman she was allowed to live on the base only because she frequently worked late for night owl von Braun, who often dictated correspondence well into the evening. 'But it was not so hard to do,' she remembered, 'because he was so committed – and he had such charisma. He was a kind boss. He never criticised.'[24]

At the time, von Braun had a girlfriend, whom he visited at the weekends in Berlin, usually getting there by piloting the speedy, sporty plane with which the Army provided him. A frequent passenger was his brother Magnus, a chemist and chemical engineer, who in 1943–4 worked as Wernher's executive assistant. Kersten recalled that her boss 'came back to work in a happy mood on Monday mornings, and everyone on the staff was happy for him'.[25]

At Peenemünde, the plane he had at his disposal was a blue Messerschmitt Typhoon, a four-seater civilian version of the Me109 fighter. During the war he made frequent use of his Messerschmitt for business trips to universities, industries and other military sites, as well as for his own personal pleasure. In addition, he used it in a few aeronautical engineering experiments of his own.

Von Braun the aviator indulged a penchant for calculated risk-taking, to the dismay of his white-knuckled passengers. He especially liked to fly in and through bad weather. His long-time chief deputy technical director, Eberhard Rees, later spoke of having flown – often reluctantly – with von Braun during those years in Germany. As another colleague revealed years later, 'Eberhard said he didn't like to fly with him because if there was a cloud in the sky, he would head right for that, because it was more of a "challenge" in flying.'[26]

In rocket development as in inclement weather conditions for flying, von Braun believed there was no substitute for first-hand observation. A colleague at Peenemünde discovered that truth the painful way. Occupied with the task of calculating the flight trajectories for the A4s soon to be test-launched, computations expert Helmut Hoelzer mentioned his heavy workload to von Braun one day and said he did not fully understand the programme objectives. The technical director immediately took Hoelzer to one of the A4 static (or hold-down) test-firing locations. The two men approached a high wooden fence surrounding the site, and von Braun found a knothole in one of the boards.

'You look through here,' he told Hoelzer, 'and in a minute you will see exactly what we are doing.' Hoelzer pressed his face against the board and saw an A4 on the test stand. The engine suddenly ignited with a roar. The sound waves ripped loose the board in front of Hoelzer and it struck him full in the face, knocking him to the ground. 'I picked myself up, bleeding at the nose,' he recalled. 'It was an indelible introduction to the practical aspects of the program.'[27]

The first two attempts to test-fly the A4 from Peenemünde were dismal failures. After the worst of the misfires, a spectacular launch-pad explosion, the only thing the usually unflappable von Braun said was, 'Well, boys, back to the drawing-board.' Was he remembering the lesson in persistence he learned from filing the cube in the locomotive factory? In any event, he became directly involved in the troubleshooting to find and fix the causes of the failure.[28]

On the third try – 3 October 1942 – the 'Cucumber', as some called the missile, roared smoothly from its pad to a height of more than 50 miles, approaching the region of airless space. The men of Peenemünde danced and wept in their happiness. Von Braun later said he couldn't recall whether he wept, but he did remember celebrating that evening with a few drinks at a party. 'I don't know about my getting wet outside,' he joked, 'but I got very wet inside.'[29]

That day Col Dornberger made his oft-quoted comment: 'Do you realize what we accomplished today? This afternoon the spaceship has been born!' He continued, however, with a less-well-known remark: 'But I warn you that our headaches are by no means over – they are just beginning!' He went on to stress that, for now, the team must focus on perfecting the rocket as a weapon.[30] Space travel and exploration would have to wait. Still, the A4 successes continued to fire the imaginations of space-minded members of the team. As test engineer Karl Heimburg remembered: 'When the V2s [A4s] worked, we knew there was no limit to what could be done. We knew then what von Braun meant when he talked about "reaching for the stars".'[31]

But by 1942–3 Nazi Germany was locked in an all-out struggle for survival, and free talk about space travel was no longer safe. The Peenemünde-East accelerometer development laboratory where Stuhlinger worked received a visit one day in 1943 from von Braun, who was checking on how its work was going. As the technical director was leaving, a member of the lab asked him if he truly believed that someday rockets would put artificial moons in Earth orbit, as Oberth and others had predicted.

'Absolutely,' von Braun assured him. He elaborated on how the Peenemünde team, assuming it remained intact after the war, might have the opportunity in peacetime to send up space satellites, then rocket humans to the moon, and later to Mars. But then he abruptly cut short any talk in that vein, warning it was 'dangerous' and 'not permitted'.[32] He quickly left the lab.[33]

A4 test shoots at Peenemünde often brought excited workers up to the rooftops of their buildings as the countdowns got under way so they would have a better view. This was not always prudent, considering the unpredictable flight paths of the missiles. Von Braun himself had a few close calls. During one such shoot he and engineer Walter Jacobi were positioned in a nearby observation trench, discussing the chances of the rocket hitting the target area. The countdown proceeded smoothly, as did the lift-off and the initial ascent.

'Then all of a sudden,' Jacobi recalled, there was 'an explosion at very low altitude. We watched the missile disintegrate, but not [until] the tail-ring with the rudder-motors hit the ground between us and the Officers' Mess did we realize the danger we were in.' Observed von Braun, 'I'd have been safer in the target area!' Then, thinking positively, he said, 'Now all we need to do is shift this impact point 180 kilometers [112 miles] farther north, and our problems will be solved!'[34]

Von Braun and Dornberger did, in fact, once station themselves dead centre of the target impact area in an A4 flight test with a live warhead. The idea was to gain a good vantage point for checking on technical problems occurring at the very end of the missile's trajectory. The way the test shots had been going lately, Dornberger said, with tongue at least partly in cheek, it 'would certainly be the safest spot'. Von Braun was standing in an open field when he spied the incoming missile's thin contrail. Then he saw, to 'my horror', that the deadly rocket was headed towards him! 'There was barely time to fall on the ground before I was hurled high into the air by a thunderous explosion, to land unhurt in a neighbouring ditch. The impact had taken place a scant 300ft away and it was a miracle that the exploding warhead did not grind me to powder.'[35]

Markedly better – and worse – days lay ahead for von Braun, Dornberger and the team at Peenemünde. Over the five years from 1937 to 1942, from peacetime to wartime, and with inconsistent support, they had carried out orders to bring the A4 ballistic missile into being. Now they had to make it reliable and move it into mass production. There were enemies against whom it must be used in surprise, bloody attacks.

Encounters with Hitler

Although progress in Army rocket development continued at Peenemünde, the support given it after the initial burst of funding waned. Hitler and other political leaders could not sustain a vision of a successful von Braun and his rockets. As early as 1939, the lengthy gestation of the A4/V2 caused Hitler to lose confidence that the project would produce anything of military usefulness. As long as the Luftwaffe remained strong, he saw that force as his best bet. Fortunately for von Braun, a few other leaders stood by him. With the pace of rocket work slowed to a crawl in the early 1940s, FM Walther von Brauchitsch stepped in. The Army's supreme commander, believing in Walter Dornberger, von Braun and their rockets, assigned 3,500 officers and enlisted men to Peenemünde under the pretext of needed 'training'. In fact, they were technicians, engineers and scientists who bolstered the sagging ballistic missile development effort.[1]

According to von Braun, most of those in his rocket group in 1932, when it signed on with the German Army's Ordnance Department, thought 'Hitler was still only a pompous fool with a Charlie Chaplin moustache'.[2] The next year, amid the misery of the economic depression and fears of a Communist take-over, the pompous fool with the moustache became chancellor of Germany. Von Braun's father, Weimar cabinet minister Baron Magnus, had warned his son privately when Hitler took office that the country would be led to disaster, including war.

Wernher von Braun later acknowledged that his feelings of patriotism were stirred by Hitler's early initiatives to free Germany from the effects of its post-First World War humiliation imposed by the Treaty of Versailles and by the ruinous inflation and economic depression.[3] After the war, when he was living and working in the United States, von Braun shared his impressions of the dictator in interviews and writings.

> I met Hitler four times. When we met in 1934, he seemed a pretty dowdy type. Later, I began to see the shape of the man – his brilliance, the tremendous force of personality. It gripped you somehow. . . . My first impression was that here was another Napoleon, another colossal figure who had upset the world. . . . At my last meetings with him, Hitler suddenly struck me as an unreligious man, a man who did not feel that he was answerable to anyone, that there was no God for him. . . . He was wholly without scruples, a godless man who thought himself the only god, the only authority he needed.[4]
>
> If Germany had won the war, der Führer would probably have lost interest in rockets. His enthusiasm would have shifted to a huge reconstruction project in the Ukraine or some such. I just know it.[5]

Hitler's first visit to a rocket R&D facility had been on 23 March 1939 – von Braun's 27th birthday – at Kummersdorf, 17 miles outside Berlin. It was planned so he could see the activities that had prompted the build-up at Peenemünde. Col Dornberger and von Braun could not guess what Hitler's reaction would be. Dornberger urged von Braun not to utter a word about space travel but to talk only of weaponry.[6]

When Hitler arrived with Gen Karl Emil Becker and others, von Braun joined the visitors at a test stand where they observed – from close range – two loud, static firings of rocket engines of different thrusts. The latter was the A5, a 21ft-tall test bed for the coming A4. A visibly unimpressed Hitler said nothing. Dornberger briefed him on the work at Kummersdorf and Peenemünde. Still Hitler said nothing. He asked questions only when shown a cut-away model of the A3 intermediate test rocket. Von Braun and Dornberger realised from Hitler's questions that the former infantry corporal did not understand the basics of rocketry, so Dornberger started the briefing again with a primer on propulsion. Hitler ultimately showed interest in plans for the larger, more powerful A4, asking how long before it would be ready for use.

At lunch it was Hitler himself who brought up the subject of space. He mentioned an early German rocketeer, Max Valier, and called him a 'dreamer'. Heeding Dornberger's advice with difficulty, von Braun for once kept his mouth shut on the subject.[7]

Von Braun returned to Peenemünde straight from the briefing and progress report he presented to Hitler and the top brass. Grasping at the few positive signs the Führer had shown, von Braun excitedly told members of his team that all had gone well and that Hitler had seemed interested and pleased with their work. There is a story that a colleague asked von Braun how he, being so young, had managed to remain calm and collected when confronting such a formidable audience. Did he perhaps have a drink beforehand to steady his nerves?

'No,' the rocket leader replied, 'I just used my imagination.'

'How do you mean?' someone asked.

'Well,' said von Braun, 'you simply imagine that all those guys sitting out there in front of you are dressed only in their underwear – and then you lose all your awe of them.'[8]

As it turned out, Hitler had not been much impressed with von Braun and Dornberger's vision of rocketry's future. He continued to rely on the firepower of his Luftwaffe bombers and fighter planes, along with his armoured and infantry ground forces. In 1939–40 he assigned a low priority to the work on rockets. It was mainly through the support of FM von Brauchitsch and other high-ranking Third Reich officials, chiefly Minister of Armaments and War Production Albert Speer, who did believe – who had caught their vision, and who foresaw bringing the operation under his control – that major resources continued to flow to Peenemünde.[9]

Rocket experiments with a nautical twist occupied at least one key member of the Peenemünde team briefly in about 1942. Scientist-engineer Ernst Steinhoff, von Braun's head of missile guidance and control R&D, was deeply involved with the firing of solid propellant rockets from a submarine in North Africa. The U-boat commander was Steinhoff's brother. The experiments were not pursued further, although 1944–5 would see an intriguing Peenemünde project to use German subs as towboats to extend the A4's field of action by hundreds and even thousands of miles.[10]

Early in the summer of 1942, before the successful test firing of the A4 'Cucumber', the once-mighty Luftwaffe was faltering as an offensive force. Dornberger and von Braun visited Hitler at his headquarters in hopes of persuading him to put his money on their 46ft-tall missile, with its 50,000lb of thrust and range of 200 miles. 'No luck,' von Braun later related. 'The next day, we received word that der Führer had dreamed during the night that our rockets would not work.'[11]

After the successful October 1942 first flight of the A4, followed by a few more good test shots mixed in with further failures, Dornberger and von Braun went back to see Hitler in July 1943. 'Der Führer looked much older, and he was wearing his first pair of glasses,' von Braun recalled. 'But when we described our accomplishments to him, his face lit up with enthusiasm. He revoked his dream.'[12]

With German air superiority now lost, and Germany's defeat in the Battle of Britain, the A4 missile was back at the top of the Third Reich's military priority list. The backing that Hitler had earlier withdrawn for several critical years from Dornberger and von Braun meant their big, revolutionary ballistic missile would not see action until months after the Allies' D-Day invasion. It would be too late to alter the course of the war.[13]

To regain Hitler's blessing for all-out development of the A4, von Braun and his team had used a piece of cinematic trickery. At the July 1943 meeting Dornberger and von Braun showed a new film of the still-developmental missile. Pictured were the first successful flight, another good test launch or two and rows of the rockets in Peenemünde hangars. Joseph C. Moquin, former US Army missile agency management whiz and board chairman of Teledyne Brown Engineering Company, recalled hearing more than one Peenemünder relate that, for the filming, a number of film camera crews had been stationed at different points around the launch site and along the expected flight path. They succeeded in getting multiple reels of good exposures of one or more of the rare successful flights. All the best footage was spliced together. The resultant film shown to Hitler and his staff pictured the A4 in perfect lift-off and glorious flight, in dramatic view after view, from every conceivable angle and position. It gave the stirring illusion of a virtual squadron of missiles flawlessly zooming off towards the enemy.[14]

The A4 progress at Peenemünde brought a stream of VIPs to see for themselves the fire, thunder and smoke at the secret base on the

Baltic. Although Hitler never visited the base, those who did included Heinrich Himmler, Hermann Göring, Josef Goebbels, Albert Speer, Grand Adm Karl Dönitz and FM Erwin 'Desert Fox' Rommel.

At Peenemünde and elsewhere, von Braun rubbed elbows with many of the highest-ranking officials in the Hitler regime. For years afterwards he retained vivid memories of them. As told to writers Drew Pearson and Jack Anderson, here are his impressions of three top Nazi leaders in whose company he found himself at times – Hermann Göring, Luftwaffe chief and First World War fighter pilot hero; Josef Goebbels, propaganda master; and Heinrich Himmler, dreaded ruler of the SS and Gestapo.

[Göring] had a kind of charm. In his way, he could even be said to have been likeable. With his perfume and his gaudy uniforms, I thought of him as a Renaissance prince who had been born into the wrong time. He was a re-incarnated Borgia, lover of food and drink, patron of the arts – with a little poison in somebody's wine at midnight, just for the fun of it.

[Goebbels] had a diabolic intelligence and an appreciation of his own power. I remember his boasting once that 'it is three days' work for me to change this nation's mind'. [Himmler] reminded me of Robespierre – a half-educated fanatic who actually believed that the extermination of the Jews was the way of truth. I'm sure he died without scruples about his crimes.[15]

For several years after Hitler came to power, von Braun had still not joined the Nazi Party. The Peenemünde base, if not the A4, was in advanced development. Von Braun was in a high position, and the dictatorship looked unfavourably on officials who were not loyal, card-carrying members of the party. Pressure grew for von Braun to sign up with the National Socialists. Earlier, he had tacitly resisted doing so. Von Braun associates cite German documents showing he became a party member on 1 December 1938. Others have pointed to a US Army statement and other papers indicating he did so on 15 May 1937. Von Braun did not address the discrepancy or cite a sign-up date in any of his known writings, and his official US Army and NASA biographies do not even mention his party membership, let alone give a date. In any event, he had been 'commanded' to join, as military historian and von Braun critic Michael J. Neufeld of the Smithsonian's National Air and Space Museum later put it.[16]

In 1940, as the war enveloped Europe, von Braun was offered a commission as a second lieutenant in the Schutzstaffel, or SS (National Socialists/Blackshirts), the military arm of the Nazi Party. It came, unsolicited, from Reichsführer Himmler through an SS colonel based in Stettin, not far from Peenemünde. He tried to dodge the offer by saying, 'I was so busy with my rocket work that I had no time to spare for any political activity.' But the colonel persisted, insisting that von Braun's 'being in the SS would cost [him] no time at all' and that it was Himmler's strong desire that he accept.[17]

The offer threw von Braun into a dilemma evidently more prickly than his deliberations on joining the Nazi Party. He turned to a small, trusted group at Peenemünde for advice. Master engineer Gerhard Reisig recalled: 'Von Braun asked us, basically, "What do I do?" In other words, "Do I accept it, or do I decline the honor?"' Uncertain, the engineers recommended he consult his military superior, Col Dornberger, of the regular German Army. Dornberger advised von Braun to accept the SS commission; rejecting it would be taken as a dangerous show of disloyalty and a personal insult to the powerful, merciless Himmler.[18]

Thirty years later, von Braun told American associate Charles Hewitt that he 'had to make a very difficult choice' in the SS commission matter. 'And he said he felt his choice was to live or not to live, and he decided to live,' recalled Hewitt, the mid-1970s executive director of the National Space Institute.[19] Less melodramatically, others said the apolitical von Braun reasoned that an SS officer's status would help him deal more authoritatively with the storm troopers present at Peenemünde in growing numbers over the years.[20] Of paramount importance to von Braun at this juncture, many have asserted, was the continuation of his team's advances in rocketry.

After he was pressed a second and third time by letter from the SS colonel in Stettin to accept the commission, von Braun consented and made the required application to the party's political army. Two weeks later a letter informed him that he had been accepted and appointed an *Untersturmführer* (lieutenant). 'From then on I received a written promotion every year,' von Braun wrote in a June 1947 affidavit. 'At war's end I had the rank of *Sturmbannführer* [major].

But nobody ever requested me to report to anyone or to do anything within the SS.' The latter assertion does not, however, jibe with later revelations that he did have duties, at least to attend or conduct periodic meetings of an SS unit at Peenemünde.

One is left to speculate whether von Braun was simply dissembling or perhaps considered his SS duties – beginning when he was a second lieutenant – too insignificant to mention. In answer to my questions, former Peenemünder Ernst Stuhlinger dismissed von Braun's SS activities at the Baltic Sea base as inconsequential – related, for example, to familiarising aircraft spotters at the base with silhouettes of Allied warplanes.

Von Braun is said not to have worn his swastika lapel pin, symbol of Nazi Party membership. In signing up he agreed to pay monthly dues of 2 Reichsmarks, which Kersten related he sometimes neglected to do. It had been thought that he publicly wore his SS uniform, with swastika armband, just once, during one of two formal visits by Himmler to Peenemünde – and then only after a mad search for it through closets.[21]

In 2002 historian Michael Neufeld reported that Ernst Kütbach, a Peenemünde worker and member of the local SS unit, had recently told the BBC that von Braun attended about half of the monthly meetings. Kütbach said that 'he would come in uniform; we also had to come in uniform. . . . And when the Doctor arrived the platoon leader would announce [the unit was] "reporting for duty". And then he would sit down and listen to [a report on] tactics and strategy.'[22]

'Apparently Wernher von Braun only wore the uniform when he had to,' Neufeld observed. 'Some colleagues were surprised to learn he was a member.' Neufeld also cited a film interview in which ex-Peenemünde rocket engineer Hartmut Küchen 'relates his astonishment when he saw the Technical Director in black'. The engineer's jaw dropped; von Braun told him to close his mouth, saying there was no way around it.[23]

Space historian Fred Ordway, a close von Braun associate from the 1950s onwards, later characterised von Braun's predicament at Peenemünde: 'Look, von Braun had to sign his letters "Heil Hitler!" He had to become a Nazi Party member because of his position. He had to accept the SS officer's commission that Himmler offered.' 'Or else?' I asked. 'Or else,' Ordway agreed.[24] Neufeld, an energetic critic of von Braun's wartime conduct, tends not to consider the dictatorship's stated wishes as tantamount to commands, and thus he does not go as far as Ordway. Nevertheless, Neufeld has acknowledged that von Braun's decision to join the Nazi Party and to accept the SS commission occurred 'under pressure'.[25] He has also said of the rocket scientist's belated sign-up with the Nazis: 'I believe he did it reluctantly', for the sake of 'convenience'.[26]

Before one of von Braun and Himmler's meetings with Hitler, Himmler had simply told von Braun, 'We are flying to Berlin,' 'Of course, in those days,' von Braun told a friend decades later, 'you didn't know if you were flying to Berlin to be killed or given a medal.'[27] (Von Braun did, eventually, receive the prestigious honorary title 'Research Professor' by Hitler's order in July 1943 for leadership in the A4/V2 development, and the War Service Cross in November 1944 for his contribution to the war effort.) His Nazi affiliations, no matter how nominal, were 'a bitter pill to swallow', von Braun later said.[28]

Roughly 80 per cent of the German rocket men who later came to the United States had joined the Nazi Party or related groups ranging from the Hitler Youth to Nazi equestrian and flying clubs, according to Maj Joseph Sestito, US Army, their security officer at Fort Bliss, Texas, and White Sands, New Mexico, in the late 1940s and at Huntsville's Redstone Arsenal in the early 1950s. 'I'm fairly sure that these men became members more or less as a matter of expediency, rather than ideology,' Sestito told Daniel Lang at the *New Yorker* in 1951. 'I believe they joined Nazi organizations primarily to hold on to their jobs. Their work is the driving force in their lives, not just a way of making a living.'[29]

In the same vein, armaments minister Albert Speer wrote in his first postwar book of enjoying 'mingling with this circle of nonpolitical young scientists and inventors headed by Wernher von Braun' during his visits to Peenemünde.[30] Similarly, rocket engineer Dieter Huzel, a conscripted German infantryman rescued from the Russian front by his mid-1943 reassignment to Peenemünde, later wrote of finding a markedly apolitical atmosphere at the missile centre when he reported for duty. Huzel observed 'the nearly complete absence of Nazi party uniforms, party lapel buttons, and party activities in general'.[31] Neufeld later acknowledged that while von Braun and other top workers at Peenemünde appeared to embrace Germany's rise under Hitler for patriotic reasons, 'none shows any sign of having been a Nazi ideologue'.[32]

Up to midsummer 1943, when Hitler decided to give the A4 project the highest national war priority, Peenemünde had escaped Allied bombing. Prototype production there of A4s had been stepped up, with forced labour in the form of both POWs and civilian detainees. Use of foreign labour at Peenemünde had been discouraged from 1939 into the early 1940s for reasons of preserving secrecy. But that factor was tossed aside as the war went against Germany and the need for labour grew critical. Against the Geneva Conventions, sizeable numbers of POWs – mostly Russian, but some Polish – were brought in and put under the strict control of the SS. By 30 April 1943, a total of more than 3,000 foreign forced labourers of all kinds were working in construction and other manual labour, as well as in skilled production at Peenemünde-East (Army) and Peenemünde-West (Air Force).[33] Von Braun evidently was not involved in the decision to use such labour for V2 production there, according to Neufeld's research.[34]

The cover story for the secret rocket centre made it out to be an experimental aircraft base. The ploy worked for years. But then film from an RAF reconnaissance plane's overflight of the island revealed upright objects looking remarkably like large rockets. British military intelligence figured it out: London was the weapons' probable future target. It took a while, but on the night of 17 August 1943, all hell broke loose on the Baltic coastal island.

A force of almost 600 RAF Lancaster, Halifax and Stirling heavy bombers, plus a hive of fighter aircraft, crossed the English Channel that night on a mission code-named 'Hydra' and struck the missile centre. It was a deadly, nearly one-hour attack intended to destroy both facilities and personnel, especially as many of the hundreds of scientists and engineers as possible.

The bombs and resulting inferno caused extensive destruction to facilities, although many buildings escaped damage. The 735 people killed – about half of them POWs and civilian forced labourers working on road-building, construction and pilot missile production – included just one key member of von Braun's team. He was Walter Thiel, chief of rocket engine design and development. Among the other team casualties were engineers, technicians and family members.

A chief instigator of the RAF raid who had pushed early and hard for it was a young intelligence officer named Duncan Sandys. His father-in-law was Prime Minister Winston Churchill.

When the British bombs and incendiaries shattered that August night, Kersten raced with others to the safety of the large underground bunker in front of the two-storey building that housed her boss's offices. The attack lasted for a seeming eternity. After the all-clear signal sounded, 'It was fire, fire, fire everywhere,' she recalled. 'Dr von Braun shouted, "Every man out, and save the documents!" He and the men saved armfuls of drawings and other secret documents.'[35]

Von Braun led his secretary by the hand into a burning building where she knew the location of, and had the key to, a cabinet safe containing vital secret papers. The two picked their way up the stairs to the safe, 'with flames all around, one wall gone completely, and the rest burning'. They found the safe overturned and blown half-open, but with its contents still intact. 'He was very courageous, leading the way and leading the other men,' remembered Kersten, herself a decorated hero for helping to salvage the irreplaceable rocket documents.[36]

The leadership displayed by von Braun in the aftermath of the devastating raid especially impressed his mentor Hermann Oberth, who later cited his protégé's 'drive for work, which is almost unbelievable for an average person'. With the heavy losses from the attack – personnel, production facilities, machinery, vehicles and roadways – makeshift transportation arrangements had to be organised, debris cleared, repairs made. 'Everyone was overtired,' Oberth recalled. 'Nevertheless, after fourteen days, operations were back to normal. Von Braun slept no more than three or four hours [a night], but he did not show it. As always, he was calm, friendly, kind, and fair to everyone.'[37]

The RAF raid, it turned out, advanced the empire-building schemes of SS and Gestapo boss Heinrich Himmler. He had shown covetous interest in Peenemünde and the A4 project well before it gained Hitler's full blessing in mid-1943, and Albert Speer had put the missile into full production at four planned, above-ground assembly plants scattered around Germany and Austria. Himmler had courted von Braun since 1939. He had visited Peenemünde twice to see test flights and to check out the place and make his presence felt. Himmler, who von Braun said lacked any technical understanding of the project, wanted control of the whole operation in order to expand his power, according to von Braun and others. 'When the military, the Nazi bosses – Himmler, Göring, and the others – came to watch the launches, Dr. von Braun hated it,' Kersten remembered.[38]

The mid-August raid on Peenemünde was followed by separate air strikes for a two-week period in September against the A4 assembly plants that were being built at the scattered surface sites. Because of the air strikes Himmler was victorious in the power struggle over control of the missile programme. Hitler ordered all A4 production to be moved literally underground. He gave the job to Himmler and to his construction chief, SS Gen Hans Kammler, who built and ran the Third Reich's concentration camps, including those used to house the workers for A4 production. Kammler used forced labour to blast and expand existing mining tunnels at the southern edge of the Harz Mountains, about 250 miles south-west of Peenemünde, into the largest subterranean factory in Germany. Located near the ancient free city of Nordhausen in the state of Thuringia, it was called Mittelwerk (Central Plant), and it was used for production of submarines, aircraft and other armaments, as well as missiles.

One of von Braun's bleakest periods, he lamented, 'began in the fall of 1943 when Himmler and his SS men wrenched control over the A-4 program out of our hands in order to enforce mass production and military deployment of the rocket long before its development and testing were completed'.[39]

But there was a war on, Germany was losing, and all-out production was the order in effect. Promoted from colonel to major-general in July 1943, Dornberger, for the time being, had managed to hold on to control of the A4's continuing developmental operations – and to von Braun as technical director of the programme. In February 1944, Himmler summoned von Braun to his headquarters in Hochwald, East Prussia. It was symptomatic of the treacherous times that the rocket expert headed to the meeting wondering, once again, if he was to receive a medal around the neck or a bullet in the head. Von Braun entered Himmler's office, he recalled years later, 'with considerable trepidation. It must be said, though, that he was really as mild-mannered a villain as ever cut a throat, for he was quite polite and rather resembled a country school teacher.'[40]

Himmler wanted von Braun to join his staff and to assist in pushing aside Dornberger and the regular German Army Ordnance Department at Peenemünde. Then SS forces could take total control of the A4 programme, including further development, testing and troop-training. 'I can do a lot more for you, Wernher, than those stuffy Army generals,' Himmler told him, by von Braun's account. The Reichsführer added that he could eliminate red tape, pour his SS personnel into Peenemünde and propel the operation ahead. Von Braun politely declined the offer, showing either great courage or foolhardiness. 'He tried to get me to go along with his idea of speeding up [A4] production on an assembly-line basis,' von Braun recalled. 'I told him that you couldn't push too fast on some things. I remembered that he was interested in horticulture, and I said that it was all right to put a plant into the ground and carefully nurture it, but that too much manure could kill it.'[41]

The earthy metaphor was not warmly received. 'Himmler smiled weakly,' recalled the missile leader. 'I could see that he was miffed, but I thought little of it – until I was arrested three weeks later by the Gestapo.'[42]

Himmler's Gestapo (an acronym for *Geheime Staats-Polizei*, or Secret State Police) had a file on von Braun. They had files on all people of any importance. That, as von Braun noted in a postwar speech, is where 'the heavy hand of dictatorship' was most keenly felt in any scientific or technical programme in Nazi Germany. He elaborated on this in 1958 in a talk given at a US military-industrial conference:

In Peenemünde, the security police kept dossiers on all of us, listing all the things we might have said about the regime or individuals of the upper hierarchy. Personal vices and weaknesses were cataloged in the files. But they left us alone so long as our usefulness, in their opinion, was greater than our debit account. Once they felt they could do without you, and you were in their way, they'd call for the dossier and destroy you. It was that simple.

I realize that this sounds quite awful to men who have never experienced it. But the sober fact is that people, whether scientists or candlemakers, learn to live with such a situation. . . . The man living under dictatorship adjusts himself to business-as-usual, whether he likes it or not, because he must, in order to survive.[43]

Von Braun's rebuff of Himmler and his 'unshakeable loyalty to [Gen] Dornberger' ('this steadfastness . . . to his immediate superior') were held up by his postwar defenders as proof of his 'moral integrity'.[44] Himmler did not see it that way. In the middle of the night of 15 March 1944, after Dornberger was lured from the Peenemünde base, Gestapo agents arrested von Braun and two prominent members of his missile team, Klaus Riedel and Helmut Gröttrup. The men were taken in darkness from their quarters to a Gestapo prison in nearby Stettin. It was a long time before Himmler's primary charges emerged. As von Braun recalled in an autobiographical account, 'I was accused of sabotaging the war effort by having my heart set on space rather than destroying London.'[45] This amounted to high treason, Himmler alleged.

In von Braun's case, the specific evidence used in what he termed Himmler's 'power play' to get him to help wrest control of the rocket programme from the Army was twofold. There was a supposedly documented remark, overheard in a train compartment, that he was interested in the A4 more as a space vehicle than as a weapon. In addition, the Gestapo accused him of making similar remarks at a party one evening in Zinnowitz, the town nearest to Peenemünde. A woman dentist – an SS spy – reported the remarks.

Himmler also cited the fact that Peenemünde's technical director always kept a fast aeroplane at his disposal at the seaside rocket centre.[46] With that aircraft 'I could be anywhere in Central Europe within two hours,' von Braun later wrote. Himmler 'accused me of keeping the plane gassed up so I could fly off to England with the [A4] secrets, which was pretty absurd'.[47] (The charge had a measure of credibility because Rudolf Hess, the third-ranking Nazi and a close friend of Hitler's, had flown to Great Britain in 1941 on a bizarre, one-man peace mission. He had been swiftly taken into custody by the authorities, interrogated and jailed for the duration of the war.) 'Well, such an accusation, trumped up as it was, was no joking matter,' added von Braun. 'It could have very easily led me to the firing squad.'[48]

Some of the evidence against von Braun was accurate. As his days in jail wore on, he 'began to learn at first hand about the realities of any totalitarian regime', he later commented. 'The men in power – and they often hold power because of machines developed by us engineers and scientists – never know when a friend may become an enemy. So their police spies build up a file on everyone, ready to be used whenever it appears expedient. They had a very complete one on me, which included some political remarks I had made years before.'[49]

Clearly, von Braun was in 'a very dangerous situation', in historian Neufeld's view.[50] His court of inquiry got under way within two weeks of his arrest and detention in Stettin prison. But this time the Army won the power play. Dornberger, who had just been promoted to major-general, dramatically interrupted the proceedings. He and Berlin's war production chief, the powerful Albert Speer, arch-enemy of Himmler's, had secured a signed order from Hitler for von Braun's release on probation. Personally intervening on von Braun's behalf, Dornberger and Speer had convinced the dictator that the A4 programme would falter without the young rocket scientist's leadership. Himmler's plot was thwarted for the time being.

As the massive war bled on, von Braun's two successive ninety-day probations were forgotten. Eventually, though, Himmler got his revenge – and his way. The attempt on Hitler's life on 20 July 1944 by a group of regular German Army officers resulted in the full SS takeover of Peenemünde. Maj-Gen Dornberger, as an officer in the regular Army, was pushed aside as overall commander of V2 development and allowed to supervise only the equipping and training of troops to fire the missiles.[51]

Years later, von Braun recalled that by the time of his release from the Stettin lock-up he had almost begun to take a perverse enjoyment in prison life. Although alarmed over his arrest, he had been allowed visitors on his 32nd birthday; he had not been starved, tortured or otherwise abused. 'I had plenty of time to think,' he added, 'and it was so quiet there.'[52] He also remembered that, fatigued by his wartime labours at Peenemünde, he was consumed by a single thought when first entering his jail cell: what a wonderful chance to sleep![53]

Comes Now the V2

Concurrent with his order in June 1943 to rush the A4 into mass production – he wanted 30,000 built, at a rate of nearly 1,000 a month – Hitler also decreed that missile attacks on London would begin that October. It did not happen. Von Braun and Walter Dornberger knew it would not happen. Developmentally, the weapon was immature. In test shots that autumn, six out of every ten missiles broke apart in the final phase of flight or otherwise failed.[1] Months of work on technical improvements to debug the weapon ensued.

Hitler had laid down another requirement. 'He wanted Peenemünde to test and train for the firing of three V2s all at once at England, Dornberger once told me,' veteran US aerospace engineer-manager David L. Christensen commented. 'He said they tried but never could get it to work.' And so Hitler's vision of waves of three-missile salvos bombarding London was not to be.[2]

The time for operational use of the missile arrived almost a year late by Hitler's timetable. Three months after the Allies' D-Day invasion of Normandy on 6 June 1944, and two months after the German senior officers' failed assassination of Hitler, the first missiles were fired at enemy targets. A mobile V2 battery in western Germany, which on 6 September had seen two attempted launchings against Paris fizzle on firing platforms, succeeded two days later in firing one missile at Paris; it struck the target area and did modest damage. Later the same day, German troops at a field site just northeast of The Hague in Holland successfully fired two rounds at London's East End.[3]

The day marked one of von Braun's blackest times of the war, he later related. 'We wanted our rockets to travel to the Moon and Mars, but not to hit our own planet.'[4] Well, yes and no. The fact that Paris was the first target may have darkened von Braun's mood, but he certainly experienced other feelings on that day too. The *Manchester Guardian*, years later, quoted him recalling that he felt little remorse when the missile attack on England began: 'I felt satisfaction. I visited London twice [between the wars], and I love the place. But I loved Berlin, and the British were bombing [the] hell out of it.'[5]

Von Braun had stronger personal and family connections with London and England than almost anyone realised. As he later wrote to contacts there, 'My mother was brought up in England, my father worked for quite a while in the London City Administration, and I myself spent some of the most unforgettable days of my life in Old England.'[6]

Within ten days of the firing of the first rounds, twenty-six more A4s struck London and the vicinity. The revolutionary missile weapons were by then in full production at Mittelwerk, the brutal, underground armaments factory that used thousands of forced labourers from concentration camps, along with its staff of skilled, paid workers. Von Braun checked on technical reliability issues – and occasional labour matters – at the plant during brief, fly-in visits from his Peenemünde home base, some 250 miles to the north-east.

When the A4 attacks began, Goebbels wasted little time in informing the German people that the Third Reich's second 'wonder weapon' – the first being the Luftwaffe's winged V1 flying bomb or 'doodlebug' – was now bombarding the enemy. He announced that this new secret weapon had been renamed the V2 (for *Vergeltungswaffe Zwei*, meaning Vengeance – or Retaliation – Weapon Two). The revenge was for the Allies' saturation bombing of the major cities of the Fatherland, including Hamburg, Cologne, Frankfurt, Nuremburg, Dresden, Darmstadt, Bremen, Hanover, Bonn, Stuttgart, Düsseldorf, Munich and Berlin. Von Braun said he learned of the new name for his rocket from a Goebbels radio broadcast.[7]

The missile developer acknowledged after the war that he and his team did indeed have mixed feelings when V2s began raining down on London. 'The Allies had bombed us several times at Peenemünde [including three US raids after the first RAF strike], but we felt a genuine regret that our missile, born of idealism, like the airplane, had joined in the business of killing.'[8] He acknowledged, however, that he would have been less than human had he not felt 'glad to be getting back at them [the British]'. These comments years later from a fellow Peenemünde team member reveal just how glad: 'Don't kid yourself: Although von Braun might have had space dust in his eyes since childhood, most of us were pretty sore about the heavy Allied bombing of Germany – the loss of German civilians, mothers, fathers, [other] relatives. When the first V2 hit London, we had champagne. Why not? Let's be honest about it. We were at war, and although we weren't Nazis, we still had a Fatherland to fight for.'[9]

An Englishman's memories of the first V2 to hit London were evoked almost thirty years later in a letter to von Braun. A.V. 'Val' Cleaver, who became a postwar friend, was one of the leading rocket engineers in Britain from the 1950s on. He told von Braun that in England, in advance of the missile bombardment, he had 'tried to help convince our authorities that the intelligence reports about your V2 from Peenemünde might be true'. Cleaver told von Braun what happened next.

> One evening in 1944, a double bang [sonic boom] over London (with no prior air-raid warning) told me I had been right – and incidentally did no good at all to the Chrysler works in the western suburbs. For some weeks, this news was still 'secret', so when I visited New York soon afterwards, dear old Willy Ley assured me he still disbelieved it. I felt unable to put him right, but did gently suggest he might be wrong. He then said that, if the rumours were true, 'a young man called von Braun' might be responsible.[10]

Despite von Braun's doctorate in physics, the consensus was that when it came to rocket-building, he was much more an engineer than a scientist – and he did not disagree. After all, he held two degrees in engineering. But just how much technical directing had Pennemünde-East's technical director been involved in with the V2 development? Postwar detractors questioned how much direct, hands-on input von Braun had had, suggesting he personally invented or discovered little or nothing. Some also contended that von Braun and his team effectively pilfered American Robert H. Goddard's rocket patents from 1914 to the 1930s in creating 'their' V2 terror weapon. (See Appendix I for a letter that von Braun later wrote elaborating on this issue.)

Maj-Gen Dornberger, himself an engineer, weighed in on the first point in an essay written in the early 1960s. He asserted that von Braun was individually or jointly responsible for an estimated twenty patentable innovations in rocket technology in the 1940s. But, because the V2 development was a secret wartime project, the documentation was locked away and the papers were never filed for patents.[11] After the war, however, von Braun filed for several patents for his wartime inventions. One example was improving the mass fraction of a rocket through increasing propellant volume by shortening the propulsion system – motor assembly and nozzle – within the same exterior dimensions. Figure 1 of his US patent application, filed in December 1959 and granted to him in January 1961 as patent number 2,967,393, consisted of cutaway drawings of a rocket in the shape of a V2. Documented accounts abounded in the 1950s and '60s of von Braun personally stepping in with the solutions to myriad aerospace engineering problems of the moment.[12]

As to the assertions about appropriating Goddard's work, von Braun contended that nothing could have been further from the truth. He wrote that, at 18, through the efforts of his mentor, Hermann Oberth, who had written to Goddard, he read a translated copy of the American rocket pioneer's trailblazing booklet, *A Method of Reaching Extreme Altitudes.* The 1919 paper, based on Goddard's experiments, postulated that rockets carrying their own oxygen supply could function quite well beyond the Earth's atmosphere and could even serve as vehicles for flights to the moon; he did not specify that he was experimenting largely with liquid-propelled rockets. Von Braun added that later on he did see various Goddard rocket concept illustrations and statements in aviation journals. 'However,' insisted von Braun, 'at no time in Germany did I or any of my associates ever see a Goddard patent.'[13] Independent accounts support his contention.[14]

In the autumn of 1944, the combination of V2 deployment and the threat of advancing Allied forces overtaking launch sites in the Low Countries led to the start of a top-secret project calling for sea launches of the missiles. Code-named Prüfstand (Test Stand) XII, the project was pursued by a special engineering group assigned from von Braun's Peenemünde team. The concept envisioned one or more U-boats towing as many as five missiles each in individual watertight containers to positions, among other locations, off the eastern coast of the United States. There, the submarines would surface, the canisters would be flooded to upright positions, and the missiles fuelled, aimed and fired at New York and other major US cities.

The idea originated with team member Klaus Riedel. His inspiration was the large, enclosed, submerged 'barges' towed behind U-boats in transporting food, ammunition and fuel supplies to German bases primarily in Norway. The team of Riedel (killed in a car accident in late 1944), Hans Hueter, Bernhard Tessmann, Georg von Tiesenhausen and Hermann Hufen performed engineering studies, executed a canister design and contracted with a shipyard in Stettin for fabrication of a test version of the container. Assisting were architects Hannes Luehrsen and Heinz Hilten. Von Braun took no direct role in the project.

Static firings of V2s inside a canister were conducted at a Peenemünde test stand. But events near the war's end in Europe led to abandonment of the effort, and the V2 never went to sea with U-boats. 'The job was not finished,' recalled von Tiesenhausen, the last surviving engineer member of the project team, who designed the canister interior. 'It was too late.'[15]

Precisely what were the wartime roles and responsibilities of von Braun and his team in regard to the V2 rocket, beyond its development and testing? Large and small distinctions surfaced in various postwar – and posthumous – investigations and studies. In addition to its R&D work, the rocket team at Peenemünde performed in-house, low-volume pilot production of the rockets, as well as training field troops to handle and fire them. The preponderant evidence shows that the team – essentially an R&D organisation that had now lost out politically to the SS for control of most of its own Peenemünde operations – did not substantially participate in the volume production at Nordhausen or the deployment and operational use of the lethal, if inaccurate, missiles. There were two major exceptions, however.

First, Arthur Rudolph, von Braun's chief prototype production engineer at Peenemünde, was detached, along with several assistants, and reassigned by the SS as operations director for V2 production at the huge underground Mittelwerk factory. Berlin put Albin Sawatski in charge as the overall director of the plant. All this was in the autumn of 1943, almost a year before the first operational firing of the missile. This subterranean plant was hidden inside Kohnstein Mountain at Niedersachswerfen, near Nordhausen, in the southern Harz Mountains of central Germany. Technically a private, but in fact a government-owned, company, Mittelwerk GmbH operated the notorious factory. In reality, the SS ran the show.

In addition to a substantial staff of paid, skilled, civilian personnel that included female office help, Mittelwerk's primary workforce in labour-starved Germany consisted of thousands of forced labourers provided by SS Gen Hans Kammler. An estimated 60,000 enslaved workers passed through Mittelwerk, the main Mittelbau-Dora concentration camp nearby and Dora's satellite labour camps in 1943–5. Some of the detainees were even rented out by the SS for work elsewhere. The forced labourers included political prisoners, POWs, criminals, Communists and a small number of Jews, Gypsies and others from several European countries as well as from Germany itself.

Multiple published accounts show they worked in grim conditions and died at a tragic rate. About one-half of the estimated 20,000 deaths occurred in the tunnel-blasting and digging to enlarge the already massive old gypsum mining complex to prepare for weapons production. The work to extend the two main tunnels and dig smaller connecting, cross tunnels was brutal, taking a horrific toll in human lives. Suspected sabotage led to several summary hangings inside one tunnel and at the Dora camp as warnings. The many other Mittelwerk deaths resulted from beatings, disease, starvation and freezing, either from sleeping overnight in the frigid tunnels or at the Dora camp and at the Ellrich and Harzungen subcamps.

Rudolph's responsibilities included assigning, scheduling and monitoring the available V2 workers. He was not responsible for procuring labour, establishing working or living conditions, or meting out punishment, according to records and his and others' assertions.[16] However, historian Michael Neufeld, who refers to Rudolph as being 'on the second level of Mittelwerk management as [A4] production director', notes several instances where he had a measure of involvement with slave-labour issues. These include his proposing premium wage incentives to harder-working prisoners and a general reduction of work shifts from twelve to eight hours – both initiatives rejected by higher management. Rudolph also attended, along with Dornberger and von Braun, a meeting on 6 May 1944 where the then-new Mittelwerk general director, Georg Rickhey, discussed the need to bring in 1,800 more forced labourers – skilled French workers

– to replace prisoners lost in the preceding severe winter. Minutes of the meeting indicate that neither Rudolph nor the other two men participated in that discussion, Neufeld acknowledges.[17]

The second exception was that von Braun made some fifteen to twenty business visits – ranging from several hours to two days in duration – to Mittelwerk from late 1943 until early 1945, as he later testified and flight logs showed. He was checking on the end-product of the full-scale production of the V2 – his 'baby', as he called it. He went primarily in his role as head of a 'final acceptance' subcommittee to monitor 'quality control' issues in manufacture and assembly,[18] he and others contended. These included implementing the debugging design changes he and his team at Peenemünde were constantly devising to enhance the unreliable missile's performance. In a broadened context, his role also included occasional screenings of new prisoners to determine levels of technical competence and to identify skilled forced labourers who could enhance the quality of the finished product. These and other such prisoner contacts constituted his involvement with slave labour. There is no evidence that von Braun ever visited the Mittelbau/Dora or Ellrich slave-labour camps that serviced Mittelwerk, or any of the Nazis' monstrous extermination or 'death' camps. However, he observed the slave labourers' primitive living conditions inside the factory tunnels before the Dora camp was built in 1944, and he did visit the notorious Buchenwald concentration camp, parent camp to Dora, to seek out imprisoned engineers and scientists on the advice of Mittelwerk's Albin Sawatski.[19]

According to Ernst Stuhlinger and Fred Ordway, in their book entitled *Crusader for Space* (published in 1994), 'some of his closest associates who still remember his words [including Gerhard Reisig]' later recalled that von Braun – against Mittelwerk standing orders not to discuss such things – said he was appalled by what he called the 'hellish' working conditions he observed in the tunnels and the forced labourers' presumed desperate living conditions at the camps.[20] Von Braun told them: 'My spontaneous reaction was to talk to one of the SS guards [at Mittelwerk], only to be told . . . that I should mind my own business. . . . I would never have believed that human beings can sink that low; but I realized that any attempt [at] reasoning on humane grounds would be utterly futile. . . . These individuals had drifted so far away from even the most basic principles of human [morality] that this scene of gigantic suffering left them entirely untouched.'[21]

The decision to use forced labourers at Mittelwerk for V2 production was that of the SS. Nominal SS officer von Braun, with what historians generally say amounted to an honorary commission, and the few other SS-connected workers at the Peenemünde base were too low-level within the Nazi military arm to have a voice in the matter. The decision was made at the Himmler-to-Hitler level. Neufeld writes that von Braun, after receiving a telephone call on 25 August 1943 from Dornberger in Berlin, chaired a meeting later that day at Peenemünde in his commander's absence in which the relocation of its V2 production operations, prisoner-labourers and some German workers to several underground factory sites in western Germany was first discussed. The historian interprets that meeting as 'the first time that von Braun was involved in decision-making about the SS prisoners'. No account is given of any decisions made or any actions taken as a result of the meeting. Previously, matters relating to the Peenemünde prisoners had been handled solely on 'the production side' – as opposed to the R&D side – from Arthur Rudolph to Godomar Schubert, army civilian head of the centre's V2 production plant, to Dornberger, according to Neufeld.[22]

'Himmler managed to obtain Hitler's approval for the employment of concentration camp prisoners in the Mittelwerk productions', Reisig wrote years later. To postwar critics of von Braun's moral failure to object, Reisig asked, 'How could W.v.Braun act directly against Himmler in the matter of forced labor?'[23] As for a related situation involving the proposed importation of more slave labourers at Mittelwerk, about which von Braun said nothing, Neufeld writes, 'Objecting would have been risky, of course', especially in view of his own recent imprisonment by the SS.[24]

Von Braun did complain, however, at one or more points about the use of forced workers – on grounds that the unskilled, disloyal, abused and sickly labourers' inferior work was resulting in high failure rates.[25] His feelings on the moral issue of using slave labour could not later be known with certainty. Close associates from that time have insisted that they did know, beyond doubt, from his shared confidences and his nature, that he was aghast at the situation but felt powerless to act.[26]

As mentioned above, von Braun's role did involve occasional screenings of new workers to determine levels of technical competence and thus to enhance the quality of the finished product. Through correspondence, visits and interviews he sought the transfer of certain foreign scientists and engineers he learned were among 'detainees' at the Buchenwald concentration camp. In a letter dated 15 August 1944 to Albin Sawatski, director of Mittelwerk, von Braun noted that on his previous visit to the underground factory Sawatski 'suggested utilizing the skilled background of various prisoners both in Mittelwerk and Buchenwald in order to accomplish additional development work [in quality control] as well as to construct a model. . . . I immediately looked into your proposal by going to Buchenwald . . . to seek out more qualified detainees. I have arranged their transfer to the Mittelwerk . . . as per your suggestion.'[27]

Von Braun noted that he had worked with both the Buchenwald camp commandant and the SS labour supply officer in effecting the transfer. Neufeld later wrote that von Braun's August 1944 letter to Sawatski may 'implicate him directly in crimes against humanity'.[28] The historian does not explain why arranging specific transfers from the notorious Buchenwald concentration camp to Mittelwerk necessarily constitutes a heinous offence, except for the 'direct involvement' with forced labour; selected detainees were assigned living quarters within Mittelwerk – an unhealthy environment but preferable to the horrific Dora labour camp.

Von Braun visited a French physicist, Charles Sadron, who had been caught as a participant in the French Resistance, in his work area near his quarters in a barrack inside one of the tunnels at Mittelwerk.[29] Sadron wrote in his 1947 memoir:

I must, however, in order to be truthful, point out one man who took an almost generous attitude towards me. That is Professor von Braun. . . . He came to see me in the shop. He is a young man, of very Germanic appearance, who speaks perfect French. He expresses to me, in measured and courteous terms, his regret at seeing a French professor in such a state of misery, then proposes that I come work in his laboratory. To be sure, there is no question of accepting. I refuse him bluntly. Von Braun excused himself, smiling as he left. I will learn later that, despite my refusal, he tried several times to better my lot, but to no avail.[30]

Von Braun's scheme, according to Stuhlinger, had been to have the French physicist (as well as several scientists at Buchenwald, whom he had transferred to Mittelwerk) eventually relocated to Peenemünde, claiming that the Mittelwerk site was unsuitable for productive work by these detainees. There, they would be assigned technical work, '[be] given decent housing, and eat the same food that we ate', Stuhlinger related. But von Braun said the French physicist felt any such transfer would make him a traitor, and he refused to be treated better than other inmates.

Von Braun's disappointment was mixed with respect for the scientist, recalled Stuhlinger.[31] Future accusers cited the case of the French professor as further evidence of von Braun's complicity in the exploitation of slave labour. Yet Neufeld terms Sadron's account 'by far the most exculpatory evidence [in von Braun's favour] that has yet been found.'[32]

Surprisingly, the percentage of forced workers within the V2 production force fell steadily at Mittelwerk during the final months of the war, as detainee deaths cut into those ranks and the proportion of regular civilian workers increased in order to meet production goals. The V2 slave-labour force there went from about 5,000 in July 1944 (of 8,400 total V2 workers) to 3,500 in October (of 7,500 total) to 2,000 in March 1945 (of 6,900 total).[33] In mid-1943, when Hitler had informed Speer that 'Himmler suggested the use of concentration camp inmates for forthcoming A-4 production to assure secrecy', Speer objected – to no avail. Hitler also ordered that 'only Germans' – including political prisoners and other Germans from concentration camps, as well as skilled, paid non-forced workers – be employed for A4 work, though this was later ignored.[34] Von Braun, who was accustomed to the presence at Peenemünde of forced labour – mostly Russian POWs – working on A4 low-volume manufacture, had expressed concern that the immature, unreliable missile was being rushed into mass production too soon. According to Stuhlinger, it was later, when he saw the 'inhuman treatment' of slave labourers at Mittelwerk, that he complained of the poor work quality and said, 'no better products could be expected' under such conditions. Official objection on purely moral grounds would have been futile at best.[35]

It should be noted that von Braun's brother Magnus was assigned to work at Mittelwerk in 1944–5 to help with quality assurance of gyroscopic components.[36] Supportive commentators, including Ordway and several Peenemünde veterans, told me that Magnus bore a strong family resemblance to Wernher and wore an identification badge that said 'von Braun'; they suggested that perhaps at least some of the Mittelwerk sightings of Wernher, claimed by surviving prisoner-labourers, were of Magnus.

During the postwar era and continuing after his death, Wernher von Braun came under allegations of 'complicity in' and 'responsibility for' the abuse and deaths of slave labourers at Mittelwerk. He gave a detailed, eight-page response to the editors of *Paris Match* in April 1966 – as background information, but not for publication – at a time when a French group of Dora-Mittelwerk survivors and their families made what he labelled 'false accusations'. In seeking a response from von Braun – by then a world-famous figure and the most tempting target of accusers – the magazine's editors had advised their New York bureau chief that the accusations 'do not appear to be based on any precise proof'.

In the *Paris Match* statement, von Braun cited his having been thoroughly investigated by US and British authorities and cleared of '[having] been in any way involved with any atrocities'; he noted that this included a review of the Dora-related and other 'open records' of the Second World War War Crimes Tribunal at Nuremberg. 'At no time did I have any authority whatever in the Mittelwerk management, in the affairs of Camp Dora, or in the setting of production goals,' he stated. That had been his testimony also in a legal deposition given in 1947 in New Orleans. He added that the findings of 'many scholars' absolving him of any atrocious conduct at Mittelwerk were 'also based on the simple fact that there are still thousands of people around (and not all of them my friends) who are intimately familiar with my own duties, the limits of my authority, the anguish to which I myself was subjected, and . . . the kind of man I really am'.[37]

He concluded his *Paris Match* statement: 'In general, I readily agree that the entire environment at Mittelwerk was repulsive, and that the treatment of prisoners was humiliating. I felt ashamed that things like this were possible in Germany, even under a war situation where national survival was at stake.' Von Braun was urged by his closest associates to issue vigorous and explicit public denials of allegations made against him by Mittelwerk/Dora survivors and their families. He responded that beyond his not wanting to drag his present employer, the US government, into controversy, these people had suffered enough without his direct challenges adding to their misery.[38]

Although 'not in the decision-making process' regarding the use of slave labour to assemble V2s, von Braun nevertheless bore a 'moral responsibility' for the 20,000 deaths at Mittelwerk/Dora, according to Neufeld in his 1995 book *The Rocket and the Reich*. (Similar views are held by others, including authors Linda Hunt and Dennis Piszkiewicz, but Neufeld is the most prominent and active of this group.) 'Clearly, there wasn't a lot he could have done to change it,' Neufeld acknowledged in a 1998 University of Alabama in Huntsville lecture, but as 'fundamentally a good human being', von Braun should at least have tried, Neufeld contended.[39] (Several members of the Peenemünde team in the audience remarked that it was easy – and naive – for Neufeld to make these judgements.) 'He was a lesser war criminal,' Neufeld alleged in another Huntsville-area college lecture two years later.[40]

Rocket-team veteran Reisig responded to the slave-labour assertions against von Braun by stating, in part, that as Peenemünde's technical director, he 'had no authority of deciding on the employment of forced labor' 250 miles away. 'The only activity required of W.v.Braun at the Mittelwerk was the inspection of the quality of the end product, the complete A-4 Rocket (propaganda boss Goebbels coined the reference 'V2').' The staff he had for this inspection work was borrowed from Peenemünde and they were 'independent of the permanent technical staff of the Mittelwerk'. Reisig added: 'Hypothetically . . . W.v.Braun could have requested manufacturing of the A-4 . . . be done by skilled German workers . . . in order to procure top-quality rockets. However, to no avail! There were no such German skilled workers available. Hitler had drafted this most valuable work force [into the military] only to have it slaughtered in the hopeless fighting conditions at the Russian front, as I witnessed from my own deployment in Russia in 1943.'[41]

The whole 'brief, sad tragedy of the enforced mass production of a still-incomplete [missile] system' was caused by 'the "leadership" and the orders of a few . . . all-powerful maniacs', Stuhlinger and other, unidentified, German- and American-born von Braun supporters wrote in a joint memo issued in protest against Neufeld's scheduled 1998 university appearance. 'Those prisoners died because of the horrible conditions in the concentration camps, and . . . where the tunnels of the underground factory were extended [to enlarge the factory], and quite generally because of the frantic and crazy effort of the Nazi government to produce weapons at all costs to win the war.' The memo stressed that all should abhor 'what Himmler, Speer, Kammler, Sauckel, and others did with their concentration camps, which will remain a dark streak in Germany's history for all times'. The statement concluded that 'to lay those atrocities at the feet of von Braun and his Peenemünde co-workers is simply not in agreement with history'.[42]

If the von Braun team of engineers and scientists held Hitler, Himmler and company fully culpable for the atrociously high death rates among the thousands of enslaved workers at the SS-run underground Mittelwerk V2 factory from late 1943 till early 1945, what moral responsibility did the Peenemünders feel for the deaths that their missiles caused among the civilian populations of wartime London and other cities? Von Braun later gave this answer: '[W]hen your country is at war, when friends are dying, when your family is in constant danger, when the bombs are bursting around you and you lose your own home, the concept of a just war becomes very vague and remote and you strive to inflict on the enemy as much or more than you and your relatives and friends have suffered.' (The preceding comments are from an early 1970s letter that von Braun wrote addressing the question of his moral responsibility for V2 casualties and the suffering inflicted on the Jews by Hitler's Germany. See Appendix II for the complete letter.)

The German Army launched the last wartime V2, at England, on 17 March 1945. Over the nearly seven months immediately preceding, some 3,255 of the missiles had been fired, out of approximately 6,400 produced at Mittelwerk. (Of the 3,255 firings, an estimated 365 did not go far enough to cause an 'incident' in the target areas.)[43] The numbers had fallen far short of Hitler's call for 30,000 missiles. The targets included London, south-east England, Paris and Antwerp. Not all the warheads reached their targets. However, more than 1,100 did so in England alone.[44]

In addition to property destruction and the disruptive terror wrought, in 1944–5 in England the missiles resulted in the deaths of

2,742 men, women and children, and seriously injured 6,467, the great majority of them civilians, according to official British figures.[45] When the victims from continental Europe are added, the total V2 death toll rises to about 5,000, not to mention the many more thousands of casualties with serious injuries.[46] In addition, at the beginning of the war, after Germany's invasion of Poland, the Luftwaffe had conducted a ruthless aerial blitz of London and its environs, intended to break the will of the British people. To put such grisly matters in some perspective, however, the total V2 casualties were a small fraction of the civilian toll inflicted by any one major bombing raid by Allied forces, out of the thousands conducted against aggressor Germany. Including RAF raids beginning early in the war, Allied bombing from 1939 until 1945 killed 593,000 Germans, mostly in night raids targeting cities with a population of 100,000 and above.[47]

By all objective accounts, the revolutionary V2 was a failure as a strategic or tactical weapon, and even as a significant instrument of terror. Also, as fate would mercifully have it, the missile found no use as a delivery vehicle for a German atomic bomb that never came about. Ironically, a number of US and Allied military experts, including Winston Churchill and Dwight Eisenhower, came to the conclusion that because Berlin spent millions of Reichsmarks and precious labour and materiel on Dornberger and von Braun's rocketry efforts, instead of building more warplanes and tanks, thousands of Allied soldiers' lives were probably saved.

Von Braun never apologised for his starring role in the creation and use of the ballistic missile in warfare, although he did confess to 'misgivings about building rockets for the Nazis'.[48] But his eyes were wide open at the start of his deal with the German Army, as he conceded in these postwar comments: 'Any moral conflict caused by the thought the rockets could be used as weapons in a war was opposed by the desire for finance for our space plans. We always considered the development of rockets for military purposes as a roundabout way to get into space.'[49]

The closest he would ever come to a *mea culpa* was expressing a 'feeling of guilt' over the civilian deaths, in remarks at a crowded news conference in Munich at the 1960 world premiere of his film biography. 'I have very deep and sincere regrets for the victims of the V2 rockets, but there were victims on both sides,' said von Braun.

'A war is a war, and when my country is at war, my duty is to help win that war.' In some press accounts he had added, 'whether or not I had sympathy for the government, which I did not'.[50]

With retreating German troops still deploying operational V2s as an offensive weapon, the von Braun team stayed busy at Peenemünde in the closing months of the war. One of their projects in those days was to work on the A9 concept, a winged version of the V2, with double its range. A variant of that design showed landing gear and a pressurised cockpit for a pilot – an astronaut, in other words.[51] The team also continued to work on the Wasserfall anti-aircraft guided missile, an effort that began in 1943 at Peenemünde and had more than forty successful test launches. Another project that was still conceptual was an A10 booster with 440,000lb of thrust. Yet another vision on paper was the A11 multi-stage rocket, with a monster first stage generating 3.5 million lb of thrust.

Paper studies were done on a long-range rocket using the winged A9 mounted atop an A10 booster. Conceived more as a space transportation vehicle than as the first intercontinental ballistic missile – or so the Peenemünde team later insisted – it would have had a range sufficient to cross the Atlantic Ocean. Hitler liked that. 'We called it the "Amerika rocket",' Ludwig Roth, a senior member of the von Braun team, noted after his relocation to the United States. 'We did not fly this rocket – but it got us here anyway!'[52]

By early 1945, as Soviet troops neared Peenemünde, US and British troops advanced in the western and southern sectors of Germany. The United States was much on the minds of von Braun and his inner circle. Conditions were fast deteriorating within the Third Reich: Allied bombings had intensified, supply trains and lorries could not get through (even to the concentration camps), communications were disrupted. The 32-year-old father of the V2 stared at contradictory orders from the German Army, the SS and local Peenemünde militia and civil defence authorities. 'I had ten orders on my desk,' von Braun later reported. 'Five promised death by a firing squad if we moved, and five said I'd be shot if we didn't move.'[53]

He and his key technical lieutenants wanted to keep their unique team together, for the future. But what to do? Where to go? To whom should they attempt to surrender? The Soviet Union seemed out of the question – it was another police state, and a hated enemy that had suffered deeply and directly at Nazi Germany's hands. They could not expect good treatment there. A targeted, battered Britain would hardly welcome the V2 team with affection, nor could it afford to finance the group's space dreams. Neither could France, a detested, perennial enemy.

At the end of January, von Braun called a secret meeting of fewer than half a dozen of his most trusted associates. During that meeting, held in a farmhouse, away from Gestapo ears, von Braun announced: 'Germany has lost the war.'[54] He then put the issue to his people: the team should keep in mind its extraordinary accomplishments, its ultimate space goals, its desire to stay intact, and the fact that both Russia and the United States would now want its know-how. According to an account in an authorised biography published two decades later, he then asked to which country the team should turn.[55]

Another account has von Braun helping along the decision-making process more forcefully. A future colleague, Col Edward D. Mohlere, US Army, recalled that one of the Peenemünde team who was present at the secret meeting, Eberhard Rees, related his recollection of what the rocket leader actually said. Von Braun had concluded that there was just one country in the world with the necessary resources for the job, 'and that's the United States. Let's go!'[56]

In any event, they took a vote. With a single dissenting voice – that of Helmut Gröttrup, who later opted to join the Russians[57] – the group chose to seek to surrender themselves and their rocket secrets to the Americans. That meant heading south, probably to Bavaria, where US forces were advancing. In the end, the SS command unwittingly aided the plan by evacuating the rocket team south, although not necessarily for surrender to the Allies. SS options included using the group as a valuable trading chip to gain the freedom of such war criminals as their Gen Kammler. But another option was, if all seemed lost, to deny this resource to the enemy by simply murdering von Braun, Dornberger and the entire core group.

Von Braun later claimed there were other, loftier factors behind his wanting to seek out US troops. 'The reason I chose America in particular is that Americans had a reputation for having an especially intense devotion to individual freedom and human rights,' von Braun later claimed. He added that the American system of checks and balances in government 'offered the highest guarantee that any knowledge we entrusted to them would not be used wantonly'.[58] This was the high-principled-sounding version. The rocket leader had earlier put it rather more bluntly: 'My country had lost two [world] wars in my rather young lifetime. The next time, I wanted to be on the winning side.'[59]

However principled or plainly pragmatic his motivations, Wernher von Braun's circuitous journey to the shores of North America began with this decision – made by him and agreed upon by his inner circle. Perhaps the odds were not good, but he hoped that from the ashes of war his one-of-a-kind rocket team might one day rise reborn like a phoenix for flight in the New World.

Bound for America

By January and February 1945 it was time to escape – escape the Russians closing in, escape the abysmal war, escape the demands of die-hard Third Reich authorities (Hitler included) that the Peenemünders remain and fight to the death. The von Braun missile team had SS orders to move south to the Harz Mountains and to await further word there.

Before heading inland, von Braun made a brief visit to the Baltic farm estate of relatives in Pomerania to say his goodbyes. He had made a point to include his cousin,[1] the lovely blonde and blue-eyed daughter of his uncle, Alexander von Quistorp, a prominent banker, landholder and aristocratic brother of his mother, Emmy. This would not be the last that 15-year-old Baroness Maria Louise von Quistorp and the almost 33-year-old bachelor Baron Wernher would see of each other.

Von Braun and his advisers hatched a scheme to carry out the retreat to the Harz Mountains, they later said, with an eye towards an eventual rendezvous with American forces. On the official SS stationery which his officer's status entitled him to use, von Braun wrote himself bogus orders to move some 5,000 civilian personnel, along with mountains of scientific equipment, machinery, vital documents and V2 missile parts, by lorry, train and car. Then he had the fictitious designation 'VABV' affixed to all the vehicles. During the dangerous exodus, whenever suspicious German guards halted the main convoy, von Braun confidently showed his 'orders' and pointed to the conspicuous VABV signs. He explained that the letters stood for *Vorhaben zur Besonderen Verwendung* (Project for Special Dispositions). He hinted that this was a top-secret effort being carried out for none other than SS Reichsführer Heinrich Himmler. Without fail, the guards waved them through.[2]

They travelled only at night, to avoid Allied aircraft attacks. During a night drive at high speed and with no headlights, von Braun's driver fell asleep. Hannes Luehrsen, a facilities master planner, was in a following car and later described the spectacular crash in reminiscing with von Braun: 'Way back, there was a moment when I feared you would not even see your [next] birthday. . . . All of a sudden your car soared like a glider from the Autobahn against the ramparts of a below-ground railroad track.'[3] In the bloody accident von Braun suffered a nasty break in his left upper arm and a facial gash the scar of which he would carry for the rest of his life. His driver was killed instantly.

First stop for the missile team was Bleicherode, near Nordhausen and the underground Mittelwerk missile factory. The retreating technical corps settled into abandoned factories and other empty buildings. A thousand lorries and dozens of trains had transported the personnel and tons of missile assemblies and parts, machinery, fixtures and documents to the interim location. Much of the V2 hardware was hidden in caves and old mines.[4]

After a month's stay in Bleicherode, new orders came in mid-March from SS Gen Hans Kammler, whom Himmler had put in overall command of Peenemünde and of V2 volume production at Mittelwerk. Army Maj-Gen Dornberger (who had retained certain administrative duties), von Braun and 500–600 of their best technical and support people were to move farther south to Oberammergau, in the northern foothills of the Bavarian Alps. They also were ordered to destroy their tons of V2 blueprints, production drawings and other secret documents before pulling out. Instead, Dornberger and von Braun sent trusted team members to hide the precious papers in nearby abandoned mines and to dynamite the entrances shut.

At Oberammergau the team of missile experts was housed with SS security men in barracks surrounded by barbed-wire fences. Von Braun had heard rumours that the SS high command was prepared to liquidate the rocket group rather than have it fall into enemy hands. He figured the notorious Kammler wanted the cream of the V2 team held under guard so he could either trade them with the Allies to save his own skin or kill them before capture.[5] In time, von

Braun was able to convince the security officers that Germany's missile brain trust could easily be destroyed by a single Allied bombing attack that would 'wipe out the last chance for the "ultimate victory"' – and how would that sit with SS higher authority? As a precaution, he argued, the rocketeers should fan out among twenty-five nearby villages in the mountains.[6]

The SS security officers bought the idea. Soon Dornberger, von Braun, his brother Magnus and two dozen colleagues, accompanied by their guards, relocated in early April to a resort hotel, Haus Ingeburg, in the village of Oberjoch. There they waited, for weeks. Von Braun later described the unusual nature of those days. 'There I was, living royally in a ski hotel on a mountain plateau. There were the French [forces] below us to the west, and the Americans to the south. But no one, of course, suspected we were there. So nothing happened. The most momentous events were being broadcast over the radio: Hitler was dead, the war was over, an armistice was signed – and the hotel service was excellent.'[7]

Gradually the SS troops guarding the dispersed V2 team members disappeared, especially after news came that Adolf Hitler had died on 30 April. Ernst Steinhoff persuaded the last SS holed up with his group to shed his uniform and pose as a missile engineer for his own safety.[8] Clearly, if the von Braun team was intent on surrendering to the Americans, now was the time to act.

And if the German missile men were looking for the Americans, the reverse was equally true. Von Braun and his fellow scientists and engineers from Peenemünde were on an extensive list of Third Reich scientists and other technical specialists sought by the US forces – and by the British, the Russians and the French, none of whom were cooperating with one another. This was a deadly serious competition for 'intellectual reparations', a quest that Winston Churchill called the 'Wizard War'.[9]

With time running short in the dying days of the war in Europe, von Braun and Dornberger decided someone must venture down from their hideout and contact the US troops they knew to be nearby. They chose von Braun's brother Magnus, then aged 26. 'I was the youngest, I spoke the best English, and I was the most expendable,' he later said.[10]

On 2 May 1945, the day after Hitler's death was made public, Magnus von Braun got dressed in ordinary clothes, climbed on a bicycle and headed down the mountain road. Two miles later, near the town of Schattwald, he encountered an advance anti-tank patrol unit of the 44th Infantry, US Army Third Armored Division. Challenged in German by Pfc Fred P. Schneiker of Sheboygan, Wisconsin, Magnus responded in alternating German and perfect English, 'We are a group of rocket specialists up in the mountains. We want to see your commander and surrender to the Americans.' The group also wanted to be 'taken to see "Ike" as soon as possible', he added.[11]

Schneiker escorted the young German straight to CIC (Counter-Intelligence Corps) headquarters in the small Austrian town of Reutte. The officer in charge, unaware of the high-priority quest for the V2 team, told Magnus to come back the next morning with the leaders of his group. At dawn the next day a fleet of cars carrying the von Braun brothers and other key missile men started down the mountain to Reutte.

What did von Braun expect? Did he fear arrest and punishment? As he told an interviewer a few years later: 'Why, no. We wouldn't have treated your atomic scientists as war criminals, and I didn't expect to be treated as one. No, I wasn't afraid. It all made sense. The V2 was something we had and you didn't have. Naturally, you wanted to know all about it. When we reached the CIC, I wasn't kicked in the teeth or anything. They immediately fried us some eggs!'[12]

Preliminary interrogation by Army intelligence officers began at Reutte. At first his questioners could hardly believe that this jolly young scientist – his shoulder and broken left arm encased in an elaborate plaster cast, his body grown pudgy from weeks of inactivity and tasty hotel food – was the main brain of Hitler's vaunted missile programme. As Bill O'Hallaren, the division's public relations sergeant on the scene, remembered, 'He seemed too young, too fat, too jovial' to have been in charge of the V2, as claimed.[13]

Von Braun mixed amiably with American soldiers, posing for snapshots, telling them about his achievements in rocketry and dreams of space flight. One GI quipped that the US Army had captured 'either the biggest scientist in the Third Reich or the biggest liar!'[14] It did not take the Army long to resolve that question. Soon, all of the more than 400 von Braun team members taken into custody at Reutte were moved to military barracks at Garmisch-Partenkirchen in Bavaria, where interrogation proceeded in earnest. Questioning by CIC technical intelligence teams composed of such first-rate US scientists as General Electric's Richard W. Porter and the California Institute of Technology's Fritz Zwicky, an astrophysicist, and rocket scientist Clark Millikan removed all doubt that the Americans had landed the real thing.

Zwicky later recalled that von Braun shared his advanced thinking on future multi-stage rockets for not only Earth satellites but also large orbiting space platforms and manned flights to the moon and planets. Von Braun found the interrogators' grilling 'extremely intelligent', and he recognised them as 'top scientists'.[15]

The US intelligence teams had each of the rocket experts write a detailed autobiography. With this and independent information, investigators began verifying their credentials, checking out their backgrounds, probing for early Nazi or other unsavoury political leanings, looking for any tendencies towards violence, and seeking clues to their true characters from teachers, former neighbours and others, recalled Maj-Gen John G. Zierdt, US Army (Ret.).[16]

The missile men waited it out, uncertain of their fate. One of them, Johann J. 'Hans' Klein, an expert in both guidance-control and propulsion systems, years later reminded von Braun of those days:

> The dark weeks of early 1945 after our move from Peenemünde began to brighten when you outlined your ideas and concepts of satellites and probes. I remember we occupied ourselves in Garmisch-Partenkirchen by making performance and trajectory calculations under the most primitive conditions. When we needed technical estimates on such strange items as life support provisions, you steered us in the right direction with the same confidence and far-sighted outlook you showed during the V2 and Wasserfall development.[17]

Within the overall American hunt for German talent, a sharp-witted West Pointer, Holger N. 'Ludy' Toftoy, had the mission of seizing V2 missiles and ferreting out and interrogating the weapon's creative corps.[18] Col Toftoy, chief of the Army Ordnance Corps' Rocket Branch in Washington, became the head of Army Ordnance Technical Intelligence in Europe, based in Paris. He was directing efforts to locate and liberate enough V2 partial assemblies and components for shipment of 100 missiles to the United States when welcome word came of the surrender of von Braun and his core group to the US Army. While ordering full interrogation of the rocketeers, Toftoy and staff were frantically spiriting the missile parts and equipment out of what would within days become the Russian and British zones of occupied Germany. They narrowly succeeded, using trains to haul the booty to Antwerp. There, it filled sixteen Liberty cargo ships bound for New Orleans. The Army team also raced against the clock to recover the hidden tons of missile documents that von Braun's group had sealed in the caves ahead of the Russians – and again succeeded.[19]

Commented one US official: 'One of the greatest scientific and technical treasures in history is now securely in American hands.'[20] The US seizure of the von Braun team reportedly infuriated Stalin. 'This is absolutely intolerable,' the Soviet chief is reported to have fumed. 'We defeated the Nazi armies; we occupied Berlin and Peenemünde; but the Americans got the rocket engineers!'[21]

'After my field intelligence teams started to interrogate the German scientists,' Toftoy recalled after the war, 'I soon felt the information on missiles was so great and so important that it should not be handled by routine field reports.'[22] He decided that a sizeable number of von Braun's associates must be brought to the United States, at least for a time, for even more extensive brain-picking – and possible temporary employment. But how many? And who? Von Braun was asked to draw up a limited list of those he thought should go.

'The first list he submitted was for five hundred – including our secretaries and everybody!' recalled Maxe Neubert. 'That did not go over too well with the Army.'[23] Von Braun, however, had convinced at least one of his primary interrogators, General Electric's Dick Porter, that the United States should import 500–600 of his team members for optimum benefit. Porter made that recommendation to Toftoy.[24]

Decades later Porter shared his first impressions and memories of his encounter with the V2 team's technical chief in May 1945 in Garmisch.

Well, we were all pretty suspicious of each other. . . . Von Braun was reserved, but he didn't refuse to answer [questions]. . . . I always liked him and got along well with him. And what I liked most was that he knew his business. Everything he said to me made sense.

First of all, we had to know where and who [his] people were. I was amazed at von Braun's ability to tell me exactly what such-and-such a man could do, what his strengths and weaknesses were, and so on. He knew his people individually. He knew at least six hundred people well enough to tell you what they could do. My greatest admiration for the guy was this ability.[25]

Porter took a leading role in tracking down many of the prime Peenemünders who had scattered throughout Germany as the fighting waned and exploring with them the idea of coming to the United States. Aerodynamics expert Rudolf Hermann and his group, found in need of food and medicine, were gathered up along with their wind tunnel and whisked out of the future Soviet sector just in time. Missile guidance specialist Wilhelm Angele, whom von Braun had put on the 'wanted persons' list, as Angele later proudly joked, was found working as a farmhand near Hanover in order to eat.[26]

Meanwhile, Toftoy knew Washington would never approve bringing as many as 500 German rocket experts to America. The colonel whittled the recommended number down to 300, and in June 1945 he sent a cable to the Army chief of ordnance at the Pentagon. Sent under the authority of the Allied Supreme Commander in Europe, Gen Dwight D. Eisenhower, the cable stated that the Army had in custody more than 400 of the top V2 R&D personnel. The 'thinking' of their technical leaders was twenty-five years ahead of US work in rocketry, the cable also stated. It urged their immediate evacuation to the United States.[27]

Toftoy later explained why he had thought the team should be imported as intact as possible. 'A lot of these men had spent their entire adult professional career in rockets, and others who were brought in were highly specialized and renowned throughout Germany in other fields of science. The team was the most complete and competent that I had ever run into, and they had years of experience in working up to that particular point. So I felt they should come as a team – not as individuals.'[28]

Toftoy believed that 'this specialized group would do the country a lot of good', saving years of rocket research and development effort, along with many millions in taxpayers' dollars.[29] Some in the US military in those pre-Hiroshima months also believed that these Germans and their V2s might be useful in defeating Japan.

After hearing no word for weeks on his request, Toftoy flew to Washington 'to personally plead the case', he recalled years later. In late June 1945, the Army and the Departments of War, State and Commerce finally approved his recommendation, provided that only a 'limited' number of Germans came over. The Army brass in Washington set the figure at 100, no more. Toftoy then returned to Germany 'to negotiate the trip and handpick those to come'.[30]

He met in August with von Braun and his top team members in an abandoned schoolhouse in the small town of Witzenhausen, where many of the mostly young missile men and their families had found temporary quarters. 'The first thing Toftoy did', von Braun recalled, 'when he met with us and saw the situation of the families, most of whom had come to Witzenhausen as refugees from all parts of Germany, was to order a generous supply of milk for the youngsters and the babies.'[31] Von Braun and his team never forgot the American colonel's compassion.[32]

The group began paring down the list of hundreds of V2 team members to just 100 who would continue their work in the United States under short-term contracts. They pored over individual files, evaluated capabilities and weighed all the factors. As tentative selections were made, more background investigations and further security screening took place. As each final selection was made, a paperclip was placed on that chosen man's file. Thus, the project came to be known as Operation Paperclip, or Project Paperclip.[33]

Try as they might, Toftoy and von Braun could not trim the list to an even 100. The operative figure became 118, which was eventually expanded to 127. 'I am really sorry, but mathematics has always been my weak spot,' a chuckling Toftoy said years afterwards. 'I often had difficulties adding and subtracting plain numbers.'[34]

No one ever challenged Toftoy over the higher number. In the end, a handful of the 127 backed out, deciding to remain in defeated, devastated, divided Germany for personal, family or career reasons.[35] Such turndowns, though few, may have prompted the notion that von Braun and Toftoy had trouble persuading enough of the rocket men to leave Germany for the recently hated United States and an uncertain future. 'Some people have the idea that we were forced to go to America,' one of the chosen, Maxe Neubert, said later. 'That's not true. Von Braun could easily have got one thousand to go.'[36]

'I saw a new life,' recalled Walter Wiesman, youngest of those selected. 'There was nothing left for us in Germany.'[37] Still, just picking up and going was not easy for some. 'It was difficult to leave our families and our homeland,' Konrad Dannenberg remembered, 'and we didn't know if the Americans would milk us dry of our knowledge and send us back as the Soviets did later with the German rocket people they got.'[38]

The Soviet Union did indeed round up its own group of V2 engineers, scientists and technicians at the war's end – some 5,000 in all (as well as abandoned trainloads of von Braun's missile equipment that was awaiting shipment southwards). A few had chosen to throw in their lot with the Russians, but most were ferreted out within the Soviet zone in Germany and elsewhere, hauled off against their will to the USSR and put to work near Moscow.[39] 'I do think the United States got the best of our group,' von Braun later said, hardly surprisingly. 'The Americans looked for brains, the Russians for hands.'[40]

Part of the US deal with the von Braun rocket experts involved setting up a temporary camp for their dependants. Most dependants had at least some money, but it was of little value in the immediate postwar period when food, medicine and other necessities were scarce at any price.

The dependants were housed in former military barracks at Landshut in Bavaria. It was to be their home while the missile men were in America – or until Washington decided what to do next with these former enemies. The men's salaries would pay for their dependants' upkeep at Landshut.

Some dependants were still trapped in Russian-controlled areas and had to be brought to Landshut 'by cloak-and-dagger methods', among them Baron Magnus and Baroness Emmy von Braun. When the Yalta agreement gave Silesia to Poland, the elder von Brauns lost their lands and home. One account had them making their way to Berlin 'on a rusty cattle train', with the help of one of Toftoy's key staff officers, Maj James P. 'Jim' Hamill, US Army, and eventually to refuge in Bavaria. Receiving word their parents had safely reached Landshut, the grateful von Braun brothers wrote to thank Hamill.[41]

Across the Channel, the British authorities and scientists understandably wanted to talk with the mastermind behind the V2. In August 1945, an uneasy von Braun was flown to London along with several other German rocket experts and kept at a military camp near Wimbledon. Von Braun was picked up daily by a military intelligence officer and driven to government offices in central London for questioning.

The man whose lethal rockets had rained down on England expected a chilly reception. 'I must admit that I thought the British might be unfriendly to me, but I found I was wrong the first day I spent at the Ministry. I was interviewed there by Sir Alwyn Douglas Crow, the man in charge of developing British rockets. I was hardly inside his office before we were engaged in friendly shoptalk. He was curious about the headaches we'd had at Peenemünde, and he gave me a good picture of the damage the V2 had done in England.'[42]

Von Braun also had a chance to see some of the V2 damage during his daily drives to and from the London interview site. One day his military driver halted the car in front of the ruins of what had perhaps been a six-storey building struck only months earlier by one of the missiles. Several minutes passed in silence as von Braun viewed the destruction out of the window. Then the car moved on.[43] After a fortnight's stay in England, von Braun was returned to US Army hands in Germany. There had been no charges of war crimes, no public trial.

Maj-Gen Dornberger did not fare so well. Von Braun, who did not hesitate to credit him for 'the better part of my success in life', had hoped his commander at Peenemünde would gain a ticket to America along with the civilian members of the team.[44] But no such luck. The US authorities acquiesced to British demands that Dornberger be handed over. He was interrogated by the British and then tried on charges of war crimes. Although he was acquitted, he spent two years in jail during the legal process before being released.[45]

The US authorities cooperated with London in another postwar request. The British sought to assemble and test-launch several V2s. More than 100 former Peenemünders – without von Braun – joined British rocket engineers temporarily at a site near Cuxhaven. The project was named Operation Backfire. Three V2s were launched out over the North Sea, all successfully, and then the Germans were returned home.

The main body of German 'paperclip' rocketeers was to leave for America in early November, aboard ships under official Army orders that concluded: 'Upon completion of this duty the civilians named below will be returned to this [European] theater.' After all, how long could it take to pick these chaps' brains, anyway, before shipping them home? Von Braun knew differently, and so did Toftoy.

An advance group – von Braun, Eberhard Rees, Maxe Neubert and four other select members of the missile team – were chosen to travel to the United States several weeks ahead of the main body. On 12 September 1945, the seven climbed aboard a covered Army lorry at Frankfurt and headed with their jeep escort towards Paris. Years later Neubert cited another example of von Braun's 'golden and never-ceasing optimism, which you were able to transplant into your friends and [in] this way [create] your devoted team . . . Riding sideways in the back of a three-quarter-ton Army lorry from Frankfurt to Paris, and while crossing the River Saar, the border between Germany and France, you said, "Fellows, take a good hard look at your country, which we are just leaving. It might be many years before you will be seeing Germany again." All we had at the time of this border crossing was a six-months contract with the United States Army and their option for a six-months extension.'[46]

In mid-September the advance group was flown from Paris to Boston. The men were taken to Fort Strong in Boston harbour for more questioning and fine-tuning of plans for the next steps. There, von Braun met Maj Hamill, the rescuer of his parents. A tall, German-speaking ordnance officer in his twenties, who had graduated from Fordham University with a degree in physics, Hamill would be the rocketeers' military overseer for the next several years.

He later shared his thoughts on meeting the V2 team's leader for the first time. 'My first impression of von Braun was that here was a very forthright man. He was perfectly willing to talk on any subject, and I saw a look of wonderment on his face at times as he perceived what the New World held. Though he was ill when we met . . . he still reflected great strength, both physically and mentally.'[47]

Hamill and von Braun travelled to Washington and the Pentagon, where they talked with Ordnance Corps top brass for five days. The rest of the advance party was taken to the Army's Aberdeen Proving Ground in Maryland to begin tackling the monumental task of sorting through the tons of seized V2 documents shipped from Germany.

The rest of the 118 German rocket men came over in three separate shiploads to Boston between November 1945 and February 1946.

Most then headed for their new, perhaps quite temporary, home, the Army's Fort Bliss, near El Paso, Texas, on the border with Mexico.

Few beyond the Pentagon and President Harry S Truman knew that these German rocket scientists were even in the United States. That fact was still a secret kept from the rest of the nation in late September 1945, when von Braun travelled incognito from Washington, DC to Fort Bliss by normal train, escorted by Maj Hamill. The first part of their trip was fairly uneventful, but in St Louis Hamill noticed that their tickets for the balance of the ride to El Paso had them in sleeper car 'O', not a numbered car, as was usual. Hamill enquired about it. 'Oh, you'll enjoy that trip, Major,' replied the St Louis stationmaster. 'We have nothing on that train but wounded veterans of the 101st and 82nd Airborne Divisions.' With great effort Hamill managed to get himself and von Braun switched to a normal passenger train.[48]

During the trip to El Paso, Hamill, who was in uniform, avoided staying constantly at von Braun's side so as not to arouse suspicion. One time when they were separated, a friendly Texan sat next to von Braun, introduced himself, and asked von Braun where he was from.

'Switzerland,' was the accented reply.

'Oh, yes,' the Texan said. 'I've travelled all over Switzerland. What's your business?'

'Uh, the steel business,' von Braun improvised.

'Why, that's mine, too!' the other man exclaimed. 'What's your product?'

'Ball bearings,' the German said, confident he was safe now. He was not. The Texan, it turned out, was an expert on the subject!

Fortunately for von Braun, his talkative new acquaintance had to leave the train then at the Texarkana stop. Departing, the Texan pumped von Braun's hand, slapped him on the back, and blurted out, 'If it hadn't been for you Swiss, I doubt if we could have beaten those Germans!'[49] It was a story von Braun took great delight in retelling.

One of his strongest first impressions of the United States was the swarms of cars he saw, especially in the larger cities. Another early impression: the vast landscapes sweeping past his window during the long train ride to west Texas. An 'overwhelmed' von Braun would later say 'it still startles me' to be able to travel for days in America 'without crossing borders or being subject to passport or custom control, such as you constantly encounter in Europe'.[50]

A Fort Called Bliss

Wernher von Braun arrived in El Paso, Texas, in late September 1945 with his Army shadow, Maj Jim Hamill. Neighbouring Fort Bliss, headquarters of the Army's new Research and Development Division Sub-office (Rocket), would be his home. Col Ludy Toftoy at the Pentagon had named Hamill as commander of that sub-office and overseer of the 118 German missile experts being relocated there under the still-secret Operation Paperclip. Von Braun was given the civilian title of project director, under Hamill.

Von Braun was ill with hepatitis when he reached Fort Bliss and was placed in an Army hospital filled with American soldiers wounded in the war. Hamill, fearful the men would harass the former enemy missile man, cautioned him not to reveal his identity. 'But I couldn't conceal my still-broken English,' von Braun recalled. 'The GIs sized me up with uncomfortable accuracy, and began calling me "The Dutchman". But they also invited me to join their blackjack poker games!'[1]

While in the hospital, von Braun was 'happily surprised' to receive a visit from Toftoy, who had flown down from Washington to discuss plans for work to be tackled in the coming months, after von Braun's recovery.[2]

The German rocketeers were under one-year contracts – unless the US government opted to cancel after the first six months and send them packing. They were in the United States officially as 'DASE' – Department of the Army Special Employees – and were, according to the government, 'wards of the Army'. They had not passed through immigration and possessed neither passports nor visas. They went to work at Fort Bliss and the nearby Connecticut-sized White Sands Proving Ground (later Missile Range) in New

Mexico. They worked first as rocket consultants to the US military and then as technicians and teachers in the assembly and launch of the captured V2s. For more than a year, their presence was kept secret from the American people.

Housed in surplus barracks, and later in a former annexe to Beaumont Army Hospital at Fort Bliss, they could not leave the post without military escort. Their post was screened and censored. While the Germans were referred to cryptically as the 'paperclip specialists' and Army 'special employees', von Braun and the others dubbed themselves 'POPs' (for 'prisoners of peace') rather than POWs.

Early on in their stay on the Army base, the Germans assumed that the nearby Rio Grande was a mighty river that lived up to its name. The men heard at night what some said must surely be 'ships' horns blowing' from that broad river, Dorette Schlidt (née Kersten) recalled being told after she arrived in April 1947 (as the spouse of scientist Rudolf Schlidt). Some time passed before the Germans saw the trickle that was the Rio Grande at that point in its course and realised they had been hearing the blaring horns of steam locomotives.[3]

Confined to the base with little money, no family members around and no contact with El Paso's residents, von Braun and his young team had few diversions to fill off-duty hours. To relieve boredom and blow off steam, they often staged night-time battles, pitting one barracks against another and using such weapons as fire hoses, sandbags, water bombs and pillows. Von Braun usually led his barracks' brigade in the skirmishes.[4]

Most kept physically active playing football, volleyball and softball, or swimming. Some gardened or built furniture. Others converted a shack into a clubhouse and lounge, complete with a bar they built and stocked with beer, whisky, gin, rum and tequila. Saturday night parties were held around the bar, 'where all "shots" were successful', as the rocket men enjoyed recalling later.[5]

After a while the Germans – travelling once a week in groups of four, each with an Army NCO as escort – 'were allowed to go shopping in El Paso, then to have dinner in a restaurant, to see a movie, and to return to the barracks', recalled Ernst Stuhlinger.[6]

Von Braun and most of the team enjoyed Sunday afternoon classical music concerts courtesy of an old phonograph. Later there were Sunday group excursions aboard military buses into the nearby mountains and desert. Stuhlinger remembered that some of the missile men would slip through a hole in the fence surrounding their barracks area and take desert walks under the stars, 'to contemplate the universe'. Maj Hamill later contended he knew all about these escapades and quietly tolerated them, but had the men 'watched'.[7]

Von Braun and his team-mates received modest pay, none of which they saw during their first year or so. Von Braun's salary, about $6,000 a year, was the highest. The salaries went to cover the Army's cost of taking care of the Germans' dependants – including wives, children, parents and others – in Bavaria, or the money was held in trust. The missile men got a small daily allowance. Out of their $6 per diem pay, recalled then-Sgt-Maj Gilbert Appler, 'the government held out $1.20 for food. The net was . . . a lousy $4.80!'[8] With part of that, they sent monthly packages of food, clothing, sewing and knitting items, and the like back to their family members at Landshut.

At Fort Bliss many of the rocketeers continued to address von Braun as *Herr Professor*, something they had begun doing in Germany after Hitler awarded him the honorary title of 'Research Professor' for the V2 successes. It was honorary because 'professor' was normally a title reserved for a university teacher, and von Braun had never formally taught. Nonetheless, he had been proud of the title – a more exalted rank than *Doktor* – and had used it on his letterhead at Peenemünde. That changed after the Germans had been at Bliss a while. Konrad Dannenberg recalled that von Braun advised him, 'Just call me "Wernher". Forget about the "professor" now that we are in the United States.'[9]

Eternal optimist von Braun also urged his colleagues to learn English, despite the short-term US commitment. Their fluency improved as contacts with Americans at the post and in El Paso widened. As Hamill put it: 'They learned English with a Texas twang.'[10] The youngest member of the group, Walter Wiesman, later joked that he learned most of his English from watching Zorro movies in El Paso, hence the 'Sí, sí, señor' flavour of his fluency. Stuhlinger had another slant on the subject: 'Those of us who had learned English at school taught those who had not, with the result that a made-in-Germany accent will prevail in the English of the "Paperclippers" for the rest of their lives.'[11]

Although von Braun had not been an enthusiastic student of foreign languages in his youth, he now studied English intensely.

Within a few months he had learned the language well, including the technical vocabulary of rocketry and many American idioms. Unfortunately, he picked up much of the slang from American soldiers in the Germans' compound. When he gave one of his first presentations to officials in Washington, he was puzzled when they occasionally laughed at the wrong places in his talk. Stuhlinger recalled that later Hamill told von Braun: 'Wernher, before you give your next talk, let me go over your manuscript. You just cannot use those GIs' slang expressions with decent people!'[12]

As the Germans began teaching their new US partners all about the V2 and how to assemble, handle and fire dozens of the rockets, von Braun strove to keep alive his team's space aspirations. Peenemünde veteran Hans Klein, in a letter written years later, reminded von Braun of his role back then:

In the early Fort Bliss days your interest and encouragement produced what I believe was the first moon-flight trajectory by hand calculations and vector diagrams. We had a lot of fun doing it and, needless to say, we widened our horizon under your tutelage. . . . I do not forget your motto for success during some of the more frustrating days at Fort Bliss: 'Man muss sein Herzüber den Graben werfen.' [Literally, 'Throw your heart over the ditch' – adapted from a First World War cavalry expression that exhorted men to be brave and aggressive in overcoming obstacles.][13]

Werner Gengelbach recalled another occasion from late 1945, when the team once again stirred their dream of space travel:

[At] a little social event . . . during our first Christmas in this country shortly after our arrival in Fort Bliss . . . some of our friends of the Paperclip group decided to brighten our otherwise lonesome holiday by some self-generated entertainment.

Your younger brother, Magnus, composed and presented a story, a reporter's account of the first take-off of man into space. He gave a very lively description of what took place. There was an old gentleman with a long white beard, supporting his somewhat weakened body by a cane, excitedly watching the space vehicle which took off from White Sands on its uncertain journey. It was Wernher von Braun in the year 2000, who had reached the biblical age of eighty-eight years.

'At this time we believed this was a keen prediction and an optimistic target, and we wondered how in a short fifty-five years we could get there.[14]

Buoyed by von Braun's optimism and enthusiasm, his countrymen put in long hours training the Americans in handling and firing the first of the captured V2s – even if the Germans yearned to be tackling a new, more challenging project instead. The team of Americans with whom the German experts worked at Fort Bliss/White Sands included some 125 employees of the General Electric (GE) Company. The contractor group was formed and managed by GE's Richard Porter, one of the main scientist-interrogators of von Braun and his key associates in Germany. Rounding out the work team at White Sands was a contingent of US soldiers – NCOs, mostly conscripts – who had engineering and scientific degrees.

When the German–US team launched the first V2 from the White Sands range in May 1946, von Braun was positioned at the 'emergency cut-off station'. With him was missile engineer Wolfgang Steurer, who recalled that the station 'consisted of a little sand dune and a single button at the end of what seemed to be an old telephone wire'.

Decades later, Steurer shared his recollections of that day: 'Upon lift-off the bird went up all right, but tumbled happily in all directions like performing a dance. I looked over to you and saw to my surprise that you obviously enjoyed the spectacle. After pressing finally – and reluctantly – the emergency button, you were all of a sudden all business again: "Herr Steurer, that was one of your lousy jet vanes! You materials people have to do everything in a destructive way!"'[15]

A few days after the V2 misfire, von Braun intercepted Steurer back at the Fort Bliss barracks: 'Herr Steurer, may I borrow your sunglasses? I have to return to White Sands.'

'I am sorry, Herr Professor, but I have apparently lost them,' Steurer lied, wary of von Braun's legendary forgetfulness about returning borrowed items. Steurer had left the sunglasses in his room, fortunately. Von Braun strode away, but returned in a few minutes – grinning and holding up Steurer's sunglasses. 'Aha!' he exclaimed. 'I have found your sunglasses, and thank you very much!'

Several days later Steurer bumped into von Braun and asked for the return of his sunglasses. Much to his surprise, von Braun pulled them out, saying, 'Here they are; thanks.' Steurer put them on. They didn't fit. 'I realized I had some sunglasses, but not mine. What I had was very likely the result of a series of exchanges.' And it was, suggested Steurer, just such a knack for 'pulling a fast one' that was 'so successfully employed by a certain individual in the ensuing years to keep our programs alive'.[16]

The first successful V2 launch at White Sands occurred in June 1946. Reaching an altitude of 67 miles, it carried a scientific instrument package for upper-atmospheric research. (Arthur C. Clarke, a veteran of RAF wartime service and future von Braun friend, had proposed in February 1945, before the war's end, that any leftover V2s be used for ionospheric research.)[17]

The Germans' first year in the United States could be summed up as follows: train the Americans, launch more V2s, potter with technical improvements, meet a few US scientists, assist in high-altitude research projects using V2s, undertake no new rocket projects, and do space thinking and planning on their own. 'Once it had them, the US hardly knew what to do with the German rocketeers,' Time magazine later said of that year.[18] Looking back in 1962, von Braun recalled that his team's first year in Texas was 'a period of adjustment and professional frustration. We were distrusted aliens living in what for us was a desolate region of a foreign land, and for the first time we had no assigned project, no real task. Nobody seemed to be much interested in work that smelled of weapons, now that the war was over, and space flight was a concept bordering on the ridiculous. We spent our time in study and teaching, and assisted with the V2 evaluation firings in White Sands.'[19]

The Army's code-name for assembling and firing the V2s at the desert range was Project Fire-Ball. The Germans had another name for it. 'We called this period "Project Icebox",' Walter Haeussermann recalled. 'The Americans had no long-range plans for us; they just kept us on ice.'[20]

This early period in Texas produced a tale that quietly persisted among surviving original German rocket team members for more than half a century. It was said that von Braun's brother Magnus had brought a quantity of platinum from Germany. One version said the undeclared metal – more precious than gold – was a simple platinum bar; another, that it had been secretly cast as a set of mechanic's tools, twice as heavy as steel. As other Peenemünders remembered it, Magnus went to an El Paso jeweller and sold the platinum, or part of his stash. Before long, the story goes, he was questioned by the FBI and later fined, perhaps $100 (£57) or so. All of that has remained unsubstantiated.[21] What is a fact, said rocket team old-timers, is that Magnus was the second German at Fort Bliss to buy a car. Wernher could not afford his own second-hand car until later.[22]

Although the Germans' one-year contracts were set to expire in November 1946, the US Army realised that these men knew a great deal more about ballistic missiles than the Americans had yet learned. New, five-year contracts were drawn up, merging the former enemies into the Army's rocket activities. No new rocket programme was begun. The new contracts gave the Germans salaries comparable to civil service pay, with annual compensation ranging from $4,300 to $6,800. Von Braun got a rise to $7,500 a year.[23]

Largely through the efforts of Col Toftoy at the Pentagon, the Army also gave the Germans the welcome word that they could bring their families to America. That December, the first of the families arrived from Bavaria. Mostly wives and children, they were housed in the former hospital annexe at Fort Bliss where the missile men had been moved two months earlier.

December also brought the first public revelation of the German rocket experts' presence in America. Although Toftoy had issued a Pentagon news release the previous May about plans to test-launch V2s from White Sands, it downplayed the von Braun team. On 4 December, however, the El Paso Times published a front-page photograph and story on the Germans, a scoop. Walther Riedel, von Braun's chief design engineer, raised eyebrows among his countrymen with his quoted gripes about Army food. For a year they had made a point of never airing complaints about anything. Von Braun brushed it off in a spirit of 'It's a free country.'

The Germans were ideal alien residents. Maj Joseph Sestito, US Army security officer to Maj Hamill, later observed: 'They seemed to have a group spirit, based on the idea that on each one's model behavior rested the glory of the Reich. Also, they may have figured they'd be sent back to Germany if they showed any resentment' of their initially sparse lifestyle.[24]

The El Paso Rotary Club invited von Braun to speak in January 1947. The Army said it was all right, and he accepted. In his talk on 'The Future Development of Rocketry' he addressed the question of rockets first being used as an instrument of war: 'It seems to be a law of nature that all novel technical inventions that have a future for civilian use start out as weapons.' He looked ahead to an era of large launch vehicles, Earth satellites, astronauts and the 'queer sensation' of floating weightlessly, manned orbiting space stations, flights to the nearest celestial bodies, and other futuristic developments. When he sat down, the El Paso Rotarians gave the former enemy scientist a standing ovation.[25]

Going public with the news that the former Peenemünders were in America had its downside, too. Democratic Congressman John D. Dingell of Detroit labelled the Army 'nuts' for bringing in the V2 villains, adding: 'I have never thought that we were so poor mentally in this country that we have to go and import those Nazi killers to help us prepare for the defense of our country. A German is a Nazi, and a Nazi is a German. The terms are synonymous.'[26]

Undeterred by such criticism, von Braun pressed on with his work at Fort Bliss. He did disappear from the base for a couple of weeks in February 1947, although there was hardly anything sinister about it. With Army approval, and escort, he had quietly returned to Germany to be married. A few months before, he had decided to ask his teenaged first cousin, Maria Louise von Quistorp, to be his wife. While most Americans regard such unions as too close for procreative comfort, marrying blood kin was anything but uncommon within the old European aristocracy. And von Braun was in love and clearly intent on marrying a noblewoman. Their age difference – Wernher was almost 35, Maria just 18 – was similarly not unusual for aristocratic European couples.

The cousins had not seen each other for almost two years, and Maria would have to leave her family, friends and homeland. Wernher had known her all her life and recalled how, aged 17, he had held Maria at her christening. But now, not sure of her answer, he decided to write to his father in Germany and ask him to enquire discreetly of Maria's feelings in the matter.

The baron, as von Braun later discovered, used 'the subtlety of a bulldozer'. After receiving his son's letter, the elder Magnus immediately confronted Maria, waved the letter, and announced: 'I am supposed to find out if you will marry Wernher. What shall I tell him?' Maria then wrote to her suitor in America: 'I told him [the elder baron] that I'd never thought of marrying anyone else.'[27]

'[Von Braun] said there were two estates of relatives his family visited' in Pomerania, recalled Dorette Schlidt, 'and that he was always glad to learn they were going to visit Maria's, rather than the other, which he didn't care for.'[28] Towards the end of the war, whenever he could briefly escape his duties at the Baltic rocket centre, Wernher had visited the von Quistorps at their nearby estate. The fact that his cousin Maria was fast becoming a lovely young woman had not gone unnoticed. 'I told her that I came for the food,' von Braun recalled years afterwards. 'And she still tells me that this was my prime reason for hanging around the house. This is one of those arguments that married people who take joy in each other shouldn't try to settle.'[29]

After their Bavarian wedding, which took place on 1 March 1947 in Landshut, the rocketeer and his 18-year-old bride anticipated honeymooning there – alone, naturally – for several days. Arriving at their small apartment, they discovered that two US military policemen had moved in to prevent a possible snatching of the scientist by 'the other side'. 'Just pretend we're not here,' one of the guards suggested. 'If you need us, we'll be in the kitchen.' The newly-weds had no choice but to accept their round-the-clock houseguests.[30]

When the time came for the young couple to sail for America aboard a military vessel, they thought: 'Ah, we'll spend our honeymoon at sea!' But all the men were quartered on one deck, the women on another.

After a year in makeshift quarters at Fort Bliss – in what had been a mental ward in the former hospital annexe – the von Brauns finally took their honeymoon. Wernher bought a second-hand Nash saloon car on credit in El Paso. The car had special seats that converted easily into beds for economy-minded tourists. After unfolding highway maps of the West, Wernher asked Maria, 'How about a honeymoon trip?' Maria replied, 'It's about time.'[31]

On his return from Germany, von Braun had brought with him not only his young bride – 'a sensitive person . . . a Dresden doll', as

Maj Hamill remembered[32] – but also his ageing parents, who had lost everything they owned in the war. Baron Magnus and Baroness Emmy adjusted to their new life as best they could. They became fascinated with American Indians, among other aspects of life in the United States.

Von Braun encouraged other single members of his team at Fort Bliss to marry and start families in what inexorably was becoming their adopted country. One ploy he used was to include selected bachelors among the guests he and his bride invited to the small dinner parties hosted in their quarters, to expose the single men to the newly-weds' domestic happiness.

One such bachelor was aerodynamics expert Werner Dahm. One day he received a telephone call inviting him to dinner. The caller did not identify himself, but Dahm was sure it was his supervisor, Ludwig 'Lutz' Roth. It wasn't. The bachelor showed up at the Roths' personal quarters on time only to learn there was no dinner party. It was von Braun who had called, and Dahm had been a no-show for dinner with Wernher, Maria, and other invited guests, he recalled years afterwards. 'How embarrassing! My older colleague Fritz Kraemer gave me sage advice, which I have never forgotten: "Get some nice flowers and apologize to Mrs. von Braun."' The peace offering worked. A second invitation led to a pleasant supper at the von Brauns' quarters.[33]

As the 1940s wore on, the Army relaxed its restrictions on the German rocketeers. They were free to mingle with the townspeople of El Paso, where they not only encountered little hostility over the war but found an easy acceptance from their former enemies. 'In America you don't seem to carry grudges, as do many Europeans who have been enemies,' von Braun said.[34]

Some of the Germans became intrigued with the West. They wore sombreros or cowboy boots and hats. They were also free to travel out of state without escorts, as long as they kept the Army posted on their itineraries. Beginning in 1948, von Braun was allowed to attend scientific conventions and present technical papers – but in the United States only. His first paper was on Earth satellites.

Proposals by von Braun and Col Toftoy to launch new, larger rocket development projects – some with space-exploration aspects – were shot down by the Pentagon. But operating on 'a shoestring budget and with makeshift facilities', the team did manage to take on several challenges beyond the repetitive V2 launchings at White Sands.[35] Their initiatives included work on a winged ramjet missile and several new versions of US Hermes missiles for anti-aircraft and other purposes. One effort involved putting a US Wac Corporal rocket on top of a V2. The two-stage result, called Bumper, scored several 'firsts' in rocketry: a record altitude of nearly 250 miles, far out into airless space; a top speed of more than 5,000 mph; the first animal trips into space, with monkeys; and the first television transmission from a high-altitude rocket.

Von Braun and his colleagues also participated in a project that finally took the V2 to sea and brought a kind of closure to the team's abortive wartime effort to have a U-boat tow the missiles across the Atlantic to bombard New York city. The postwar project was a test launch of a V2 from the deck of an aircraft carrier, the USS *Midway*, in 1947 near Bermuda. All did not go well after lift-off, as Capt William C. 'Bill' Fortune, US Navy, who was present, reminded his friend von Braun many years later. 'The launch was a success, giving credence to Admiral Dan Gallery's premise that the Navy should get into space. Some weren't so sure immediately thereafter, when the V2 veered toward the bridge, with admirals, generals and the rest of us trying to dig foxholes in the steel decks.'[36]

The seagoing V2, nicknamed 'Sandy', broke apart early in its wayward flight. Still, it helped prove the feasibility, after a fashion, of naval launches of large ballistic missiles – from future giant Polaris nuclear submarines, for instance.

In all, the von Braun team and its American partners launched some seventy V2s from White Sands between 1946 and 1951 for training and research purposes. More than two-thirds of the shots were rated successful. Some of the failures were spectacular. Survivors of that era will always remember one especially wayward missile. Shortly before seven o'clock one evening, the V2 blasted off on a trajectory that should have taken it north, over the desert. It rose beautifully but turned smoothly, to everyone's horror, in a southerly direction, towards Mexico. The missile was responding perfectly to the guidance commands of its gyroscope, which had been installed backwards!

One of the top scientists in the group, Ernst Steinhoff, forcibly restrained an engineer who wanted to push the 'destruct' button before the errant rocket had exhausted its explosive fuel supply. Fifty

miles away, it roared over the heads of fiesta dancers in Juarez –
and crashed harmlessly at the edge of a nearby cemetery. Fort Bliss
and White Sands commanders made immediate apologies to the
Mexican authorities.

'But by the time American officials got over there,' Steinhoff
revelled in telling, 'some enterprising Mexicans had set up a souvenir
stand near the cemetery and were selling the missile pieces to
tourists!' The missile engineers who went from White Sands to
Juarez to inspect the impact crater, the story goes, found that the
total weight of the rocket parts offered as 'genuine' souvenirs
equalled that of at least three V2s.[37]

The years at Fort Bliss held a combination of happiness and
disgust, of hope and despair. In 1947 and 1948, for example, von
Braun wrote a story of the first manned, large-scale, international
mission to Mars as an essay with voluminous technical appendices.
Dorette Schlidt typed *The Mars Project* manuscript. It was first sent to
a New York publisher in 1948. No sale. Eventually, eighteen US
publishers rejected it. Only parts of it were ever published in America
and elsewhere, beginning in 1952. Another writer turned it into a
novel, with which von Braun disavowed any connection.

On 9 December 1948 came happier news: the joyful birth of the
von Brauns' first child, daughter Iris Careen.

The following summer, in a surprising move so soon after the war,
the British Interplanetary Society invited von Braun to become an
honorary fellow 'in recognition of your great pioneering activities in
the field of rocket engineering'. Unable to appear in London in
person, von Braun nevertheless sent his grateful acceptance. He
noted both his 'regret' that his life's devotion to the 'noble cause' of
rocketry had increasingly found 'military application', and his little-
known ties to England, adding, 'You may imagine that the wartime
abuse of our V2 baby against the country I am connected with by so
many links has been one of the most disappointing experiences in
my life.'[38]

It was several years after he began his new life in the American
south-west before von Braun experienced the land of cowboys,
horses and ranches. He had enjoyed horseback riding in his
homeland as a youth, and he had continued riding as a college
student in November 1933 with – purely because of the convenience
of the stables, he later explained – the SS equestrian organisation,

SS-Reitersturm I, at Berlin-Halensee. In 1949 he seized the chance to play cowpoke on a Texas ranch. That year, Dr Hubertus Strughold, an aviation medicine specialist at an Air Force school in San Antonio, invited von Braun to lecture there on the past and future of rocketry. Von Braun accepted. Professor Strughold, ultimately hailed as the 'Father of Space Medicine', reminded the rocket leader years later about what happened next: 'At the end of our dinner, you said to me: 'I have seen so many Western movies and heard so much about Texas ranches – but I never have been on a Texas ranch!' The next day, late in the afternoon, we rode in my car to a famous ranch near New Braunfels. At the ranch we were shown some dozen horses wearing the colorful Western-style saddles. You even made a tour on a horse around the ranch and were very pleased.' Then the young von Braun returned to work on his rockets and what Strughold noted as a very different breed of 'horse power'.[39]

Von Braun continued to be thwarted by the lack of any major new project to advance rocketry and space exploration, and by the lack of scientific research facilities and other resources for his team. Stuhlinger recalled that an impatient von Braun in those days 'used to say [to his German associates] it would be a mistake to think only about space. He told us, "Space isn't our problem. Our problem is time."'[40] The team was not getting any younger, and its goals were no closer to realisation.

Von Braun later referred to that period as their years of wandering in the wilderness. The less than blissful El Paso/White Sands era was 'irretrievably lost' in the postwar development of more potent rockets.[41]

Another problem at Fort Bliss, as von Braun and his team saw it, was the relationship with Jim Hamill. The gangly officer, still in his twenties when given the command, held the difficult job of military commander of these recent enemies throughout their stay in Texas. Although they respected his military rank and position, they cared little for his command style. They found him too authoritative and short-tempered, and often cool to their concerns. Moreover, Hamill tended to ignore von Braun's written complaints, requests and even threats to resign. Some attributed it to Hamill's hammering home the point that he was in charge, that the German team was under him, not von Braun.[42]

At low points during this period, von Braun – the natural optimist, the tireless cheerleader – did consider quitting and joining industry.

But encouragement from several key people, including then-Capt William E. Winterstein, US Army, coupled with his conviction that only government could undertake a space programme, led him to stick it out.

Reflecting von Braun's deep frustration, one letter of resignation in January 1948 stood out in a collection of his personal papers more than two decades later. In the four-page handwritten letter he complained of being bypassed and ignored by Hamill to such an extent that it forced him to conclude, 'I do not have your full confidence anymore.' He cited several specific incidents. Unless Hamill could show 'unlimited confidence' in him, von Braun would step aside for a lesser role. He suggested that Ludwig Roth replace him as head of the German group.[43]

Von Braun later recalled being 'irritated' then and acknowledged having written the letter, but added he may never have given it to Hamill. 'I apparently wrote the letter, slept on it and then decided not to send it,' he speculated to me.[44] (Or just as easily, it could have been his handwritten duplicate of a letter he wrote and delivered; or Hamill could have returned the original to von Braun in rejecting his ultimatum.)

'Whether he wrote it and then put it in his pocket and never sent it, or whether he did send it, I really can't recall,' Hamill said nearly a quarter-century later. 'Anyway, he did that [threatened to resign] every two weeks. We parted company many times – not really, of course. But, as Dr von Braun himself used to say, "You don't make progress without friction." Well, we made progress, and of course there was some friction.'[45]

Ever the diplomat, von Braun took pains even decades later to tell me: 'Jim Hamill handled his very difficult job in a splendid fashion. I would not want this letter to indicate that I felt he had done his job clumsily or that he had almost broken up our team, because that certainly was not the case.' Nevertheless, the letter clearly reflects that such was the case from von Braun's perspective at the time.[46]

If the years from 1945 to 1948 had largely been a period of professional gloom for von Braun and his team, a series of events in 1949 changed everything. The Soviet Union tested atomic weapons, and intelligence reports revealed significant progress in its ballistic missile development efforts. It became clear that what Winston Churchill two years earlier termed an 'Iron Curtain' had indeed fallen across Europe, delineating the Soviet-dominated Communist bloc and launching what came to be called the Cold War. At the same time, very real war clouds loomed over a politically divided Korea.

That year, a Pentagon that had vetoed Col Toftoy's earlier proposal to develop a missile with nuclear capability and a range of 500 to 1,000 miles suddenly embraced a form of the idea. It ordered the Army and its German rocketeers to invent a 200-mile, nuclear-capable missile on a high-priority basis, to be followed by longer-range ballistic weapons.

On 1 August 1949, Toftoy got word that his request for expanded facilities at Fort Bliss for the escalated rocket programme had been denied. More of the base was now needed for the threatened Korean conflict. Two weeks later, Toftoy visited North Alabama to check out a pair of mothballed Second World War Army arsenals. Overcoming early opposition, he won the Pentagon's approval to combine the Huntsville and Redstone Arsenals as a home for the newly minted Army Ordnance Rocket Center and a corps of several hundred US-born and imported German rocket specialists from out West.

Word of the impending relocation reached the von Braun team in an offhand way. Hannes Luehrsen, chief architect at Peenemünde and the rocket team's lead architect-planner at Fort Bliss, recalled that he was at work one day when Maj Hamill 'came in my office, tossed some drawings and maps on my desk, and said: "Here's where we're going!"' Hamill explained to a startled Luehrsen all about Redstone Arsenal. 'Does von Braun know this?' the planner asked. 'No,' replied Hamill. 'I'll tell him tomorrow.'[47] Although von Braun may well have already been told of the plans by Col Toftoy, Hamill's nonchalance further illustrated how things stood between the two men.

The US government had begun thinking the year before about the Germans' irregular resident status in America and what to do about it. Matters came to a head in 1949. With their multi-year work contracts and the plans to move to Alabama the following year, von Braun, his Peenemünde veterans and their families heeded advice to start the US citizenship process. A small problem developed with the INS (Immigration and Naturalisation Service) bureaucracy: because the Germans had entered the country in 1945 as 'special employees' of the Army and had bypassed the INS, they were technically illegal immigrants.

The government decided that the alien group must physically re-enter the United States as legal immigrants so that the five-year waiting period for citizenship could commence. The solution: walk across El Paso's Rio Grande Bridge into Juarez, Mexico, then turn around and immediately return to US soil – a distance of two blocks.

'On my immigration papers,' von Braun enjoyed later recounting, 'where the "Vessel of Entry" column normally would have a romantic name such as "Queen Mary" or "Isle de France" or "Mayflower", it states: "Entered at El Paso, Texas – via Streetcar"! . . . We called it our "Streetcar named Desire". The fare was 5 cents, and it was the most valuable nickel I have ever spent.'[48]

In northern Alabama, the German–US rocket team would have a new workshop and a secluded, spacious playground. Dorette Kersten Schlidt remembered von Braun's return to arid Fort Bliss after a first visit to Huntsville and the nearby 40,000-acre Redstone Arsenal in the Appalachian foothills in Tennessee river country. She recalled his excitedly telling his German compatriots: 'Oh, it looks like home! So green, green, everything is so green, with mountains all around!'[49]

One of the American civilians who had signed on with the Army's expanded rocket team at White Sands, Ramon Samaniego, remembered von Braun's enthusiasm over the move. 'Dr. von Braun came and personally talked to me about going [to Alabama]. "We are going to make history," he told me.'[50]

New Home Alabama

Two adjoining, shuttered Second World War Army Ordnance and Chemical Corps weapons-producing arsenals in agricultural northern Alabama were combined to form Redstone Arsenal, the Army's new rocket development centre. Along with the existing buildings and wide-open spaces came cheap, plentiful electric power from the Tennessee Valley Authority, access to the navigable Tennessee river, and thus a connecting water route clear to the Gulf of Mexico.

Beginning in the spring of 1950 and continuing through that year, von Braun and his team of peripatetic rocketeers moved with their families to their new verdant 'homeland'. The contingent consisted of several hundred General Electric Company contractor employees; a small corps of young US Army draftees with engineering, science and maths degrees; Army civilian workers; and about 115 German missile men under contract. They set up shop at the sprawling base of almost 40,000 acres. Of von Braun and his team, future Redstone public information chief David G. Harris later observed: 'They arrived with the US space programme in their briefcases, only we didn't know it then.'[1]

Huntsville, with a population of 15,000, was then the self-proclaimed 'Watercress Capital of the World'. It was also known as the birthplace of actress Tallulah Bankhead and the Madison County seat from which the fateful Confederate order to fire on Fort Sumter had emanated in the American Civil War. But by the mid-twentieth century it was primarily a community where King Cotton reigned, both in the surviving textile mills and in the white-dappled fields covering the countryside. During the war, a workforce of more than 14,000 had earned paycheques by making poisonous gases and other 'chemical munitions' at one arsenal, while at the other they loaded the stuff into shells and also prepared conventional artillery shells. Then, after peace broke out, the jobs nosedived to zero. Although economically deflated in the postwar period, Huntsville's response to the late-1949 news of the German rocket team's imminent coming was mixed. Not all the smattering of negativism reflected hostility towards the wartime enemies of just four years earlier. Before the announcement of the reopening of 'the Arsenal' for rocketry research and development, Huntsville had been considered for a big new Air Force wind-tunnel facility. That plum would have required a workforce of 3,500 for construction and another 3,500 for operation. A hungry Huntsville lusted after the facility. Alabama's junior US senator, John Sparkman, who lived in Huntsville and who had been born in Hartselle in the next county, had been lobbying hard for it, as had Alabama's senior senator, Lister Hill.

But the senior – very senior – senator from Tennessee, Kenneth D. McKellar, had more clout. In the autumn of 1949, the 80-year-old McKellar, serving in his sixth six-year term, was chairman of the omnipotent Senate Appropriations Committee. He was not requesting, he was demanding that the wind-tunnel facility go to Tennessee rather than to the Air Force's preferred Huntsville location.[2] Tennessee's junior senator, Estes Kefauver, carried some pretty fair clout of his own as well.

So the coveted wind tunnel, which became the US Air Force Arnold Engineering Development Center, went to a remote wooded site near Tullahoma in the Tennessee hills a little more than an hour's drive north-east of Huntsville. Soon afterwards, it was announced that the Germans were coming to Redstone. The development was widely seen as a consoling bone tossed to the Alabama interests. Huntsville native Patrick Richardson, then a law student at the University of Alabama in Tuscaloosa but later von Braun's personal attorney, summarised half a century later the reaction of the disappointed community: 'They're getting 7,000 workers [in Tennessee], and we're getting a hundred Krauts.'[3]

Whether because of the town's economic hunger or simple good manners, or both, open enmity towards von Braun and his band of German missile men was minimal. Isolated instances included a petrol station owner, still grieving over the loss of a loved one in the war, who posted a sign announcing that Germans were not welcome as customers. That sentiment was understandable. 'The last time our boys had seen Germans, they were shooting at them,' one townsman later remarked of the coolness.[4]

But in general the von Braun team found a wait-and-see attitude among the people of the area. 'The local natives', von Braun soon observed, 'at first were as sceptical about us as we were curious about them.'[5] One of his top men, Karl Heimburg, could smile nearly two decades later in a wry recollection of the townspeople's period of adjustment: 'It took them a while to accept us. About five or ten years.'[6] The city's popular mayor at the time, Robert B. 'Speck' Searcy, later admitted, 'At first we thought they were a bunch of crazy rocket men.' That applied in spades to their leader. 'I thought Dr. von Braun was a nice fellow when I first met him, but I thought he was crazy,' Searcy confessed in 1962. 'Now I'm willing to believe anything he tells me, and if he says we're going to visit the planets and the Moon, then that's what we're going to do.'[7]

Moving to north Alabama was much less of an adjustment for the Germans than coming to El Paso had been. 'As Germans, we felt much more at home here [in Huntsville],' remembered von Braun's head architect-planner Hannes Luehrsen. 'Everything was so green and beautiful. I remember smelling the fresh-mown hay at Redstone. And the trees! At Bliss we had had to drive two hundred miles to see five trees together!'[8]

But what of the Deep South's history of racial troubles? Especially Alabama, the self-proclaimed 'Heart of Dixie', its capital the proud 'Cradle of the Confederacy'? Would the Germans find cross-burnings and lynchings still being perpetrated?

Scientist Werner Dahm and his German compatriots were not far removed in time from the Third Reich's genocidal Holocaust. 'We had some concerns here,' he recalled, 'not so much about segregation' as about possible open strife between whites and 'coloureds'. He said, 'We were quite relieved when we found it wasn't that bad here, at least the signs we could see.' He remembered heading to work on one of his first mornings at Redstone Arsenal in 1950 and seeing a pick-up lorry with a white worker and a black worker in the back engaged in a lively discussion. Dahm thought it was a positive sign.[9]

Von Braun was 38 when he moved to Alabama. He wasted little time carrying his gospel of space exploration to the people of

Huntsville, Madison County and the surrounding area. His first speech to the local Kiwanis Club included a slide presentation showing his step-by-step plans for putting a man on the moon, the space shuttle and other things he had in mind, the then-general manager of the local Chamber of Commerce later recalled. He added, 'I remember one of our well-known farmers remarking to me: "The day a man lands on the Moon, I'll fly a bale of cotton to Washington and back with wings strapped to my back!"'[10]

Another community leader, lawyer Louis Salmon, recalled a presentation the scientist gave on past and future rocketry at a current affairs seminar for the fledgling University of Alabama-Huntsville Extension Center in 1951. The attorney found the review of the past 'fascinating', but the futuristic part struck him as completely 'incredible'.

He later reminisced with von Braun: 'You spoke of such senseless things as weightlessness, walking in space, space stations, and concluded with the forecast that with funding we could have a man on the Moon in ten years. Charlie Shaver [a lawyer friend] and I walked out of the meeting together and agreed you belonged in Tuscaloosa – Bryce Hospital [the state mental asylum], not the University.'[11]

With equal candour, M. Beirne Spragins, then-chairman of the board of Huntsville's dominant bank, later talked with von Braun about their early days together. 'I was always interested in missiles, but when you were quoted to me as saying that you were going to the Moon, I said "the damn man is crazy!" Later, when I got to know you better, I told you what I had said and we both enjoyed it.'[12]

In those early years in Huntsville many thought von Braun was not only 'touched' in the head but godless, too. His file of 'religious mail' included one note from a woman beseeching him to forget all this irreverent space talk and just 'stay home and watch television like the Lord intended!'[13]

Few then knew that the man of science and technology had a strong spiritual side – and extensive knowledge of the world's religions. Von Braun used his familiarity with the Old Testament to good advantage one evening in the 1950s when he faced a fundamentalist religious group at a dinner meeting. A church deacon there challenged him: 'This drought we've had in Alabama these past two years has ruined our crops! When are you going to stop punching holes in the clouds with those rockets and drying up the rain?'

Scattered applause broke out, and von Braun stood to respond to the deacon. 'I know you are familiar, sir, with the Bible and with the story of Jacob's ladder. The angels are ascending and descending the ladder. So are we. If the good Lord does not want us to go up and down His creation, all He has to do is tip over the ladder.'

'The applause was deafening,' recalled Army information officer Reavis O'Neal Jr. 'That day Wernher von Braun became Huntsville's own personal angel!'[14]

It was banker Spragins who helped von Braun build his first home in America. Not long after the rocket team's arrival, the banker promoted a meeting at the burgeoning town's Russel Erskine Hotel to seek Federal Housing Administration (FHA) financing for much-needed homes for the new arrivals. At the head table with Spragins were Ludy Toftoy (now a brigadier-general at Redstone), von Braun, an FHA official and several other luminaries.

'Things were going so good', the banker later reminded the rocket leader, 'that you leaned over and asked if it would be okay if you asked about financing your residence. This was immediately agreed to during the meeting.' But a small hitch developed for von Braun, who was renting a home for his family and a nearby flat for his parents: he lacked the required cash down-payment for the home loan. No problem. Five new friends made him an unsecured loan so he could qualify for the FHA-backed mortgage. 'Needless to say,' remembered his banker friend, 'the house was built, the loan repaid', and all ended well.[15] The von Brauns' mortgage payment for the three-bedroom home at 907 McClung Street, a mile from the town centre, was $61 a month, taxes and insurance included.[16]

Several other German families soon pooled their resources, passed the cash around from account to account, obtained FHA loans and built starter homes near the von Brauns' on a knoll quickly dubbed 'Sauerkraut Hill'. Other ex-Peenemünders jointly bought a sizeable parcel of land on top of Monte Sano (Healthful Mountain), a mini-sized alp rising 1,000ft above the city on its eastern edge. They subdivided the parcel into lots and built homes in a cluster there, which some townspeople called 'the German Colony'.

Hans F. Gruene, one of von Braun's Sauerkraut Hill neighbours and colleagues, later shared with his boss the story of an incident involving a workman who apparently didn't realise Gruene was one of those crazy German rocket scientists. 'When you had your first paint job done on your house at McClung,' Gruene related, 'the painter came to me in the backyard and told me point-blank that he was not too happy to work for "a nut that wants to go to the Moon despite the fact that the Bible says we could [not] and should not go". I did not try to match . . . Bible quotations with an Alabama handyman because I knew I would have lost, hands down.' Instead, the German scientist tried to use 'pure logic' in defending notions of space travel with the workman. Gruene said he lost at that, too.[17]

And then there were the yelping beagles. Robert and Frances Gates Moore's home stood on a 3-acre site directly across from the von Braun residence. Engineer Bob Moore had gone to work in 1951 at Redstone in Hans Hueter's laboratory in the Guided Missile Development Group, of which von Braun was technical chief. The Moores at one point had a pack of six young beagles with a habit of leaving the house in the small hours of the morning to chase rabbits through the nearby fields. The yelping pups were waking up von Braun early in the morning, and he was not an early riser.

'Dr von Braun called Mr Hueter', Frances Moore remembered years later, 'and asked him to please ask Bob to please keep the dogs penned up until seven o'clock in the morning, because they were coming under his bedroom window about four o'clock in the morning, hot after a rabbit. He didn't want to call Bob himself because he was afraid it would scare a young engineer to death! He was extremely considerate.'[18]

Von Braun's feelings for his graceful wife matched her devotion to him, friends and associates observed. As their family grew (they eventually had three children) he encouraged Maria to avoid the hausfrau trap. While remaining, by all accounts, a devoted young mother who put her children's interests ahead of her own, she earned a pilot's licence in Huntsville, attended horse shows in Tennessee and elsewhere, enjoyed water sports and indulged her love of travel, attending concerts and touring museums whenever possible.

Maria von Braun had let her husband know at the outset of their move to Huntsville that as a big-city girl, raised mostly in Berlin, she had no desire to become a rocket-and-space widow stuck in small-town Alabama. Friends said her husband kept his promise to make at least one trip each year to New York for Broadway shows, symphony concerts, art museums, fine dining, Christmas shopping and just being with each other. In 1951 he wrote to a British rocketry colleague in London, in part to explain a lapse in correspondence: 'I spent two weeks on a vacation . . . at Lake Wisconsin, and I had promised my wife that there would be no rockets, no letters and no nothing even remotely reminding of business. As to the moon, her role was reduced to a device for the stimulation of romance.'[19]

Recognising early on that his German comrades tended naturally to be absorbed in their work and to associate only with one another away from the job, von Braun urged them to get involved in the community. Many responded by doing just that, in both large and small ways. They became active as musicians and board members with the Community Concert Association that led to the Huntsville Symphony Orchestra. They physically helped build a Lutheran church and a community astronomical observatory and planetarium. They took active roles in museum and arts groups, in school PTAs and in scout troops with their children. They taught night classes in technical courses at the new University of Alabama-Huntsville Extension Center.

'It is essential', von Braun preached, 'that the team members understand the system of government under which they work and that they maintain contact with the unscientific members of the community. . . . We must meet the people and learn to share their problems and their activities. Only in this way can we become real Americans instead of transplanted Germans.'[20]

The gregarious von Braun was quick to practise what he preached. He mixed and mingled. He became president of the local astronomical society that he, team members and new friends had founded. He went hunting and fishing with men of the city's business, professional and political leadership circles. He formed close, valuable friendships with such influential people as Reese T. Amis, erudite long-time editor of the *Huntsville Times*, who had been an artillery captain fighting Germans in the First World War; Milton K. Cummings, a wealthy cotton broker, stock trader and Democratic Party insider; and the town's own US senator, John Sparkman. The von Brauns socialised with them and their wives. They judiciously accepted social invitations from others within the community, although they tried to avoid hosts who merely wanted to 'show us off', as Maria complained on occasion. Von Braun spoke to countless organisations. Maria took part in many of their children's school and extra-curricular activities, as did Wernher to the extent he could.

At parties and receptions in Huntsville-area homes, Wernher was an outgoing guest, an engaging conversationalist. His wife, although naturally reserved, was regarded as thoroughly charming in her own way. To attorney Patrick Richardson she seemed like a princess.[21] He also remembered a small dinner party he and his wife hosted at which the von Brauns were among the guests. The mellow, after-dinner conversation turned to the question of what was truly important in life, what had lasting value, aside from love of family. Von Braun spoke of how generations of his family were raised to believe in conserving their ancestral lands, then passing them on to the next generation. They were taught that the land would sustain them, always, and that it was the most important thing in the life of the family. But then von Braun and his brothers saw their ancestral lands lost to war and political events, 'and we came to realise that all one can be sure of leaving one's children is what's inside their heads. Education, and not earthly possessions, is the ultimate legacy.'[22]

In the Huntsville workplace, von Braun enthusiastically employed the same approach with the development of the Redstone missile as he had used with the V2 in Peenemünde. Akin to the US Army's traditional 'arsenal system', it called for building strong internal R&D capabilities up to and including prototype production. Pilot manufacturing of the first missiles 'in-house' strengthened the agency's hand later when they were overseeing and evaluating the production contractor – in the Redstone's case, Chrysler Corporation.

Among von Braun's most notable skills, observed Col Edward D. Mohlere, US Army (Ret.), in 1998, was his ability to develop a high-quality technical workforce from the available labour supply in Huntsville.

Just picture the existing technical capabilities of an extremely rural area. It's very low on the scale. One of the remarkable things about von Braun was his taking training programmes and using them to train people who had small farms in the area and who rode around in pickup lorries, and making of them a very competent, highly qualified technical workforce. They were working in dimensions and tolerances that are almost unbelievable. A millionth of an inch here and there was quite important.[23]

Starting in 1953, the Redstone missile, patterned in part after the V2, was flight-tested at Cape Canaveral on Florida's east coast. When von Braun first saw the cape he was struck by its similarity, in its 'isolation and inaccessibility', to his former seaside missile base at Peenemünde.[24] An early test there produced a classic example of von Braun's promotion of honesty in the pursuit of perfection. The rocket soared flawlessly until mid-flight and then suddenly failed. Telemetry readings confirmed the 'bird' had performed well until a precise point. That enabled troubleshooters to localise the likely source. The suspected area had been checked and rechecked during many lab tests, but none of the possible explanations seemed to ring true. Finally, one was accepted as the likeliest, and corrective action was ordered. Several versions of what happened next circulated in conversation and print over the years. Here is what von Braun himself wrote:

At this point an engineer who was a member of the firing group called and said he wanted to see me. He came up to my office and told me that during pre-launching preparation he had tightened a certain connection just to make sure that there would be good contact.

While so doing, he had touched a contact with a screwdriver and drawn a spark. Since the system checked out well after this incident, he hadn't paid any attention to the matter. But now that everybody was talking about a possible failure in that particular apparatus, he just wanted to tell me the story for whatever it was worth. A quick study indicated that here was the answer. Needless to say, the 'remedial action' was called off and no changes were made.

I sent the engineer a bottle of champagne because I wanted everybody to know that honesty pays off, even if someone may run the risk of incriminating himself. Absolute honesty is something you simply cannot dispense with in a team effort as difficult as that of missile development.[25]

It was von Braun's 'rare brand of leadership' that Col James K. Hoey, US Army (Ret.), saw as equal in importance to his 'imagination and technical brilliance'. Hoey recalled that around 1951 von Braun had decided to tackle 'that most traumatic of events in bureaucratic life – a reorganization – and a far-reaching one at that'. It especially affected the director's German-born colleagues. As Hoey elaborated to von Braun on his 60th birthday:

> You and one or two others had worked out the structure that you felt would do the job and had hung names on the various blocks. At this point, you divorced yourself from day-to-day operations for about two weeks. You could be seen slipping in or out of your office with one of the group in tow; or engaged in long and earnest conversation with some other individual in the corner of the coffee shop; or wandering for hours among the trees outside . . . the old headquarters building with one or another of the more difficult personalities.
>
> The result was that when the reorganization was finally announced, there were no surprises; the scars [were] minimal, and were exactly as predicted. The team moved on without missing a step. It was a quietly magnificent performance.[26]

The colonel added that the validity of von Braun's approach was evidenced by the fact that 'the resultant internal structure stood for a good many years without change'.

Von Braun's 'skill as a director and a manager of people' was the quality that George C. Bucher, an Army manager at Redstone (and later with NASA), found most memorable. As a newcomer to the organisation in the 1950s, Bucher got his first exposure to the rocket leader's handling of his characteristically marathon meetings. While 'entranced by the technical discussions and decision-making in [the] meetings with the [German] laboratory directors, I sometimes thought, "Gosh, this is getting long-winded. Every single person feels he has to speak. Why doesn't Dr. von Braun take a strong stand, cut off discussions, and make a decision on his own?"'

Years later, Bucher told von Braun that he failed to realise then that he was

> purposely directing the meeting in your own unique and masterful way. After giving everyone an opportunity to speak, you then skillfully synthesized the contributions so that everyone felt he had contributed to, and hence was committed to, the objective that you defined and the action you outlined. You then turned to

those who had supported a somewhat different approach and asked, 'How does this sound to you? Can you live with it?' Invariably, the response was, 'I'll do my best to support it.' When the meeting ended, everyone knew what was to be done . . . and knew every strength and every weakness, every bright spot and every dim spot, in every laboratory that was to contribute.[27]

The team's liquid-fuelled rockets, with their fantastic mazes of pipes and valves and pumps and tubing, required maximum engineering – von Braun's first love. He would get out into the shops and labs, seeing the detail work up close, getting his hands dirty and schmoozing with his 'co-workers', as he referred to employees. Engineer Donald Bowden was in his early twenties when he was put in charge of operations at a test stand at Redstone in the 1950s. 'I'd look up, and out of the blue there'd be von Braun coming around the corner,' Bowden remembered. 'And he'd call us all by name; he'd learned everybody's name. . . . He didn't have any qualms about talking to everybody.'[28]

Around lunchtime one day in April 1952, Larkin Davis was busy at his workbench in the jig-boring room of the machine shop in Building 4711 at Redstone. He was startled to see the missile centre's technical director and another man walking towards him. Von Braun, whom Davis had not met, introduced himself and then his companion, an engineer from the Bulova watch company. The director asked Davis to brief the visitor on the work being done. Davis explained a portion of the work that involved machining extremely 'close-tolerance, precision parts'. About then, 'Dr. von Braun turned to the engineer and said, "I'm just a plumber myself."' Davis added, 'That engineer and I almost broke up laughing.'[29]

No one in Huntsville worked more closely with von Braun during this period than a small-town northern Alabama woman named Bonnie Holmes. Two years after the rocket team arrived, she was 21, a high school and business college graduate, and a civil service clerk-stenographer working at Redstone Arsenal. She answered a call for workers willing to change jobs at the Army post. Twenty years later she wrote a letter of reminiscence to her long-time boss:

In early 1952 while I was working in the Army Post Engineer Office your intriguing article 'Crossing the Last Frontier' appeared in *Collier's* magazine. It was to me a captivating and imaginative article, although there were some who didn't think so.

Shortly after, the Personnel Office called to ask if I would be interested in being interviewed for a secretarial position in the Guided Missile Development [Division], specifically to work for you. At that time, a move from a sure position was a big decision for me, but recalling that inspirational article made it easy. My Post Engineer Office associates didn't [agree]. . . . With looks of shock and disbelief there came the remarks 'You must be out of your mind. Surely you can't really want to work for those crazy Krauts. There is no future for you in such a business.'[30]

The negative talk did not deter the young clerk-stenographer. 'My interest in space had been sparked by your writings and I was determined to go and accept the challenge,' she continued. 'Ever since I have thanked my lucky stars for being selected for the position – for the opportunity to work for you and with you and with all the "greats" in the dedicated team.' She called the eighteen eventful years that she was his secretary 'a great adventure – and the thrill persists to this day.' She could not resist adding in her 1972 letter, 'I wonder what they think of "those crazy Krauts" now!'[31]

Von Braun liked to introduce visitors to his office, famous and otherwise, to Bonnie Holmes, 'the lady I work for – my real boss'.[32] As the secretary he called 'a phenomenon', she handled most of his correspondence, served as office gatekeeper, scheduled all his appointments and travel, and carried out countless other duties. The rocket scientist and his wife sometimes travelled abroad under the names of the secretary and her husband for the sake of privacy and security. 'Von Braun relied upon her heavily,' observed Joseph M. Jones, long-time Army and later NASA public affairs officer at Redstone. 'Dealing with her was tantamount to dealing with von Braun himself.'[33]

Of course, what brought the von Braun team and its American partners to Redstone Arsenal and Huntsville in the first place – aside from their barely suppressed space dreams – was developing new Army guided missile systems. By 1951, work was under way to create the Redstone as a nuclear-capable ballistic missile, designed for a 200-mile range and maximum 6,500lb payload capacity, and the Nike-Ajax anti-aircraft rocket.

As the Redstone missile's development progressed, first came static, or hold-down, test firings on a primitive steel-and-concrete stand built on the cheap at the arsenal by von Braun's team – 'the poor man's test stand', he called it. Then followed modifications based on the test results. After just three years in development, in August 1953 the Redstone underwent the first in a series of test flights from Cape Canaveral. Some were successful, some not.

Witnessing one catastrophic failure on the launch pad, von Braun's Army boss, then-Maj Gen Toftoy, turned to him and asked, 'Wernher, why did that rocket explode?' The scientist said he didn't know, but that just as soon as he and others could check production, test and launch data, he would try to have the answer. The general, however, continued to press von Braun for an immediate explanation – and von Braun continued to say he had none. Persisting, Toftoy asked, 'Don't you have any idea why it exploded?' His patience exhausted, von Braun fired back: 'Yes. It exploded because the sonuvabitch blew up!'[34]

Sometimes it was what the loquacious von Braun did not say that carried a strong message. The launch – or attempted launch – of the Army's third Redstone came on 5 May 1954. Viewing the proceedings from on top of an old lighthouse a mile or so from the pad were von Braun, Toftoy, Sam Hoffman of North American Aviation's Rocketdyne Division, and several others. The missile 'rose only a few feet, faltered and exploded in a white ball of flame due to a Rocketdyne engine failure,' Hoffman painfully recalled years later. The Rocketdyne chief noted that Toftoy had looked at him and asked pointedly: 'You will be here for the next launch, won't you, Sam?'

Hoffman noted that von Braun did not say a word. But his 'silence spoke volumes and impressed me more than anything [he] could have said. I returned to Rocketdyne instilled with a new sense of dedication and purpose.'[35]

It was during such trying times that von Braun pressed his agency and contractor team even harder for the highest quality of workmanship and reliability of product. Twenty years later, Arlie R. Trahern Jr, a Chrysler Space Division executive, recalled that von Braun had disarmingly promoted missile reliability as something 'which would make the target area more dangerous than the launch area'.[36]

In time, successful flight tests of the Redstone came with increasing frequency, eventually earning it the nickname 'Old

Reliable'. But even its being declared operational, put into production and deployed in the mid-1950s did not slow engineering modifications to the weapon. The multiple design changes required the issue of a series of field 'mod kits' to soldiers already deployed with the missiles. A meeting was called at Redstone Arsenal to deal with the changes continuing to emanate from the weapon's development office, headed by veteran rocket team member Arthur Rudolph. With von Braun, Redstone production chief John C. Goodrum Sr and others in attendance, Rudolph persisted in defending the design changes as necessary further improvements in reliability.

'Finally,' recalled Goodrum, 'Dr von Braun reached over, put his hand on Rudolph's shoulder and said, "Art, you remind me of the fellow who married a virgin and kept telling her how good it was going to be once he got it perfected."' A design freeze soon went into effect.[37]

In the early days of American missile development, Gen James M. Gavin, US Army, and a technical group in Washington were engrossed in evaluating Soviet missile programmes, mainly via extensive covert photography. But the experts at the Pentagon were 'at a loss to understand much of the information' about the new Soviet missile sites. Gavin suggested that von Braun be brought up to examine the pictures. That idea was vetoed. 'I was told that he couldn't be allowed to see them because he didn't have the necessary special security clearance!' recalled the general. The group was making no progress, however, so it ultimately relented and had the German scientist come up from Huntsville to see if he could help. 'He came to my office, where we showed him all the photographs,' Gavin related. 'At once he described the complete system, what the plumbing was, what kind of fuel they were using, where they were storing it, where the missile was fueled, and gave us the general characteristics of the missile.'[38]

The young metallurgist William R. Lucas, who had early on joined in the development of the Redstone missile and who worked with von Braun for nearly all of the latter's twenty years in Huntsville, quickly recognised the rocket scientist's brilliance in technical leadership. But he also soon grasped other realities about the man. For one thing, his 'tremendous breadth of intellect' was more impressive than his 'depth [in] any one area'. For another, when he had an idea, goal or fixed view, 'you couldn't dissuade him from it. . . . You had to get his attention.'[39]

One of the scientific challenges that had drawn Lucas to rocket work was corrosion of lightweight missile materials in the environment of a subtropical sea-coast launch site and corrosion caused by liquid propellants. Lucas and his boss at the time, Karl Hager, were seeking equipment and funding for a study of such corrosion. For von Braun's 60th birthday book Lucas wrote:

Since you had just come from White Sands where corrosion was one of the least problems, we were meeting very little success in convincing you. Then you spent a week in Florida in preparation for a launch. Upon your return, you directed us to begin a program on corrosion immediately because in one week at Cocoa Beach, the bottom of your tooth-powder can had rusted and the salt air had attacked the chrome on your automobile bumper. White Sands was never like that. You 'convinced' us that Florida was much worse and that we must do something immediately. We were delighted to implement your idea.[40]

Although von Braun's work occupied much of his time and attention, he did not let it crowd out personal and family life. Wernher and Maria's family grew during the early 1950s. For a time, three generations of von Brauns resided in Huntsville. Brother Magnus also lived there until 1955, when he left the Army to work for missile contractor Chrysler Corporation in Detroit. The brothers' mother, Baroness Emmy, began teaching English classes for the wives of the other German rocket experts. Their genial, walrus-moustached father, Baron Magnus, regularly took walks in the community, acknowledging the greetings of residents along the way. In 1953, von Braun's parents decided to return to Germany, where the baron had qualified for a liveable government pension for his years of pre-Hitler public service. There he wrote a memoir, *From East Prussia to Texas*.[41]

In the beginning, Wernher and Maria were almost as accessible as the elder von Brauns to their fellow Huntsvillians. Their home telephone number was listed in the city's directory – until one New Year's Eve, when a drunk called to request 'a ticket to the moon, Doctor'. Then the von Brauns went ex-directory.

On 8 May 1952, a second daughter, Margrit Cecile, was born to
Maria and Wernher, joining big sister Iris. Both girls had dual US
and German citizenship at birth. Wernher and Maria had begun the
process of naturalisation as American citizens in 1949, when long-
term employment with the Army seemed assured. That step had
triggered a Department of Justice action and an FBI 'investigation
and surveillance that would last . . . decades and fill thousands of
pages with facts [Nazi Party membership, SS commission, and the
like], but also gossip and often frivolous innuendo'.[42]

On 14 April 1955, Wernher and Maria joined with thirty-eight of
the German rocket experts and some seventy family members and
others in renouncing their German citizenship, which was required
of them to adopt US citizenship. They became naturalised US citizens
onstage in the Huntsville High School auditorium before a crowd of
1,000 townspeople and students. A broadly smiling Wernher von
Braun told reporters it was 'one of the proudest and most significant
days of my life – almost like getting married. I am very, very happy.'

Maj-Gen Toftoy was there. He assured the new citizens that, in the
ten years since he had met them in a schoolhouse in Germany and
invited them to come to the United States, even though he 'promised
you no future', this 'great team of rocket engineers and scientists'
had achieved much. 'I am sure that the future holds even greater
things for you.'[43]

Early Media Trail

For Wernher von Braun, the early 1950s was also a busy time for promoting his vision of space flight to the American public. In wartime Germany, expressing interest publicly in space flight could – and did – get him into serious trouble. But now he was free from Hitler's dictatorship with its secrecy and its controlled news media. He faced no such restraints from his new employer, the United States Army, as long as it didn't interfere with his weapons work. To spread the gospel of space exploration, he began giving press interviews, making speeches and writing articles. 'In those early years, Wernher von Braun was a good storyteller in his own right,' space scientist Rick Chappell observed.[1] And he quickly became savvy about using the mass media, understanding that their help would be needed to tell the story widely and to gain public and political support.

All through the 1950s, von Braun cultivated relationships, some close and long lasting, with journalists and other communicators, including writer Arthur C. Clarke, television broadcaster Walter Cronkite, Walt and Roy Disney, writer-publisher Erik Bergaust, Associated Press missile/space correspondent Howard Benedict, author and magazine writer-editor Cornelius 'Connie' Ryan and, later, US television's Hugh Downs.

The first major breakthrough in spreading the space gospel to the American people came via the popular magazine *Collier's*. Editorial appetites had been whetted in October 1951 when representatives of the magazine attended a symposium on space travel sponsored by New York City's Hayden Planetarium. Plans for a magazine series were finalised the following month over drinks and dinner in San Antonio during the Symposium on Physics and Medicine of the Aeropause.

'Four of us sat at the table through cocktails, dinner, and long into the evening,' recalled Fred L. Whipple, chairman of Harvard University's Department of Astronomy. With him and von Braun were University of California physicist Joseph Kaplan and former war correspondent Connie Ryan, then a writer-editor at *Collier's*.

'Our conversation from the beginning . . . was aimed at convincing a highly skeptical . . . Ryan that space travel was possible and desirable,' said Whipple. 'The fact that we succeeded in this tour de force is evidenced by the subsequent series of articles in *Collier's* magazine and the two volumes *Across the Space Frontier* (1952) and *Conquest of the Moon* (1953).'[2]

The eight articles, written by von Braun, Whipple, Kaplan, Willy Ley and others and published by the magazine between 1952 and 1954 with spectacular illustrations, created a sensation. They made a cogent case that launching Earth satellites and sending flights to the moon would soon be realities – with few advances in existing technology required. Ryan, the erstwhile space sceptic before von Braun and friends made a convert of him, was the lead editorial writer on the articles as well as editor of the books spun out of the magazine series.

Also present at the Texas symposium was the leading space illustrator Chesley Bonestell. After hearing von Braun's formal presentation, the painter had turned to Ryan and said, 'There is the man to send our rocket to the moon.'[3] Bonestell recalled the exchange he had with von Braun when he and Ryan met him in the crowded hotel restaurant:

'Dr von Braun, you ever thought of going to the moon?'

'Call me Wernher. Yes, indeed I have.' (He started to draw a rocket on the paper napkin.)

'What, not streamlined?'

'Do you want it streamlined?'

'No, you design it the way it should be.' Pause. 'Well, never mind at present – we'll go into it later,' the artist added, suddenly aware of 'the craning necks of the rival magazine editors at the next table from *Look* and *Saturday Evening Post*'. They met the next day, at another hotel, away from eavesdropping competitors.[4] The attention-grabbing paintings created by Bonestell and fellow illustrators Fred Freeman and Rolf Klep to accompany the *Collier's* articles became space-art classics.

In reminiscences years later, Connie Ryan remembered the first issue, which came out the day before von Braun turned 40. 'Remember March 22, 1952, when we waited for the reaction to a special *Collier's* issue called "Man Will Conquer Space Soon", urging an all-out US space program? As the editor in charge of the project I shall never forget the controversy that followed. We were both praised and damned for what *Collier's* called "one of the most important symposiums ever published by a national magazine".'[5]

Ryan further observed: 'It is amusing now to read the vehemence of those critics who, at the time, considered your proposals far-fetched.' He cited 'the well-known science writer who called you "the high priest of space" . . . and who summarized *Collier's* issue this way: "If we seriously accept von Braun's plans, then surely he will, scientifically and economically, bankrupt the US."'[6]

The *Collier's* series was also praised highly by British rocketeer Val Cleaver, who wrote to von Braun after the opening issue in March 1952. 'Congratulations to all concerned on that magnificent effort, quite the best piece of popular publicity the cause has ever had. Your hand still retains its cunning – you are not only the world's leading rocket engineer, but a superb "political engineer" as well, a technical salesman without equal!'[7]

Von Braun understood that controversy helped the cause, as illustrated by the second annual Symposium on Space Travel, held at the Hayden Planetarium in New York City in October 1952. Von Braun, Milton W. Rosen and several other scientists had been invited to present papers. Rosen planned to take issue with von Braun's aggressive, 'great leap forward' approach as spelled out in *Collier's* and the spin-off book, *Across the Space Frontier*. Von Braun had proposed what Rosen characterised as a 'mighty three-stage rocket that could be used to assemble in orbit a large, permanent space-station'. His own paper, 'A Down-to-Earth View of Space Flight', argued for a gradual, research-based, let's-not-go-too-fast mode.[8]

'The conflict in opinions was going to be obvious,' Rosen remembered, 'and Willy Ley, who was arranging the program, wanted me to modify or withdraw my remarks in fear that they might do damage to the cause of space flight.' The Hayden Planetarium took the position that Rosen could say what he wanted, and a clash loomed. Rosen reminded von Braun in a letter of reminiscence: 'But you were not concerned. Indeed, you pointed out that if all of us sang the praises of space flight, the press would take little note of it; however, if we presented a strong difference of opinion, it could be newsworthy. We did do that and your prediction was proven correct. "Experts Differ on Future of Space Flight" was the headline on front pages of New York newspapers. Moreover, the conflict was featured in *Time* magazine.'[9]

Milt Rosen continued to have serious differences with von Braun, but eventually friction gave way to a friendlier relationship. He added in his correspondence with von Braun 'what everybody knows – that you have always looked ahead with optimism, and that you have been tolerant and gracious to those who differed with you'.[10]

The magazine series,[11] plus other coverage in the early 1950s, enhanced the public image of von Braun as a space visionary – or crackpot. Science and technology author and *Time* science editor Jonathan Norton Leonard, in a 1953 book, cited him as 'an unusual example of the sub-species homo tecnologicus'. Ambivalently noting the rocket scientist's 'fantastically successful' and 'horrendous' V2 missile, Leonard reflected:

> There is no doubt about his engineering competence. . . . But von Braun is more than a mere technician; he is also something of a prophet and something of a mystic. He is regarded by the more conventional rocket men with the mixture of suspicion and admiration that must have been felt by cozily established clerics toward Saint Francis of Assisi or Peter the Hermit. He worries and frightens them with his technological visions. When he talks to the lay public about his confident plans for voyaging into space, they accuse him of preaching to the birds. When they observe the following that has gathered around him of little boys in toy-shop space suits and teen-age enthusiasts with space dust in their eyes, they accuse him of leading a children's crusade toward sure disappointment.[12]

Von Braun met such criticism, wrote Leonard, with a reply 'much like that which Peter the Hermit must have given to the abbots and bishops who deplored his disquieting influence'. The rocket pioneer commented: 'Enthusiasm and faith are necessary ingredients of every great project. Prophets have always been laughed at, deplored and opposed, but some prophets have proved to be following the true course of history.'[13]

Not all of von Braun's space visions were benign. He claimed he was consistently driven by a primary desire for peaceful exploration of the cosmos. Yet, in his eagerness to see the space age commence, in the early 1950s he proposed that the United States be the first to orbit a 'satellite station' or series of satellites armed with atomic missiles. The idea was to give America permanent military control of the entire globe – but only for good, of course. From such a station, or stations, he postulated, the United States could observe, and if necessary, punish errant nations by bombarding them from the ultimate high ground of space. Leonard satirically summed up the envisioned potential import of von Braun's militaristic, yet idealised, Cold War proposal: 'No nation will challenge the power that looks down upon it from an artificial moon. No nation will attempt to challenge it; the earth will enjoy a pax Americana and can beat its radars into television sets.'[14] Critics assailed the startling proposal to militarise space and, thus, invite a cosmic arms race. Von Braun quietly withdrew the scheme and never put it forward again.[15]

Von Braun's high-profile salesmanship offended the conservative tastes of many scientific colleagues. Not even all of the original members of his rocket team approved of von Braun's spreading the space dream to the public and to political leaders. A major dissenter was Adolf K. Thiel, veteran of the V2 days at Peenemünde. First at Fort Bliss/White Sands Proving Ground, scientist Thiel remembered, 'We analyzed, designed, calculated, and dreamed about bigger rockets to the moon. [But] not much response from our bosses!'[16] Thiel, who left the team in early 1955 to join the TRW Corporation in California, reminisced years later with von Braun: 'You started to get your thoughts into magazines. Some of us longhairs and purists didn't think very highly of that. And one day I opened my big mouth about it and still remember your answer: 'We can dream about rockets and the moon till hell freezes over. Unless the people understand it and the man who pays the bill is behind it, no dice. You worry about your damned calculations and I'll talk to the people.' You did, and succeeded; we listened, and learned.'[17]

As a master salesman and promoter,[18] von Braun made maximum use of his most outstanding trait, Thiel said. 'He was so charming. He charmed the pants off you! And he was absolutely convinced what he was doing was right.'[19] And, despite any appearances to the contrary, von Braun 'was not a glory hound', rocket-team member Werner Dahm emphasised. 'He didn't seek that [for himself]. He wanted to advance space exploration.'[20]

Close on the heels of the major exposure in *Collier's*, von Braun began learning to master the young medium of television as well. The opportunity came via a series of Walt Disney programmes on space that featured von Braun and his chief scientist, Ernst Stuhlinger, among other experts. The shows, introduced by Disney himself, began airing in 1955 with *Man in Space* and continued with *Man and the Moon* and *Mars and Beyond*. They kept alive the *Collier's* momentum to educate Americans about, and stimulate their interest in, the coming space age.

A Walt Disney Productions associate, William R. Bosche, worked with von Braun in California on the series. Calling himself a 'charter member of the Mickey Mouse Chapter of the Wernher von Braun Fan Club', Bosche cited one especially memorable workday.

You came in about 5 PM, rumpled but ready after a day at [a contractor's plant in] Santa Susana. Along with Willy Ley and Ken O'Connor, we were working out the details of a 'bottle' type space suit for maintenance and repair of the orbiting space station you had designed.

As Ken and I were asking questions and making sketches, you and Willy were carrying on a running argument about the number of cells in a monkey's brain and just how much they could be programmed to do.

As midnight approached we had things fairly well nailed down, so you looked at your watch, commented that you had no appointments until 7 AM, and asked if we would like to work straight through until then on the next problem.

Unfortunately, we all found that we had pressing engagements at that moment that forced us to decline your kind offer. However, we were impressed by your eagerness and predicted then, as we do now, that if you maintain such an ambitious attitude you may make a name for yourself yet![21]

In addition to wooing and working with the mass media, von Braun hit the speech-making trail across the United States in the

early 1950s, addressing any semi-respectable group that would
invite him. On early speaking engagements outside the South, he
would occasionally soften up audiences by saying, 'I want first to
apologise for my accent. I'm from Alabama.' He combined, as one
observer put it, 'a German lisp, Texas twang, and honey-soft
Alabama drawl that enthralls his listeners'. His composite accent
prompted one wag to dub him 'Sam Houston Cornpone T. von
Braun'.

On occasion he was unable to deliver his speeches in person. In
1951 he was invited to present his technical paper on 'satellite
vehicles' to the second annual congress of the International
Astronautical Federation (IAF) in London. That March he explained
to a British Interplanetary Society (BIS) official why his prospects of
attending the IAF gathering were rather dim. 'Old Joe [Stalin] from
behind that Curtain keeps us pretty busy and my personal finances
aren't too rosy, either. A couple of months ago I broke ground for a
new home and now I'm broke myself. But let's wait. It is still six
months hence, so Old Joe may meanwhile become a monk and I
may hit the jackpot. Who can tell?'[22]

Von Braun neglected to cite a third reason he would probably not
be going to London any time soon: national security. The US Army
feared kidnapping or other harm to its ace rocket scientist – not yet
an American citizen – by the Soviet Union if he ventured back to
Europe. Von Braun made a further reference to the sad state of his
own finances in a follow-up letter: 'Ah, if my pocket would be as full
as my head I would most certainly come.'[23]

The London gathering led to his meeting a man who would figure
prominently in his life – Frederick C. Durant III, an American
Second World War naval aviator, test pilot and early space activist
(and future aerospace industry executive, author and sometime CIA
agent). Durant, an IAF officer, learned that von Braun had prepared
a paper for the conference but could not attend, so he wrote and
offered to read the paper for him. The situation prompted von Braun
to write to a colleague in London:

I greatly enjoyed your suggestion that it would give the meeting
the right cosmopolitan flavor if an American would read a
German's paper at an Astronautical Congress held in London.
By God, isn't all this a business between one planet and another,

after all? When, during the war, I made my business trips through Germany in a little Messerschmidt, I always felt like a house-fly trapped between a double-glazed window. In the 'cosmic age' we are entering now, there is no room for frontiers, at least not for spiritual ones.[24]

With Durant's letter in hand, von Braun arranged to meet him over coffee in a New York café so he could decide whether he wanted Durant to represent him, and so Durant could decide whether he wished to associate himself publicly with this German rocketeer's views.

'It took us about three minutes only, and we became great friends. It was the first of many meetings through the years,' recalled Durant, now the retired deputy director of the Smithsonian's National Air and Space Museum and founding head of its Astronautics Department. The von Braun paper on space satellites was 'very well received' by the IAF delegates from nearly a dozen countries, Durant remembered, despite 'some negative comment in the British press'. The Allies' VE (Victory in Europe) Day was a scant six years in the past, and some damage from V1s and V2s could still be seen in London.[25]

In an implausible occurrence, two years earlier (1949) von Braun had been voted, in absentia, an honorary fellow of the BIS. He responded that this action, 'despite the grief the work of me and my associates brought to the British people, is the most encouraging proof that the noble enthusiasm in the future of rocketry is stronger than national sentiments which in the past so often have hampered scientific progress to the benefits of all mankind'.[26]

Von Braun, in postwar correspondence with another BIS officer in England, both deplored the recent past and held out the pacifistic hope of a missile-free future filled with cosmic travel and exploration: 'Is it not a shame that people with the same star-inspired ideals had to stand on two opposite sides of the fence? Let's hope this was the last holocaust, and that henceforth rockets will be used for their ultimate destiny only – space flight.'[27]

11

Towards the Cosmos

By the mid-1950s the Korean War was over, but the perilous, thermonuclear-tipped Cold War persisted. Now the Soviet Union had the hydrogen bomb. It was also suspected of leading in the development of large ballistic missiles that could accurately hurl nuclear warheads from one continent to the other. Despite the 'star-inspired ideals' that von Braun shared with scientists and a portion of the general public, it seemed he would always be tied to developing rockets as potential weapons of war.

In early 1956 a new chapter opened for the von Braun rocketeers at Redstone Arsenal and for their industrial partners. Their top-priority mission was to develop the Jupiter intermediate-range ballistic missile (IRBM), with a range of 1,200 miles, while perfecting their shorter-range, nuclear-capable Redstone rocket. They were placed under a new command, the Army Ballistic Missile Agency (ABMA), headed by soon-to-be Maj-Gen John Bruce Medaris. Von Braun, moving administratively with his civilian team of experts from the existing missile command which was partially located at the Huntsville base, was named the director of the ABMA's Development Operations Division.

The official birth date of the agency was 1 February 1956. The swashbuckling Medaris arrived brandishing a riding crop, an impressive intellect and photographic memory, and an ego to match. He soon showed himself to be a hands-on, spit-and-polish commander. In two of his first informal edicts he ordered that civilian workers' shirt-tails be tucked inside their trousers, and that Confederate flag images be removed from pick-up lorries and other vehicles driven to the post. Early in the civil rights revolution, he opposed the display of that flag, considered a symbol of defiance of

Washington, by federal employees or government contractor workers on a military base in the 'Heart of Dixie' state.[1]

Medaris quickly assembled his top people – von Braun included – for a meeting to get things moving on the Jupiter Project, and to make sure everyone knew who was in charge. Charles A. Lundquist recalled that Medaris gave the group an opening 'pep talk' and emphasised that he 'would be available to fix all kinds of problems'. 'If there are any problems you can't resolve, you let me know.'

With only the briefest pause, von Braun responded: 'Ah, yes, General, there is a problem that we've been working on. On the date you have selected for the next launch, there is a full moon, and this will make optical tracking of the vehicle very difficult.'

'Well,' remembered Lundquist, 'Medaris understood he had sort of been challenged: "What are you going to do about that full Moon?!" He took it in good spirit, and some bantering went on. But it was clear that von Braun was testing him with the whole crowd there. Von Braun didn't let Medaris just get away with making a statement like that.'[2]

Medaris, who gained the nickname 'the Big M' and a second star (shoulder-pip) at Redstone, had begun his military career as an NCO in the US Marine Corps. He was just what the Army missile programme needed. It was locked in a struggle with the Air Force over control of all US medium- and long-range guided missiles and stood to lose that fight. Von Braun soon came to appreciate his strong-willed commander: 'Our survival as a rocket-building team is at stake. Only a tough fighter in command of the Army Ballistic Missile Agency has a chance to keep it alive, and General Medaris is such a man!'[3]

A mutual respect between the two developed into friendship. Thirty years later, Medaris told his and von Braun's old rocket team: 'Wernher and I were about the most perfect possible match between two men who wish to pursue great projects together. . . . He called me "boss", but I am sure that I was as much under his guidance as he may have been under mine. . . . My life was so much richer because his path crossed mine.'[4]

Medaris headed the ABMA for four productive years. When he left, von Braun worked under one other military commander, and then only briefly. But even under a supporter like Medaris, von Braun remained frustrated in his efforts to persuade the US government to undertake space projects. 'Galileo, the Wright Brothers and Thomas Edison wouldn't have a Chinaman's chance here today,' the space advocate lamented to a journalist in 1956. 'They'd be thrown right out of the Pentagon on their ears!'[5]

In the years before the autumn of 1957 'von Braun was considered by many in the military as something of a flake . . . with this space business', recalled Col Edward D. Mohlere, then a US Army Ordnance Corps commander in Detroit and later a member of von Braun's team in Huntsville. 'But he was resolute, and I was captivated by his enthusiasm. I thought, "Why not? It's going to be the wave of the future."'[6]

Despite Defense Department resistance to his ideas, von Braun tried to maintain his characteristic optimism. In 1955, an interviewer asked him whether men could be trained to adjust to sleeping and working over long periods in the alien weightlessness of 'outer space'. 'My personal feeling', the scientist replied, 'is that the time will come when old space men will complain that they can't sleep in a bed.'[7]

In the mid-1950s plans were being formulated and then announced in advance for the International Geophysical Year (IGY) for 1957–8. The IGY was to be a cooperative global research programme with emphasis on the physical sciences to advance knowledge of the Earth, the atmosphere, near-space, solar flares and various other energy emissions from the sun, through the use of sounding rockets, astronomical observatories and other means. With his Redstone rocket now performing well in test flights, von Braun and others in US military and scientific circles began seriously thinking of mounting Earth satellites.

One catalyst for action in 1954 was Frederick C. Durant III, the president of the IAF, whom von Braun had befriended at the time of the 1951 IAF congress in London. Ex-naval aviator Durant arranged a private conference on 25 June for von Braun and several other key leaders from academia, industry and government with Cdr George Hoover, chief of the Office of Naval Research in Washington. The subject was the possibility of launching an Earth-orbiting satellite in the near future.

Hoover later remarked: 'By the end of the meeting, they were all quite surprised by von Braun. He had been doing so much writing that they had almost forgotten what a very practical scientist he is. When he discussed the combination of the Loki-Redstone configuration as a possible solution to [launching] the first satellite and went into the nuts and bolts of the thing, they were really impressed.'[8]

The ad hoc meeting produced an Army–Navy plan using existing hardware to lift a 5lb payload into an orbit of at least 200 miles' altitude as early as autumn 1956 and no later than November 1957. It would employ a souped-up Redstone with lengthened fuel tanks as the booster rocket; upper stages of clustered small solids such as the Air Force's Loki, refined by Cal Tech's Jet Propulsion Laboratory; a satellite designed by the Navy; and Navy tracking facilities.

Could the Army be counted on to provide one of its Redstones for the proposed project? Von Braun took Hoover and his boss, Capt Bill Fortune, US Navy, to see his own Army boss at the time, Maj-Gen Ludy Toftoy. 'Don't worry about it,' Toftoy said, by von Braun's account. 'There's only one method for getting ahead in this game. That's by action. I promise you all the cooperation you need.'[9]

The plan was first code-named 'Project Slug', for reasons of military politics, and later renamed the less obtuse 'Project Orbiter'. Von Braun put key people on his team to work on the project, which had so far been authorised only at lower levels, in a clandestine so-called 'skunk works' (an innovatory R&D Unit) that was 'quietly tolerated by his superiors' at Redstone, recalled Ernst Stuhlinger.[10]

Gerhard B. Heller, von Braun's scientific assistant from 1953 to 1958 and a colleague since 1940, reminded him years later: 'You wrote a letter to me [in Huntsville] from the Silver Sands Motel in Cocoa Beach [Florida] in August 1954 which shows essentially the concept of the first US satellite [vehicle]. This letter caused me to become awfully busy.' Heller was working on solving the problem of atmospheric re-entry of longer-range ballistic missiles by launching a test nose cone on a multi-stage rocket using the Redstone as the first stage. 'Your satellite vehicle configuration in the letter used the same first three stages as your re-entry vehicle and the fourth stage as the orbiter,' remembered Heller, noting the 'amazing' coincidence.[11]

American officials and scientists had known since the early 1950s of Soviet intentions to launch Earth satellites. In the autumn of 1954 Gen Toftoy submitted the Project Orbiter proposal – written by von Braun and several team members – to a higher authority within the Army. About the same time, the American Rocket Society's Space Flight Committee, chaired by Milton Rosen, submitted an independent report to the National Science Foundation suggesting a study on satellites. Also, the Air Force was working on a satellite project that would use its future Atlas missile as a launch vehicle.

No formal response to any of those initiatives had been heard by 29 July 1955, when the Eisenhower administration announced that it had opted for a high-level government committee's choice of the Naval Research Laboratory's Project Vanguard to launch the first American satellites sometime in 1957. The chosen launch vehicle would combine a modified General Electric Viking liquid-fuelled first stage, an Aerojet liquid-rocket second stage and another contractor's solid-propellant third stage. The rocket was to be capable of boosting research satellites of up to 40lb into 300-mile-high orbits. The White House, rejecting the use of a military missile – such as the proven Redstone – for the job, lauded the Vanguard's composite launch vehicle as a new, elegantly engineered and wholly 'scientific' rocket befitting an IGY undertaking. It was also an untested design.

'This is not a design contest,' von Braun protested. 'It is a contest to get a satellite into orbit, and we [the Army] are way ahead on this.' To the claim that the Vanguard possessed 'more dignity' than the Army–Navy Project Orbiter proposal, a frustrated von Braun retorted: 'I'm all for dignity. But this is a Cold War tool. How dignified would our position really be if a man-made star of unknown origin suddenly appeared in our skies?'[12]

The Soviet Union followed the Vanguard announcement with a public statement that it, too, planned to launch an orbiter during the IGY. Its top space officials could hardly believe that Washington had chosen the Vanguard concept with its unproven launch vehicle over an existing, reliable rocket such as the von Braun team's upgraded Redstone.[13]

'The United States has no time to lose if it wants to be first in orbit,' von Braun told reporters in 1956. He had pleaded in the early 1950s that an 'American star' rising in the West would greatly impress the peoples of Asia and the rest of the world – and that it could be done with little difficulty and at no great cost. Now, he warned, Russia was 'working hard' to launch its satellites.

The White House did not view this as a 'space race' and did not share von Braun's sense of urgency. Presidential assistant Sherman Adams had advised Eisenhower that no one would get excited about a trifle like a little man-made object circling Earth.

Von Braun also had a tough time 'convincing Congress that launching a satellite was possible', Gen James M. Gavin, US Army, later recalled.[14] So, as chief of Army Research and Development at the time, he brought von Braun before a Senate committee. 'Dr von Braun began to talk about the Soviet capabilities' for satellite launching, remembered Gavin. 'After listening a while, Senator [Allen Joseph] Ellender [Democrat – Louisiana] said that we must be out of our minds, that the Soviets couldn't possibly launch a missile or a satellite. He had just come from a visit to the Soviet Union, and after seeing the ancient automobiles, and very few of them, on the streets, he was convinced that we were entirely wrong.'

As von Braun listened, nodding his head in polite acknowledgement that he was at least hearing the nonsense coming from the senator, Gavin grew concerned that the nods were being misconstrued and recorded as agreement. So he handed von Braun a note suggesting that he be careful not to give the impression of concurrence with the senator, 'since I knew that neither of us did agree with him'.

That did it, Gavin recalled. 'The chairman of the committee brought the hearing to an end and then called me before him and threatened to throw me out . . . for attempting to influence a witness. It brought the hearings to an end and no one was convinced that the Soviets could possibly launch a satellite.'[15]

Despite the Eisenhower administration's decision, and the attitude in Congress, the never-say-die Wernher von Braun and company persisted with Project Orbiter. They promoted it as a back-up to Vanguard, in case the untried launch vehicle ran into problems. In April 1956, a congressional committee shot down that proposal by a split vote. 'Wernher von Braun and his rocket team, the world's most experienced, were specifically ordered to forget about satellite work', *Time* magazine later reported. 'They did no such thing, and neither did their Army bosses.'[16]

Von Braun showed a similar resolve a year later, when Secretary of the Army Wilber Brucker, acting on orders from Eisenhower's defence secretary, visited him. Brucker said there was much current talk of satellites and warned von Braun to confine his work to weapons development. The secretary asked, 'Do you hear me?' Stuhlinger reminded von Braun: 'You said, "Yes, sir", because you had heard him.'[17] While immediate military commanders looked the other way, the von Braun team continued to prepare a satellite and booster rocket for the day when a call would come for its services. Shrugging off its frustration over being officially left out of the satellite picture, the Army missile team pressed on with its primary job, the Jupiter Project.

During this time, von Braun continued to take a hands-on interest in his co-workers. Young Army officer R.P. 'Hap' Hazzard, who served from 1956 to 1959 with the ABMA at Redstone Arsenal, reminded von Braun years later of 'one early incident in those traumatic years [that] served to set my own mind at ease that I was working among a dedicated and understanding group'.

'To the young officers of the [Jupiter] Project Office you were . . . a rather unknown quantity,' then-Brig-Gen Hazzard told von Braun. Assigned the task of comparing the technical features of the Air Force's Thor IRBM with those of the Army's Jupiter, he had first examined 'the relative re-entry wind-drift error of the opposing nose cones. Lacking today's computer support, I had addressed the problem with a slide rule, the enthusiasm of youth, a shaky data base, and the inexperienced knowledge of a recent graduate student.' After Hazzard's presentation, von Braun had complimented his conclusion as basically correct – 'but commented with a degree of understanding not usually attributable to technical directors, that you felt that the indicated size of the error was about an order of magnitude too small (which it was)'. With this 'helpful redirection, all of us were convinced that you were not only human, but a human with empathy'. The incident had a lasting favourable influence on the group, recalled Hazzard.[18]

The Huntsville-based rocket team's work on the Jupiter IRBM, in competition with the Air Force's Thor, actually gave the Army the chance it needed in the satellite sweepstakes. Saying he must have test vehicles to work on nose cone and other developmental problems with Jupiter, von Braun won permission to build twelve Jupiter-Cs. They bore an uncanny resemblance to the juiced-up Redstones he had proposed for Project Orbiter in partnership with the Navy.[19]

By 20 September 1956, von Braun's team had readied the first Jupiter-C for launch from the Air Force missile test range at Cape Canaveral. The four-stage rocket contained a dummy satellite package in its nose. Pentagon bosses, suspicious that von Braun might try to pull a fast one and beat Vanguard into orbit, sent specific 'Don't you dare!' orders to Gen Medaris, who had no choice but to comply. 'Wernher,' said Medaris in a telephone call to the cape, 'I must put you under direct orders personally to inspect that fourth stage to make sure it is not live.'[20]

The Jupiter-C, with an inert fourth stage and no satellite aboard, flew to an altitude of 600 miles and covered a distance of 3,300 miles – a record then for any US rocket. As reports confirming the success reached him in the blockhouse, von Braun did a little dance. Any doubts that they could put a satellite into orbit evaporated.

The elation was short-lived. The Jupiter-C success was rewarded two months later with an order from Defense Secretary Charles E. Wilson limiting Army missiles to 200 miles – regressing the German-born rocketeers to the range of their V2 days. Wilson was burned in effigy in Huntsville's Courthouse Square. The US Air Force and its savvy Maj-Gen Bernard 'Bennie' Schriever had temporarily won the missile political wars. (Wilson's edict was later rescinded.) A measure of perceived justice came in 1957. The defense secretary pinned the Civilian Service Award, the Pentagon's highest non-military citation, on von Braun's chest for the ABMA's successful first re-entry by a nose cone using the agency's innovative ablative design.

The Vanguard Project's top leadership included older and newer acquaintances of von Braun's. John P. Hagen, a new acquaintance, was project director; Milton Rosen was chief engineer (and a persistent critic of the gung-ho von Braun's early ideas on space); and Richard Porter (a key interrogator of von Braun and his team in Germany after their surrender) headed General Electric's work on the Viking booster. All were able professionals, yet von Braun and the Army knew that Vanguard was bound to hit technical snags because of the complexity of the new, unproven (if beautifully conceived) launch vehicle. Sure enough, setbacks arose with the start of Vanguard testing in 1956. The von Braun group offered to partner the Vanguard team and substitute the Old Reliable Redstone (Jupiter-C) for the Viking-based rocket but keep the Vanguard name and payload. The Navy spurned the offer.

Von Braun had one more card to play. As a contingency plan for the future, he decided to hang on to one of the Jupiter-Cs for what he officially termed, with tongue firmly in cheek, a 'long-term storage test'. Under the same cover story, Jet Propulsion Lab officials did likewise with several Sergeant solid-propellant, upper-stage substitutes for the previously chosen Loki rockets. Just as a back-up.[21]

In November 1956, von Braun talked about Vanguard with visiting astronomer John O'Keefe in his office at Redstone Arsenal. O'Keefe worked for the map section of the Army Corps of Engineers and had come to ask von Braun for precise geographic data on Moscow for missile targeting – and to discuss satellites. Just weeks earlier the von Braun team had been forbidden to allow the fourth stage of its upgraded Redstone to achieve orbit. He told O'Keefe that radar scans showed the Soviets were firing ballistic missiles capable of putting up satellites.

Early signs revealed Vanguard was in trouble. It all meant an open path to a Russian coup in orbit, von Braun warned. An upset O'Keefe asked what he could do. The hyperconfident von Braun answered: 'I want you to go see John Hagen . . . and I want you to tell him that if he wants to, he can paint 'Vanguard' right up the side of my rocket. He can do anything he wants to, but he is to use my rocket, not his, because my rocket will work and his won't.'

Escorting O'Keefe out, von Braun said Hagen would probably respond that it doesn't matter who reaches orbit first. If that happens, von Braun implored, 'Will you say to him, if that's what he really thinks, will he for Christ's sake get out of the way of the people who think it makes a hell of a lot of difference!'[22]

In the early autumn of 1957, members of the von Braun team had detected various clues that the Soviet Union was on the brink of launching an artificial moon around planet Earth. Most US officials and Army brass were unbelieving: not those technologically backward, peasant Russians! On 1 October, Radio Moscow broadcast the radio frequencies to which people could tune in for transmissions from a forthcoming USSR object in space.

Then, on 4 October, the 'man-made . . . Red Star' that von Braun had warned about appeared in the heavens. It was spherical, weighed 184lb and emitted beeping signals from orbit. It was named Sputnik ('Little Moon'). It stunned the United States and other nations of the West. It gave birth to the space age.

Von Braun got word of it during cocktails before a private dinner party at Redstone with visiting senior Army and Defense Department officials from Washington. His response: 'I'll be damned!'[23]

Nobody's Perfect

Along with an uncommon intellect and extraordinary range of talents and interests, Wernher von Braun had his share of personal failings. For example, he tried to quit smoking but failed, so he cadged cigarettes from other people. When travelling he liked to stay up late talking and drinking good whisky and then sleeping late. He used profanity when he was with the boys. 'He was the best cusser of all the Germans,' remembered one close associate, senior information officer Foster Haley. 'He had a tremendous sense of humor – profane, like most of the Germans.'[1]

He was charitably described as a 'hearty eater'. (His favourite foods were steak, potatoes and fresh vegetables.) A maid at a dinner party returned to the kitchen shaking her head, 'That man has the worst table manners I've ever laid eyes on!' He spent so much time discoursing at the dinner table that he would eat in frenetic bursts, shovelling it in, holding the fork not in his right hand, American style, but in his left.[2]

On trips he rarely carried any money, leaving it for others to pay the bills and tips. At times he fit the stereotype of the absent-minded professor, forgetting to wear a belt, wearing mismatched socks to the office, neglecting to pick up the waiting associate with whom he car-pooled. He had little use for speed limits or for red lights and stop signs, if the way seemed clear. He got pulled over by police officers, highway patrol units and the US Army's military police (MPs).

The engineer-scientist was hopeless when dealing with everyday gadgets and machines. 'He doesn't like mechanical things', noted his long-time secretary Bonnie Holmes. He never used a dictating machine in the eighteen years she worked for him. 'He doesn't trust them,' she revealed. And he wasn't any good at working the buttons and switches, so he invariably called her in to take dictation. It was easier that way – for him.[3]

Engineer-manager Thomas L. 'Tom' Shaner, the civilian aide-de-camp to von Braun 1969–70, observed, 'Dr von Braun . . . could do anything he set his mind to. He flew jets. He loved to fly any kind of aircraft. But it was funny to me that he couldn't, or wouldn't, operate a simple thing like a VCR. Those things would frustrate him, and he would tear up more videocassette recorders just out of anger and frustration. He would beat on it and throw it on the floor.'[4]

One time Shaner and a dozen associates were gathered in von Braun's office to watch the televised splashdown of a space mission. The reception was terrible, with a sepia tone. Shaner casually got up, walked over to the TV set, flipped open the little door covering the controls and turned the colour knob. Instantly the true colours of the blue sky, shimmering water and bright sun came in beautifully.

'Tom!' exclaimed von Braun, 'Vot did you do? Vot did you do?'

'I just adjusted the colour on the TV,' answered an apprehensive Shaner, certain he had upset the boss. 'Do you want me to put it back the way it was?'

'No, no, I didn't know you could do that!' von Braun answered. 'It's been that way for five years!'[5]

Barber David Hinkle, who cut the scientist's hair for most of von Braun's twenty years in Huntsville, recalled the day von Braun walked into the shop with his son, Peter, aged 6. While his father got a haircut and hid behind a newspaper, the boy played on the floor with a large toy aeroplane. It was a battery-operated tin plane whose engines emitted sparks when the toy was scooted along the floor. After a few minutes of vigorous play, the extra-long batteries fell out. The boy handed the plane and batteries to his father in the barber's chair. The scientist tried to fix it, but to no avail. Then, his haircut finished, he moved to a seat in the corner of the shop and continued to wrestle with the batteries – still without success.

Then Peter's two sisters arrived, sized up the situation, and one of them implored, 'Oh, Daddy, let me do it.' As Hinkle looked on, struggling to suppress a smile, von Braun's young daughter replaced the batteries in the toy plane in no time.[6]

He wasn't very handy around the house, either. Ernst Stuhlinger and Fred Ordway wrote about a party at which Maria mentioned that she had repaired a broken picture frame by drilling a hole and driving a screw into the wood.

'Do you have a drill?' a surprised von Braun asked.

'Oh, yes,' his wife replied. 'I have an electric hand drill, and a whole box full of other tools – and I even use them quite frequently.'

'I did not know that,' her husband admitted.

'It makes no difference, Wernher,' continued Maria, 'because you would not know how to operate the drill, anyway, and you would certainly hurt yourself.' When others called von Braun a genius, he would jovially protest: 'I've never considered myself a genius – and my wife is always ready to attest to this fact!'[7]

Writer Arthur C. Clarke caught a glimpse of his friend's minimal prowess around the house during a Huntsville visit in 1958. Maria sent the two men out to fetch ice for drinks. Von Braun inserted a quarter in an ice-dispensing machine and out came a bag of ice – and then a second bag, and then another, and another.

'Lest a new Ice Age descend upon Huntsville,' Clarke later reminded von Braun, 'you took the approved remedial action. You kicked the deranged robot in the guts. It promptly returned your quarter. That was the moment that I decided that you were, indeed, a man who had Power over Machines.' Unfortunately, Clarke recalled, 'this respect lasted only ten minutes'. When the two men returned home and deposited their complimentary ice into the kitchen sink, 'you attacked it with a pick. You made such a mess that Maria ordered you out of the kitchen, and did the job herself.'[8]

Von Braun was not above taking advantage of free assistance. His personal attorney in Huntsville, Patrick Richardson, dropped by one Saturday morning in the early 1960s. He found von Braun holding a ladder in his garage for the chief engineer of radio telemetry from the space centre at Redstone Arsenal. He was fixing the electric garage door-opener. The rocket scientist explained that every time a radio-equipped cement-mixing lorry passed by, his garage door flew open. It was always helpful to have expert resources one could call on.[9]

Von Braun's driving habits followed the pattern set in his rocket-propelled youth. In Berlin one day in 1938, he drove the wrong way around the Wittenberg-Platz – a busy roundabout – just to see, he claimed, whether he would be arrested. He was.[10]

Driving to and from work at Redstone Arsenal, he was sometimes pulled over by MPs and booked, or given warnings, for transgressions ranging from speeding to driving without due care and attention. A long-time associate at Redstone said: 'He usually drives through the gate here at about sixty miles an hour – reading a speech!'[11]

Ed Mohlere, a NASA associate, described him as 'a wild driver. He didn't know the meaning of a stop sign.'[12] Another colleague said that in the early years at Redstone von Braun tended 'to drive on worn tyres till they blew out'. J.N. 'Jay' Foster, who also worked closely with the rocket scientist on both the Army and NASA sides, and in Washington, reminisced with him about 'the days I accompanied you to the [Capitol] Hill. Why is it you can get away with parking in a No Parking zone and the rest of us can't?!'[13]

William R. Lucas, an eventual successor to von Braun as space centre director in Huntsville, recalled the time he rented a car, met von Braun and planned to brief his boss while driving from Baltimore to Washington. 'Well, von Braun wanted to drive,' Lucas said. 'If you were flying an airplane, he wanted to fly it. If you were driving a car, he wanted to drive. So he was driving and I was looking at my material and briefing him. He was breezing along down the road. I was not conscious of how fast we were going, but he said, "Look at all those cars stopped up there. What's wrong?"' The heavy-footed von Braun began to get the picture when police pulled him over to join the line of cars. He had been speeding and got caught. It was the one time Bill Lucas was glad his take-charge boss was at the controls.[14]

Von Braun was a night owl, especially when travelling on business. His associates suspected he preferred to rise just before the crack of noon. 'He was not an early riser, that was for sure,' Lucas remembered. 'I travelled with him several times, and one of the things you were always told is that you have got to be responsible for waking him up. That wasn't easy, and you had better wake him up more than once. He would stay up all night, and he didn't like to get up early in the morning.' Lucas never let von Braun sleep through a morning meeting in Washington or anywhere else, but it took some doing. 'I would go to his hotel room and bang on his door the next morning. You would have to get yourself up and plan plenty of time to get him awake.'[15]

Assistant Tom Shaner endured many such experiences when on the road with von Braun. 'He was definitely a night person and notorious for not getting up early,' he said. And in Washington 'if you are scheduled to go before a congressional subcommittee hearing at nine o'clock in the morning, you have to be there.' He would get a firm agreement with von Braun the night before: 'Okay, we are going to meet for breakfast downstairs at eight o'clock.'

'Eight o'clock would come, and no Dr von Braun,' recalled Shaner. 'I would call his room, and he would answer and say, "Good morning. I'm not dead. I'll be right down." He would hang up the phone and go right back to sleep.'

So his travelling companion would go up to his room and pound on the door. Von Braun would eventually answer, 'Yes, yes. I will be right down.' Shaner said, 'I would make him walk to the door and open it. He would be standing there in his shorts [underwear] right out of bed. He hadn't taken a shower, he hadn't shaved. . . . So I learned not to trust Dr von Braun when he would say he was up.'[16]

Von Braun said, on more than one occasion, at least half-seriously: 'I'm convinced that none of the important achievements in human history were accomplished before ten-thirty or eleven in the morning!'[17]

Von Braun was no better at handling money than at getting an early start on the day. 'He never carried cash,' Shaner recalled. 'He never had any money with him.' So the aide carried the cash and credit card and kept up with all the travel expenses and receipts. After each trip he would settle accounts with Bonnie Holmes and Maria von Braun, 'the family bookkeeper, accountant, and banker'.[18]

Although von Braun didn't pursue the big bucks personally – at least not until joining private industry in his later years – he worked the system to his advantage. In his twenty-seven years on the US government payroll, he found ways to supplement his income. Beginning in the late 1950s, his fame as 'Mr Space' brought him abundant opportunities for paid speech-making, consulting work for Walt Disney (on Tomorrow Land at the original Disneyland, then on Walt Disney World) and others, and writing articles and books. He gave many more unpaid speeches than paid ones, and he often shared his honoraria and other outside payments fifty-fifty with the civil-servant ghostwriters who had helped with his articles and speeches, staff members said. Still, the paid activities became so lucrative that NASA placed a limit on his outside income.[19] But it allowed a substantial sum per year, knowing that he could command a fortune with private industry.

'He'd mix business with personal business,' recalled Shaner. 'He gave a lot of speeches back then, and he got paid for many of the big ones. They'd give him $5,000 or whatever as an honorarium. . . . He would work all these social things in with NASA business. He'd have a trip somewhere on business, and then that evening he'd be the guest speaker at some big banquet. That way he didn't have to buy a ticket to fly commercial to do the dinner. If he worked it in with business, he could fly the NASA [aeroplane].'[20]

Government travel also allowed von Braun to indulge – free of charge – what biographers Stuhlinger and Ordway described as his 'strange predilection for pills'. During his travels he tended to carry a pocketful of pills he just might need for various maladies. 'If he had to visit a military installation,' the pair noted, 'he rarely missed the opportunity to make a quick call at the infirmary and ask for some pills that might help to subdue an upset stomach, a sore throat, a little sinus trouble, or some slight pain in the chest.'[21]

On the road von Braun also took full advantage of the supplies of complimentary alcohol provided for him and his staff or guests in the hotels where he would speak at large affairs. Before leaving, he made sure his aide packed up the booze for transport home, Shaner recalled. Von Braun, who considered the alcohol simply part of the deal, always offered to split the loot with the assistant, based on who needed what for their home drinks cabinets.[22]

Despite his brilliance, von Braun's mind could go AWOL at times. Unless his secretary intervened, he often lost track of the time at work and let meetings run long, which threw his schedule off. He tended to forget to stop by the supermarket when Maria asked him to pick up a few items on his way home. On a sailing outing in Chesapeake Bay with associates Frank Williams, Jay Foster and others, he dropped anchor, only to realise he'd forgotten to secure the other end of the rope to the boat.[23]

A similar side of von Braun came into view on a trip to Los Angeles, as aerospace official Lee B. James reminded him: 'When we checked in to the hotel lobby, we set our bags down there, and you said, "Lee, would you watch my bags a minute while I make a call?" That was the last time that I saw you that day.'[24]

As for his forgetfulness, missile and space agency translator and personal friend Ruth von Saurma once remarked, 'In this respect he is very much the typical scientist. I suppose there is only so much room in [such] a man's mind, and he eliminates the unimportant things.'[25] Similarly, his daughter Iris wrote in a schoolgirl essay about her father that, if he forgets a few everyday things, 'that's because he always has so much on his mind. He's always thinking about some problem.' (When the local newspaper learned of the revelatory essay and wanted to publish it in its entirety, Wernher and Maria gave their blessing.)[26]

On business in Washington, von Braun once left his wallet in the back-seat of a taxicab. Later realising his loss, he could not remember the name of the cab company, let alone the driver or the cab number. His wallet contained his 'Secret' clearance ID cards and other valuable documents. He telephoned Bonnie Holmes back home and gave her the impossible task of recovering the lost wallet. It was his (and her) good fortune that the very next passenger in the cab in question was the widely read Washington investigative reporter and syndicated columnist Jack Anderson. He had noticed the wallet, quickly discovered the famous name within and, knowing how to reach von Braun's office, had taken prompt steps to reunite wallet and owner.[27]

Von Braun had a natural flair for getting what he wanted from others. 'He was a master psychologist,' observed Shaner. 'He was an expert at manipulating people to get them to do what he wanted without being domineering. He was always making it seem like it was your idea and not his. . . . Many times he knew the answers to his dilemma or problem, but he wanted you to come to that conclusion and offer the solution.'[28]

One day in the 1960s, the custom-built 'Pregnant Guppy' jet transport, designed for hauling large rocket upper stages to Cape Canaveral and other oversized cargo here and there, made its first visit to Redstone. Von Braun asked one question after another about the bulbous plane's flight characteristics. The owner-pilot got the picture: did von Braun wish to go up for a spin and try his hand at the controls? Yes, he did, and so it came about.[29]

For all his charm and charisma, von Braun the media star managed to alienate more than a few of his fellow scientists. Some saw him as an undignified huckster who hogged the limelight. Aerospace scientist Herbert Friedman found his colleague 'an affable person with a friendly personality [who] certainly had the capability to charm people'. However, 'many of us resented him because he would appear [at meetings] . . . and monopolize the stage. He was always followed by media people, and he could disrupt a meeting very thoroughly just by being there. . . . I was put off by his public image.'[30]

Given his personal magnetism and good looks, not to mention his fame, von Braun had ample opportunities to be a skirt-chaser. Late in life he cheerfully confessed publicly that 'I am a sinner, worse still . . . I sometimes even enjoy being one.'[31] But he did not specify those sins, so one is left to wonder. Von Braun 'certainly admired women', recalled management associate James Daniels, who worked with him in Huntsville and Washington and accompanied him on trips. He enjoyed women's company, their friendship, their attention and their flirtations at dinner parties and receptions. But he was not, to Daniels's knowledge, a philanderer. In travels with von Braun, he witnessed the handsome scientist's 'No, thanks' response to colleagues' late-night offers of arranged female companionship.[32]

'He always maintained the high road,' Shaner concurred. 'There were people who didn't know him or respect the way he was, but he would never get involved in things that were out of line. He was above all that. A lot of contractors would approach me about women for him, and all that sort of thing. I set them real straight on that: he didn't do those kinds of things.'[33] For whatever reasons, when it came to extramarital indiscretions, von Braun evidently didn't engage in them, despite the odd rumour in Huntsville.

By all accounts, von Braun shunned racial epithets and ethnic slurs, except for rare lapses. The scientist never made 'any kind of off-hand remarks about any ethnic groups or any person', aerospace engineer-manager Leland Belew asserted. 'He had an affinity for life to the point that he had a very high tolerance for all people.'[34] However, in a private letter written in the 1950s he used the then-common vulgar expression 'nigger in the woodpile' in alluding to some problem or potential trouble.[35]

In satellite and other work in the 1950s and beyond, von Braun and his German rocket team-mates had frequent association with Jewish scientists and engineers. 'We had many contacts with . . . people of the Jewish religion, and there was never a problem between us in meetings,' engineer Rudolf Schlidt recalled. 'Everybody respected each other. These guys we were working with – at JPL and so on – were terrific, absolutely terrific. There was never, never any feeling or remark of "Are you one of them?"' added Schlidt, meaning ideological, anti-Semitic Nazis. 'The issue didn't exist. We had very professional working relationships. During the day we had conferences and decision-making processes.'[36]

Perhaps the truest test of the personal relationships came after hours. 'We were in the homes of many of the people out there,' noted Schlidt. 'I remember Dr. Joseph Kaplan [who was Jewish] of Cal Tech, for example. He invited us in his home in the evening – von Braun, myself, and Stuhlinger. Never, never was there any problem between us.'[37]

Physicist Stuhlinger concurred: 'Wernher von Braun and I had many scientist friends who were Jewish. We did in Germany, too.'[38] Until the Hitler regime's increasingly anti-Semitic policies made such relationships virtually impossible, that is. In postwar America, despite Schlidt and Stuhlinger's rosy recollections, more than a few Jewish members of the US aerospace community did not always feel so friendly towards von Braun and his team. The same could be said, of course, for a number of non-Jewish figures.

And what of the Prussian arrogance that some in the aerospace world – especially competitors – saw in the rocket pioneer? 'Sometimes you think of people of nobility as haughty individuals,' observed senior aerospace manager Robert Lindstrom, who worked closely with von Braun for years. 'I don't recall him being that way. I'm sure that on occasion he could be. He was a very strong individual when he wanted to be.'[39]

Was von Braun prone to lose his temper? Did he ever go ballistic? Shaner and other associates recalled that he did have a fiery side but kept it suppressed, at least publicly. After an occasional run-in with what he called a 'hard head' outside his team, an angry von Braun would privately unload to Shaner about 'that damned so-and-so!' the aide euphemised.[40]

Von Braun cursed intrusive news reporters in late 1969 while on a family holiday. And he exploded during the filming of one of the many NASA promotions he was asked to do. An over-eager TelePrompTer operator sped up the lines of script von Braun was reading. The rocket scientist began speaking faster and faster, soon sounding like a speeded-up voice tape. 'Dammit!' he finally shouted. 'Will you slow the damn thing down?' In a moment he regained his composure and was ready to continue. The shaken machine operator could not.[41]

Bill Lucas learned the hard way that von Braun 'didn't have lunch with someone just to have lunch' – that there was method and purpose in virtually everything he did. In the Skylab development days of the late 1960s, Lucas managed a portion of the design work. The project was running behind. 'We were having a Management Council meeting in Washington,' he recalled. 'We had a lunch break, and von Braun asked me to have lunch with him. He chewed me out – up one side and down the other – for something that came up during the meeting about Skylab, something where we were not doing all we should be doing.' Other attendees returning from lunch asked Lucas how his lunch had gone. 'Well,' he told them, 'when I went to lunch I didn't know I was going to be lunch.'[42]

Von Braun was notorious for allowing endless, even if usually productive, meetings. But at times during them an unplugged, glazed look would cover his eyes. Engineer-manager James Kingsbury learned precisely what was happening with him at such moments. 'One of the things people [on the team] looked forward to was making technical presentations to von Braun, because he always appeared to be totally immersed in what they were saying,' Kingsbury recalled.

Little did some of them know. 'In a lighter moment, he [confided that] he had mastered the art of drifting off with his eyes open. When the presentation ended and he came back to the real world, there were a few key questions he had memorised' to show how alert he had been – generic questions about 'vibration' or 'rough combustion'. 'And then,' recalled Kingsbury, 'von Braun concluded [in his revelation], the presenters were delighted, because he obviously had paid careful attention.'[43]

While known for showing great patience with briefers in meetings – unless they tried to bluff their way through or otherwise put something over on him – von Braun could be impetuous on other occasions in the workplace. It happened when Tom Shaner showed up as a nervous, 29-year-old aerospace engineer in mid-1969 for an interview for the job of assistant to the director.

Von Braun had not wanted an aide in the first place (other than his brother Magnus, who had served for a time in that capacity in Germany), but in the late 1950s his closest associates had convinced him that he needed one to 'manage' him, especially when he was on the road for up to three weeks of every month. He tended to make commitments and cut deals during his travels but did not always remember to share the details with staff back home. For what was typically a two-year assignment, the aide would stick to von Braun's side, attending briefings and other private meetings and all public appearances with him; monitoring conversations and any decisions reached; and taking care of whatever the boss needed done. Von Braun insisted that the aide be a young person with a technical background who could later take professional advantage of the insider experience. Jerry McCall, mathematician; Frank Williams, engineer; and Jay Foster, technical manager, had filled the post earlier.

In 1969 Shaner was the first of three finalists – chosen from a dozen candidates within the centre – to be interviewed by von Braun. The young engineer had seen the director mostly at a distance during his six years in the Test Laboratory. Shaner recalled that he 'was in total awe of him and scared to death when I went into his office. But he put me at ease. . . . Von Braun asked me why I wanted the job. I told him I really didn't want it but that Karl Heimburg [Test Lab chief] was my boss and he told me I should try for it and that I would always have a job with him afterward.' Von Braun apparently liked the answer. 'He told me this was not his idea, either, but the idea of these other people because they got tired of putting up with him and all the trouble he caused when travelling.'

After determining that Shaner was committed to the agency for the long haul, von Braun asked him if he drank – alcohol, that is. 'Of course, that took me aback,' the engineer remembered. 'I didn't know what he was feeling for there. I said, "Well, I like to drink socially. I can take it or leave it, but I don't have any objections to it." Evidently I gave him the right answer. He just said, "I never trusted a man who wouldn't have a drink."' Then von Braun smiled and cancelled the two other interviews. Shaner was in.[44]

A devilish von Braun delighted in having fun at his associates' expense. Shaner was one such victim. From the start, the young assistant had let his boss know he preferred to fade into the wallpaper at big events where von Braun was speaking. The director would have none of it. 'Before Dr von Braun would give a speech, he always felt like he had to introduce me to the audience,' Shaner recalled. 'There might be 5,000 people sitting out there at this big, fancy, black-tie dinner, and he'd introduce me. I didn't like him to do that. It embarrassed me. A lot of times he'd have them put the spotlight on me and I would have to stand up. Later I would go to him and say, "Dr von Braun, don't do that. Please don't do that."' But the more Shaner protested, the more von Braun would embellish his introduction: 'If he was introduced as the most important man in the space agency, then he'd say I was the second most important man in the agency, because without me, he couldn't do anything. He got a big thrill out of doing that. And the more he could embarrass me, the better he liked it.'[45]

It wasn't only new employees who got teased, however. A long-time colleague from Germany also fell victim to his needling. Von Braun was an early and eager practitioner of MBWA – management by walking around. One day in the late 1960s he strolled into his centre's Space Sciences Laboratory. With him was a small group including Ernst Stuhlinger, the lab's director, and Hermann Weidner, the overlord of all the labs. Harry Atkins, a young research physicist in the SSL, was at an optical workbench, performing infrared astronomy research and using a special soldering iron. As Atkins recalled:

Dr von Braun and the group were . . . seeing what people were doing. He had this extreme interest in science. He used to go down into the labs and hide from his secretary. . . . With von Braun, all you had to do was mention something scientific you were working on and he'd just get into it. I happened to mention that this infrared detector I was working on was made out of a compound called lead sulphide. Well, he knew about it from their World War II work. He was very interested and asked me all kinds of questions.

During the discussion Hermann Weidner backed up against my bench and his suit coat came in contact with my pencil-thin soldering iron. During all the talking someone smelled smoke. Weidner looks around, jumps forward, starts beating his smoking coat, and they're all helping him. Here I am, thinking my career has ended very fast!

Von Braun started laughing and said to Weidner, 'No, Hermann, this is good! It's time you bought a new suit! You've been wearing that one since the first day I saw you in Peenemünde!'

'I will never forget it,' added Atkins. 'Weidner, a big, burly guy, was just stunned. . . . You could tell it was an older suit. It was funny. Von Braun's comeback was just fabulous.'[46]

An odd thing about von Braun, remembered Bonnie Holmes, was his noisiness. When not talking, he hummed, whistled and otherwise made sounds. 'You can hear him before he comes in,' she said. 'He just naturally makes noise.' He was also a man gifted with an 'almost photographic memory', virtual total retention and recall of everything he ever read or heard, Holmes said.[47]

Although he was said to have been an exceptional listener, von Braun was also a world-class talker, an adroit conversationalist, polished presenter of reports and testimony, and spellbinding speech-maker. 'He would take information, he would tap knowledge no matter where he found it, and he would enter into extensive conversations,' remembered I.M. Levitt, astronomer and early space advocate. Von Braun 'was like my wife', observed Levitt in 1999, when he was 90. 'She visits with people – the cashier at the supermarket, a woman in line, whomever. Wernher liked to visit with people. He was a talker.'[48]

Others noted that von Braun tended to dominate small-group conversations. He was often guilty of interrupting other, faltering speakers before they had finished. Occasionally, he oversold his ideas in face-to-face meetings with higher authorities. 'The problem with Wernher sometimes', Maj-Gen John G. Zierdt, US Army (Ret.), observed, 'was getting him to shut up.'[49]

He once contracted a severe throat infection and, after hospitalisation and treatment in Washington, DC, was ordered to take an additional two weeks of complete voice rest, away from work at Redstone. Von Braun normally held daily informal lunch gatherings with members of his management staff. And normally, recalled staff member David Newby, he would talk a blue streak (very rapidly) during these sessions, whether trying out parts of a speech or congressional testimony he was soon to present or bouncing new ideas off staff. 'His absence with the throat infection prompted a lot of conversation – and wisecracks,' Newby remembered. 'People said things like . . . "Has anybody noticed how quiet it is around here lately?" and "This [two-week silence] has set back the space programme a whole month!"'[50]

The scientist enjoyed telling jokes, and he was good at it – in English or German. American- and German-born associates said he often livened up his all-male staff meetings at Redstone with a risqué joke or two. And hardly surprisingly, those with a space angle held a special appeal for him. Ordway recalled this 'favourite von Braun joke that he'd tell with delight':

Two American astronauts are launched to Mars. When they reach the Martian surface they see a beautiful red-skinned, red-haired woman. Using their language-translation machine, they tell her they are from Earth. They see she is stirring a big pot, and every now and then she pulls a newborn baby from the pot. So they ask her, 'What are you doing?' She replies, 'Making babies', and she continues to stir. 'How strange,' the astronaut commander says. 'Well,' the beautiful woman asks, 'how do you make babies on Earth?' So the commander says, 'Let me show you', and they go behind the Martian rocks and bushes. After a while, they come back out and the woman asks, 'Well, where is the baby?' And the commander says, 'Oh, back on Earth it takes nine months to have the baby.' Perplexed, she asks, 'Then why did you stop stirring?'[51]

At one staff meeting, playing off the rivalry between his Huntsville space center and the Manned Space Center in Houston, von Braun told this one, to rousing laughter. 'One day a plane landed at the Dallas Airport and out stepped a dozen or so male dwarfs. Someone asked, "Who are those little people?" The other fellow there says, "Oh, they used to be big tall Texans – before they got the shit beat out of them!"'[52]

Von Braun was also capable of making, or repeating, the occasional ill-considered, politically incorrect remark in public. Asked by reporters early in the space age where he stood on the question of future women astronauts, von Braun wisecracked: 'Well, all I can say is that the male astronauts are all for it. And, as my friend Bob Gilruth [in Houston] says, "We're reserving 110 pounds of payload for recreational equipment."'

Aerospace manager Lindstrom was present when von Braun was fielding questions after making a presentation on large launch vehicles to a technical society gathering in Dallas. 'A man stood up and asked von Braun, "What do you think about automation?"' Lindstrom recalled. 'And his response was, "Well, automation is much like having a wife: she helps you solve some of the problems you wouldn't have if you hadn't got married in the first place!"'[53]

It is clear that while von Braun demanded technical perfection in rocketry, he did not pursue it in his life. Along with his many positive personal qualities, a rich complement of imperfections only showed how very human this extraordinary man was.

New Age of Space

If anyone did, Wernher von Braun knew the value of timing. In a coincidence worthy of a Hollywood film, on the evening of 4 October 1957 he was enjoying drinks before dinner at the Redstone Arsenal Officers Club with a roomful of Defense Department and Army top brass visiting from Washington. The main guest of honour was President Dwight D. ('Ike') Eisenhower's new Secretary of Defense-designate, Neil McElroy. Community leader W.L. 'Will' Halsey Jr, who was present, recalled that the room was so heavy with top brass that 'it seemed like the two-star generals were serving drinks to the three-star generals'.[1]

From his nearby Huntsville home, Gordon Harris, public information chief at the Army Ballistic Missile Agency at Redstone, telephoned the group with the sensational news he had received a moment earlier from a London *Times* reporter seeking reaction from von Braun and others. The Soviet Union had launched an artificial moon into Earth orbit! It was named Sputnik, and it was broadcasting beeping signals for people the world over to hear.[2]

'Those damn bastards!' was Army Maj-Gen John Bruce Medaris's response (to Ernst Stuhlinger a little later).[3] Von Braun seized the moment with his visitors: 'We could have done it with our Redstone two years ago!' He began talking 'as if he had suddenly been vaccinated with a Victrola [gramophone] needle', recalled Medaris. 'In his driving urgency to unburden his feelings, the words tumbled over one another.'[4]

'We knew they were going to do it!' von Braun told McElroy. 'Vanguard [the anointed US satellite project] will never make it! We have the hardware on the shelf! For God's sake, turn us loose and let us do something!'[5] Then, a bit more calmly, he pressed on: 'Sir, when you get to Washington you'll find all hell has broken loose. I wish you would keep one thought in mind through all the noise and confusion: we can fire a satellite into orbit sixty days from the moment you give us the green light.'[6]

'Not sixty days!' objected an incredulous Secretary of the Army Wilber Brucker, who had accompanied McElroy to Huntsville.

'Sixty days,' von Braun insisted.

'Wernher, make it ninety days, will you?' interjected Medaris, not quite so impetuous.

As von Braun remembered it: '"Okay," I said, "ninety it is." But I really meant sixty.'[7]

Sputnik shocked America. It was the nation's Pearl Harbor of space, a national call to action. It also threw into serious question the presumed technological superiority of the United States and its cherished public education system. It heightened fears about national defence; the same powerful Russian rocket that hurled a satellite into orbit could also boost nuclear payloads great distances. Editorial writers railed. The public worried.

I.M. Levitt, the prominent astronomer and director of the Fels Planetarium in Philadelphia, soon went public with the charge that an 'astonishing piece of stupidity' by Washington had led to this 'astounding propaganda victory' by the Soviets. Levitt revealed that the Army's von Braun team could have launched a satellite earlier with a rocket already in hand but had been reined in by the Pentagon.[8]

Much to von Braun's chagrin, President Eisenhower dismissed Sputnik as 'one small ball in the air . . . something which does not raise my apprehensions – not one iota'.[9] The President's closest aide, Sherman Adams, also pooh-poohed the orbiter's importance. He said America's purpose in planning to launch, eventually, its own satellite was to serve science, not to win 'an outer-space basketball game'.

Texas Democrat Lyndon B. ('LBJ') Johnson, then majority leader of the US Senate, disputed the Republican administration's benign assessment. 'The Roman Empire controlled the world because it could build roads,' LBJ observed to the press. The Second World War Navy veteran added, 'Later, when men moved to the sea, the British Empire was dominant because it had ships. Now the Communists have established a foothold in outer space.' Referring to Eisenhower's selection of the Navy's elegantly designed Vanguard over von Braun and the Army's improvised Jupiter-C, Johnson needled: 'It is not very reassuring to be told that next year we will put a "better" satellite into the air. Perhaps it will even have chrome trim and automatic windshield wipers.'[10]

The morning after Sputnik's debut, von Braun and Medaris escorted Defense Secretary-designate McElroy and his party of visiting Washington officials around the Army Ballistic Missile Agency (ABMA) facilities at Redstone. Next came an impromptu briefing. Army officer Truman F. Cook attended both the reception and the follow-up briefing. Reminiscing years later with von Braun, retired Col Cook recalled 'being impressed by the confidence of your unrehearsed presentation the following day regarding the Jupiter-C capability to rescue the reputation of the United States in the space race'.[11]

Out on the speech-making trail, a frustrated von Braun declared that what America's various rocket and space projects needed most was freedom from obstruction and interference by government scientific committees and the like. In other, more upbeat remarks, he predicted that the space-age obstacles the nation faced – the medical barrier, the propulsion barrier, the atmospheric entry heat barrier – would soon fade away. 'Right now, however,' the rocket scientist added, 'we're up against the "cash barrier" – and that one doesn't always fade away so quickly.'[12]

To those who wondered, and worried, in late 1957 about the Soviets' obvious lead in rocket launch power, von Braun explained: 'There was no ballistic missile development in the United States between 1945 and 1951 because there was no obvious need for it, no interest in it, no money for it.' After the Second World War, 'public interest in the United States turned away from weapons toward consumer goods. Few people in this country realized that the rulers of the Soviet Union were in no mood to lose their wartime gains through hasty disarmament.'[13]

Von Braun elaborated on that point in a late 1957 magazine interview: 'The six years between 1945 and 1951 are irretrievably lost. The Russians are turning out more scientists than we are, and good ones too. We could have done what they did if we had started in 1946 to integrate the space flight and missile programs. Our lot has been one crash program after another. One fine day we suddenly decided we had to have an ICBM. It was like telling the Wright Brothers to build a B-29.'[14]

But he labelled as 'nonsense' much of the self-criticism that swept the United States after Sputnik I. He cautioned Americans against overreacting and beating themselves up over real and imagined national shortcomings.

People often asked why the United States lagged behind the Soviet Union in space, and what it would take for America to pull ahead. Von Braun usually asked his audiences, 'How much would you be willing to sacrifice for us to do it?' 'The classic answer I got', he recalled, 'was from a fellow who said, "We have two automobiles. I'll sacrifice my wife's."'

The rocket team at the ABMA was still waiting – and hoping – for the green light from Washington in late October 1957 when the Association of the US Army invited von Braun to share his post-Sputnik thoughts in a talk in the nation's capital. His address was entitled 'The Lessons of Sputnik'. With an audience that included Army Secretary Brucker, Chief of Staff Gen Maxwell D. Taylor and other Army heavyweights, he pulled few punches. 'October 4, 1957, the day when Sputnik appeared in the skies, will be remembered on this planet as the day on which the Age of Space Flight was ushered in. . . . For the United States, the failure to be the first in orbit is a national tragedy that has damaged American prestige around the globe.'

Furthermore, he said the 'Soviet success' taught America no real lessons in science, technology or project management. Rather, this country 'made some grave errors in judgment. We failed to recognize the tremendous psychological impact of an omnipresent artificial moon, visible to anyone with a pair of good eyes, and audible to anyone with a simple radio receiver.' He warned of other Russian space 'firsts' to come.[15]

The wait was brief. On 3 November, a month after the first Sputnik, the Soviets launched Sputnik II. The orbiting 1,120lb payload included a dog named Laika. America's Vanguard was nowhere near a launch pad yet.

A few days later the Pentagon said it was ready to direct the Army's Jupiter-C team to 'prepare' a satellite payload for launching – if needed. It intended to stop short of giving full authority to launch. Von Braun, Medaris and their Jet Propulsion Lab partner in California, William H. 'Bill' Pickering, all sent word threatening to resign if they did not receive an unqualified go-ahead.[16]

Fortunately, Defense Secretary McElroy had taken office less than a week before. He remembered the confident sales pitches by von Braun and Medaris at Redstone. On 8 November the news agencies carried the story: 'The Secretary of Defense today directed the Department of the Army to proceed with launching an Earth satellite.' Under heavy public and congressional pressure, the Eisenhower administration had put aside whatever misgivings and biases it held regarding the Army's von Braun team and its plan. McElroy flashed the green light to Medaris, who immediately gave von Braun the good news over the office intercom: 'Wernher, let's go!'[17]

With the 'Go!' now in hand, the ABMA team followed up on earlier steps to secure a scientific payload superior to the 5lb chunk of hardware that von Braun had first proposed in Project Orbiter. Arrangements were made with James A. Van Allen, head of the Physics Department at the University of Iowa, whom Stuhlinger had first visited in 1954, concerning possible satellite experiment packages for Jupiter-C.[18]

Van Allen had prepared payload instrumentation – cosmic radiation counters and other sensors – that would conveniently fit in the nose compartment of either Vanguard's Viking composite vehicle or the Jupiter-C. The Iowa scientist recalled decades later that during ABMA–JPL talks, Bill Pickering had pointed out that Van Allen's cosmic-ray instrument package was the only scientific package planned for America's International Geophysical Year satellite project that had also been configured for Jupiter-C, as a backup to Vanguard. Van Allen noted that von Braun, 'who had endorsed and fostered this decision', had replied to Pickering, with feigned innocence, 'Isn't that interesting?'[19]

In the frenetic post-Sputnik days, the typically forward-looking von Braun quietly developed a plan to use a standard Redstone missile to put a man in space. The ABMA's Gen Medaris liked the idea. He dispatched von Braun to Washington to sell the idea to higher authorities. Teamed with him were then-Col John G. Zierdt, chief of the ABMA's Control Office, and Ted Hardeman, head of financial planning. The men called on Herbert York, then a member of the President's Science Advisory Committee (and later the head of the Defense Department's Advanced Research Projects Agency). Maj-Gen Zierdt recalled the scene years later, and a memorable von Braun ploy:

Your presentation was lucid and to the point. You answered all the technical questions perfectly and then Dr. York said, 'Now let's talk about money.' At this point you threw up your hands, said, 'That is for somebody else to worry about', and walked out.

So I was left holding the bag and trying to explain some dollar figures I had never seen before. I muddled through somehow, but I have always remembered how neatly you dumped the problem in my lap.[20]

Von Braun's plan ultimately gained approval – and the necessary funding. It was also in the Sputnik aftermath that von Braun telephoned an old colleague in Boston, Joachim Kuettner, who was not then working with the ABMA missile/space team.

'Would you want to head up a project to put the first man in space?'[21]

'Of course,' Kuettner answered, 'but I think I don't know enough about it.'

'Neither does anybody else,' replied von Braun.

Kuettner recalled: 'That was all, and I was off for Huntsville.'[22]

The Navy's Vanguard/Viking had already suffered test failures. On 6 December 1957, carrying the nation's intended first satellite payload, it was readied for flight from Cape Canaveral. The Navy gave it the downplayed designation of TV3, for Test Vehicle 3, in case of failure. It exploded at lift-off, tracking the pattern of almost all untried rockets on their first few attempted launchings. The fiery failure, pictures of which was splashed all over television screens, the front pages of daily papers and the covers of news magazines, shook the national psyche. Labeled 'Kaputnik' and 'Stayputnik' by the press, the disaster turned the spotlight brighter on the experienced, unsurprised von Braun group, who were busy working offstage.

Within a week of having gained the green light from the Pentagon on 8 November, von Braun had looked ahead and made a launch-pad reservation at Cape Canaveral for 29 January 1958. The Number 29 super-Redstone (Jupiter-C) vehicle – set aside in 1956 under the subterfuge of a 'long-term storage test' – was hauled out and made ready. 'All she needed was a good dusting,' von Braun later remarked.[23]

On the night of 29 January, the Army's rocket and its satellite payload were 'Go' on the pad. Weather conditions were not. Strong winds at high altitudes forced a postponement until the following night. The same thing happened then. The evening of 31 January arrived. For a change, the weather cooperated. It was time to launch. First-stage ignition came at 10.55 p.m. Eastern Standard Time (EST). The rocket rose on cue, its 'spinning tub' of assemblies of small rockets whirling at the top for added stability in flight. The launch vehicle roared upwards.

Gen Medaris was at the cape for the launch, but von Braun was waiting out the countdown and lift-off in Washington, along with Pickering and Van Allen. They had been told 'in so many polite words', von Braun recalled, 'that we had to sweat it out' in the Pentagon's Communications Center. He had worn a dark suit that evening because 'we had been told that if it was successful we had to go meet the press' at the National Academy of Science in Washington. 'Nobody had said what we should do if it missed. But just in case things didn't come off so well, I had a pair of dark sunglasses with me and was determined to sneak away to a still darker movie theater.'[24]

Von Braun had calculated that the tracking station east of Pasadena should begin receiving radio signals from the orbiter 106 minutes after lift-off, or at precisely 12.41 a.m. JPL's Pickering got an open phone line to the California station. At 12.41 he asked if there was any signal. The answer was no. More minutes passed.

'None of the stations had heard a thing' from the overdue satellite, von Braun recalled. 'That went on for what appeared to be hours. Meanwhile, we had to keep up appearances, and had to smile and convince everybody that things were in perfect shape.' At one point Army Secretary Brucker probed, 'Wernher, what happened?' Anxious generals looked at one another and asked, 'What's wrong? What happened?' Von Braun said, 'I heard Bill [Pickering] shouting into the telephone, "Why the hell don't you hear anything?!"' Then, in quick succession, four receiving stations 'came in and said they had a clear signal. At this moment we knew that we were in the satellite and space business.'[25]

The delay was later attributed to a slightly greater velocity at fourth-stage burn-out and, thus, a higher orbital path than expected. What had seemed an eternity to von Braun and the others had lasted 8 minutes. 'Those moments were the most exciting 8 minutes of my life!' he recalled.

President Eisenhower made the public announcement: 'The United States has successfully placed a scientific satellite in orbit around the Earth' as part of the nation's IGY participation. He named the 18lb, 4ft-long satellite Explorer. It was in an elliptical orbit measuring 223 by 1,580 miles.[26]

At the full-scale, 2 a.m. press conference in Washington, a joyful von Braun told reporters: 'We have firmly established our foothold in space. We will never give it up again.' He noted that there was 'dancing in the streets of Huntsville' – a spontaneous, late-night celebration had erupted in the centre of town in the Courthouse Square – 'and jubilation all over the country'. Of the Explorer feat and his team of German-born rocket experts, now 'Americans by choice', von Braun told an interviewer: 'It makes us feel that we paid back part of the debt of gratitude we owed this country.'[27]

A news photo showed von Braun, Pickering and Van Allen holding a full-scale model of the satellite over their heads in triumph. Van Allen's comment on first seeing the picture was, 'Wernher, as usual, carries the brunt of the load!'

With Explorer aloft and America at last in space, von Braun became an instant national hero in the eyes of many and a household name in many lands. Among the stacks of congratulatory messages was a telegram that read: 'Please accept my warmest congratulations on your great achievement which has thrilled and delighted us here in Britain. You and I had some differences during the war. I am so glad we are now working together for the same cause. I hope we may meet personally one day. Best wishes.'[28]

The well-wisher was Duncan Sandys, the British intelligence officer who had helped plan the devastating 1943 RAF strike against the Peenemünde rocket centre. His specific task had been to target the quarters of von Braun and his key team members and blow them to oblivion. Sandys, son-in-law of wartime Prime Minister Winston Churchill, had become the minister of defence in 1957.

News of the launching of Explorer also elated von Braun's parents, then living in Oberaudorf, West Germany. 'We have been waiting for a long time for Wernher to get a chance to show what he can do,' a happy Baron Magnus von Braun told reporters. 'Two years ago, Wernher warned over television that sooner or later we would see a "Red star" rising over the horizon. Few people believed him then, but now they have turned to him to match that Red star.' The elder von

Brauns cabled their middle son with this happy message: 'BEEP,
BEEP, BEEP!'

The space scientist had waxed philosophical – and poetic – to the
press when the Soviets orbited the first space satellite. He observed
that Sputnik had launched an era that 'will free man from his
remaining chains, the chains of gravity which still tie him to this
planet. It will open to him the gates of heaven.' After Explorer 1 he
philosophised in a speech about this 'small beginning' and the
unpredictable, intangible benefits of man's leaving Earth and
exploring other worlds:

> We have stepped into a new, high road from which there can be
> no turning back. As we probe farther into the area beyond our
> sensible atmosphere, man will learn more about his
> environment; he will understand better the order and beauty of
> creation. He may then come to realize that war, as we know it,
> will avail him nothing but catastrophe. He may grasp the truth
> that there is something much bigger than his one little world.
> Before the majesty of what he will find out there, he must stand
> in reverential awe. This, then, is the acid test as man moves into
> the unknown.[29]

Von Braun could not resist pointing to Explorer 1 as an affirmation
of his earlier exhortations that his team must stay together and not
defect to industry, if it was to make history, accomplish the greatest
good and achieve collective glory. 'What corporation', he emphasised, 'would have sent up a satellite?'[30]

Less than a week after their satellite triumph, President
Eisenhower invited von Braun, Pickering and Van Allen, among
other guests of honour, to a White House dinner saluting Explorer
and other recent national achievements. The JPL director, von
Braun, and their wives were staying at Washington's DuPont Plaza
Hotel and planned to share a cab that evening to 1,600
Pennsylvania Avenue.

'About ten minutes before the taxi was due,' Pickering recalled,
'Wernher called up to my room and said, "Have you got a spare
white tie?" I said, "What do you mean? I rented my suit and I got
one white tie with it." Wernher said, "I rented my suit and I got zero
white ties with it. What are we going to do?"' Pickering suggested

that von Braun ask the hotel manager for help. Von Braun's young Army aide, Lt Fred Kleis, had already phoned formal-wear rental shops in the capital but found all had closed. 'After a while,' Pickering recalled, 'Wernher called back and said, "No, the hotel manager was no help. But I have decided that maybe we ought to just go to the White House, and when we get there they should be able to handle this problem."'[31]

Presidential Press Secretary James Hagerty was telephoned and told about the problem. He told von Braun not to worry, that a white tie would be awaiting him at the White House. 'We drive up to the front door, get out, and somebody whips Wernher aside,' Pickering recalled. 'He's gone for about ten seconds, and he comes back to the line – with a white tie on!' After a delay, and with the guests already seated in the dining-room, to von Braun's shock the President entered the room wearing a black tie and complaining to his press secretary, 'Jim, I can't seem to find my white one anywhere.' Ike apologised both for being late and for wearing a black tie, Pickering recalled.[32] 'We looked all over the White House,' said the President, 'but we could not find my white tie.'[33]

Soon after Explorer's ascent, a class of primary school children in New York wrote to von Braun and sent their drawings of suggested satellite designs. When he answered them and sent an autographed picture, the children's teacher proudly hung a sign in their classroom proclaiming 'Wernher's Little Learners'. And when a group of children in Illinois posted him $50 they had collected to boost America's second-place space programme, he returned it with the suggestion the class 'might wish to apply the money in some way to reward deserving boys and girls who do well in science or mathematics'.[34]

In February 1958, a second Vanguard satellite launch attempt fizzled out within 14 seconds of ignition. By March, when von Braun's team prepared to launch its second Explorer, only two man-made moons – Explorer 1 and Sputnik II – circled the planet. Sputnik I's orbit had decayed fast, and the trailblazing satellite had burned up on re-entering the atmosphere. The ABMA team fell short with Explorer 2 when a JPL upper stage failed to ignite. Later in March the beleaguered Vanguard team succeeded on its third try and put a satellite into orbit. The Army soon recovered, achieving back-to-back successes with Explorers 3 and 4 that year.

Earlier in 1958, the von Braun/ABMA team had received a new directive saying it could return to developing and test-launching rockets of greater range than the standard 200-mile Redstone. The United States had the Army's Jupiter IRBM, first fired successfully in 1957; the comparable Air Force Thor; and the Air Force's much more powerful Atlas ICBM in the works. Still, the Soviet Union held a commanding lead in launch power.

Not widely known is the fact that the US Navy had shown interest in the Jupiter back in late 1955 as a shipboard-launched IRBM. Design changes and engineering support from the Army ensued with that in mind, laboratory director Hans Hueter related. His Systems Support Equipment Lab within the von Braun organisation at Redstone Arsenal proceeded to develop 'a support equipment and [on]board launch concept' that examined such aspects as 'ship motions, missile training, emergency missile dumping, missile tail grab and release mechanisms', Hueter reported in 1960. But after several months of development, a contractor's study found that 'the handling of large quantities of liquid propellant on board a ship was . . . too hazardous'. The result: 'the Navy cancelled the project and turned to solid propellants'. Still, the Army's Jupiter, which was later turned over to the Air Force, retained design features that resulted from the Navy's interest.[35]

Also in 1955, a senior engineer on von Braun's team, Georg von Tiesenhausen, conducted a secret study for the Navy on a submarine-launched Jupiter IRBM. It was called 'Project Navy: A Study on Submersible Launching Containers for Guided Missiles'. The investigation harked back to a concept pursued at Peenemünde near the war's end that envisioned floating V2s long distances before firing them at New York and other targets.[36]

Several months into the space age, von Braun detected certain assumptions being made by a sizeable segment of the US public and its national leadership – assumptions that perturbed him. He spoke out with what was fast becoming characteristic candour, and was widely quoted.

I am getting hot under the collar in a most unscientific fashion. Like it or not, the United States is in a space race, yet our science program lags. Apparently a large section of the public believes that we can sit on our hands until Soviet

science falls apart as it is bound to do, they assume, under a
dictatorship. That is wrong.

The second assumption is that we can catch up if we spend
enough money. Wrong again. The third is that we have no
business in space. That is the worst mistake of all.[37]

Von Braun also warned against the US preoccupation with reacting
to whatever its chief competitor did, rather than pursuing a
maximum effort of its own design. The von Braun group had done
preliminary work at ABMA on a space launch vehicle with a
booster stage of more than 1 million lb of thrust – for a possible
lunar project. The team submitted a proposal to conduct
developmental studies of clustering engines for the first stage of the
Nova heavy-lift rocket to the Defense Department's Advanced
Research Projects Agency (ARPA). The ARPA had been created in
1957 to referee among the military branches and their competing
space proposals. It had yes-or-no authority over all space proposals
with defence applications.

Representatives of the ARPA visited Redstone in early 1958 for a
meeting on the ABMA's Nova booster proposal. After a briefing by
von Braun, an ARPA official enquired, 'How much funding will you
require?'

'How much do you have?' asked von Braun, no fool.

The official replied that $10 million was all the agency could
spare.

'That's wonderful!' exclaimed von Braun. He'd take it, he said, and
the project would get under way.[38] It did. The big-booster work
would lead directly to the team's Saturn family of heavy-duty space
launch vehicles.

Donald R. Bowden was a young project engineer with the ABMA
in the late 1950s when he gave one of his first briefings to von
Braun. He discovered a surprising trait that made him agree with
Hermann Oberth's assessment of his former protégé; of the qualities
he 'admired most' in von Braun – 'your organizational talent . . .
relentless drive . . . self-discipline . . . even-tempered fairness and
engaging ways . . . most of all I enjoyed your frankness and easy
comportment, never putting on airs'.[39]

Bowden was working in California on the Jupiter missile's S-3D
engine with North American Aviation's Rocketdyne group.

A technical problem with a critical valve had arisen, and the contractor came up with a solution. Back in Huntsville, he went to von Braun's office seeking approval of the proposed change. 'I was explaining to him how this new valve worked,' remembered Bowden. 'And he said, "Wait a minute, wait a minute! You're on Lesson Two here, and I need Lesson One. Tell me how this valve works, and what does it do, and how does the gas generator feed the turbine here, and . . ."'[40]

The young engineer was 'astounded that a world-famous rocket scientist would admit he needed some "Lesson One" instructions'. Bowden gave von Braun a detailed A-to-Z explanation of the new valve and how it would solve the problem. Von Braun quickly grasped the specifics. But the briefer had seen 'humility', he said, in his leader that day. Bowden, who stayed for years with the von Braun team as project manager and later succeeded as an entrepreneur and corporate CEO, added: 'The lesson I learned from von Braun is that no matter how important or even famous you are, if you don't understand something, you don't let your ego get in the way of your truly learning the details. . . . Don't try to fake it and go on to make a wrong decision.'[41]

Around 1958–9, von Braun became involved with the ABMA's foreign missile intelligence unit at Redstone, recalled Rankin A. 'Randy' Clinton, then the intelligence unit's technical chief. He needed to know what he could and could not say about Soviet missiles and space capabilities, from a classified-information standpoint, as he prepared to testify before Congress or to give a major speech. 'He would come over to our office for a session,' Clinton remembered, 'and he'd always say, "Randy, make me smart!"' In other words, he had to learn what information was cleared for "white world" use', as opposed to the 'black world' of military secrecy. 'He was very sensitive to the security aspects of it.'[42]

Von Braun soon became directly engaged in the analysis of photographs taken by America's high-flying U2 spy planes. 'Many times he would come over and look . . . at the film of Soviet missile emplacements,' Clinton revealed. 'We bought a stereo microscope for his use, so we wouldn't have to set up a screen display every time. He was certainly helpful to us. He could spot the "fingerprints" of old [Peenemünde] colleagues' in the U2 photos.[43] 'It took a presidential waiver to clear von Braun for this work with us,' added Clinton, because the newly naturalised US citizen's security rating was not high enough.[44]

Meanwhile, with the Soviet Union continuing to boast of its space victories, US leaders, including Senate Majority Leader Lyndon Johnson, urged that space exploration be given a high national priority. In April 1958, a still-lukewarm President Eisenhower sent Congress a proposal to establish a national space agency. A month later the Russians sent Sputnik III into orbit – a record-setting 3,000lb satellite. Congress passed the Space Act in mid-July to create NASA (the National Aeronautics and Space Administration), largely from the old NACA (National Advisory Committee for Aeronautics) headquartered at Langley, Virginia. Ike signed the measure into law later that month and appointed T. Keith Glennan as NASA's first administrator.

Von Braun went to Washington late that summer for a private meeting with Glennan. He was aware of discussions between the new NASA chief and Defense Secretary McElroy about the possible transfer of von Braun and the bulk of his 4,800-person Development Operations Division to NASA. He was also aware of Secretary of the Army Brucker's and Gen Medaris's vigorous opposition to that move. Glennan reminded von Braun years later that the scientist was, understandably, trying to learn more about how his team and operation would fare if the transfer to NASA went through. Glennan reminisced: 'While I would not accuse you of it, I think I detected an effort to secure a preferred status for your group. After a discussion that lasted more than two hours in which you were unable to gain your desired objective, you rose to depart. Your final remark, delivered with that wonderful smile of yours, was one I shall never forget. It went, "Dr. Glennan, all we really want is a very rich and generous uncle!" That, you did get!'[45]

Von Braun also got an early taste of intra-agency intrigue and in-fighting in the late 1950s during the transition to NASA's control over US space activities. And some of his Huntsville teammates got a good look at a von Braun not easily shut out of the game. As plans for America's Project Mercury and its one-man space 'capsules' were taking shape, von Braun took a planeload of ABMA people up to Langley, later called 'the first Houston'. Elements of what was to become NASA's Manned Spacecraft Center (later the Lyndon B. Johnson Space Center) in Texas were then still housed at the Virginia centre.

The discussions dealt with various aspects of implementing Mercury. NACA's Robert R. 'Bob' Gilruth, soon to become director of the Houston field centre, took a leading role in the talks. They went on all day and covered the von Braun team's plan for using its standard Redstone rockets to lift the first US astronauts into space on arcing, 15-minute, suborbital flights over the Atlantic. The plan was deemed the quickest way to get Americans technically 'into space', albeit short of orbit.

Charles A. Lundquist, a member of the Huntsville contingent, recalled that the dialogue at the meeting eventually took a testy turn. 'Toward the end of the meeting von Braun said to Gilruth, "I hear reports that you're planning some kind of two-man capsule and you're talking to the Air Force about it. Would you care to give us a briefing on it?" And Gilruth said, "No, I wouldn't." And von Braun said, "Well, we'll see about that."'

And on that combative note, Lundquist remembered, the meeting ended. 'It was the Gemini programme they [Gilruth and company] were working on, and they were trying to exclude Huntsville,' he recalled. 'When von Braun went to headquarters he got the assignment for us to be involved in the [launch] vehicle end of it.'[46]

At the outset of their relationship, master aeronautical engineer Bob Gilruth cared little for von Braun, according to Houston space centre's Christopher Columbus 'Chris' Kraft and scientist Charles Lundquist of the von Braun team. Gilruth had been engaged in warplane design in the Second World War, worked for a time in England and came to detest Nazi Germany's V2 missile effort and those behind it, Kraft wrote.

His early distrust of von Braun eventually developed into mutual professional respect, von Braun colleagues maintained. 'In the pre-Apollo days, when roles and responsibilities were not yet decided, there was some early resentment between the centres' and their directors, Frank Williams recalled. 'It was not the most harmonious of relationships.'

But that antipathy eased as soon as Project Apollo roles were established and other major issues settled. 'The two shook hands. They didn't go out drinking together, but there was definitely a mutual respect, admiration and friendliness between the two,' Williams added. Von Braun gained further appreciation from Gilruth 'by giving him more payload [weight] than promised [through increased launch vehicle thrust], because all three Apollo spacecraft modules came in overweight'. From the Houston end, Gilruth's chief engineer, Maxime 'Max' Faget, concurred, 'They got along very well together.'[47]

As NASA field centre chiefs, the two men followed different programme management styles and remained cut-throat competitors for new roles and funding. For public consumption, however, the pair sought to present 'an image of full cooperation', for the good of the overall programme.[48] Von Braun, in particular, worked hard to preserve that perception and never spoke ill of Gilruth, in private or otherwise, according to both Stuhlinger and von Braun management associate James T. Shepherd. But the 'hard feelings' between Houston and Huntsville were the reality, and the 'image of unity' was a fabrication, Shepherd recalled.[49]

Through the years Gilruth gave every sign of believing that all von Braun wanted was everything. He was not far off the mark. Von Braun's organisation had long controlled Army rocket launch operations at Cape Canaveral, as well as testing at the White Sands Missile Range in New Mexico, and von Braun lobbied hard for basing the astronauts and their training activities at Huntsville. As the NASA organisational battles came and went, von Braun won some and lost some. While Gilruth had Senate Majority Leader Lyndon Johnson of Texas in his corner, von Braun could count on help from savvy veteran Senator John Sparkman of Huntsville. Before long the von Braun-led space centre had facilities and operations at five locations in four southern states, including support activities at Cape Canaveral and a coast-to-coast network of project branch offices.

The first of the ABMA's Redstone/Jupiter-C rockets carrying space payloads – the Explorer satellites – were rechristened 'Juno I' vehicles. When the agency's more muscular Jupiter IRBMs were pressed into service as space launchers, they gained the unmilitary-sounding designation 'Juno II' – Jupiter plus three upper stages. In early March 1959, it was an Army Juno II that launched the unmanned Pioneer 4 spacecraft on the first intentional fly-by of the moon. The spacecraft went on to become the first US satellite to go into permanent planetary orbit around the sun. (Two months earlier, the Soviets' Luna 1 – in Russian, Lunik – had missed its planned lunar landing and gone into solar orbit.)

Less than three months later, America's first space passengers of significance – the small rhesus monkey 'Able' and squirrel monkey

'Baker' – took a 1,600-mile, suborbital ride in the nose cone of a Juno I rocket as a warm-up for Mercury astronauts. Both primates survived the test flight and were safely recovered, although 'Able' died while Navy surgeons were removing biomedical probes. 'Miss Baker' went on to live a long and healthy life as the star of the show at Huntsville's US Space and Rocket Center.

The Army's von Braun group had a string of three straight Explorer successes going when, in the summer of 1959, it prepared to launch Explorer 6 at Cape Canaveral. The satellite had a planned orbit with a perigee, or low point in altitude, of about 100 miles. But at launch the Jupiter-C rose briefly, veered abruptly and then dived back low over the launch site. The range safety officer pulled the 'destruct' switch, and the rocket and its satellite payload were blown up before anyone got hurt.

The payload designer, team veteran Josef Boehm, was distraught when his two years of work were obliterated in seconds. He managed at least a smile a few days later, though, when von Braun handed him a photograph. It showed the errant rocket zooming horizontally among the launch towers at a height of only 200 to 300ft. The picture bore an inscription:

> To Josef Boehm –
> Nothing wrong with it,
> just perigee a little low
> Wernher von Braun

'I will always treasure your inscription,' Boehm reminisced years later. 'Many of your associates will remember situations where you eased the impact caused by mishaps or frustrations with your generous attitude and exuberant humor.'[50]

It was also in 1959 that von Braun began mentoring a young Associated Press reporter. The arrangement evolved into 'a lasting friendship', remembered Howard Benedict, who went on to become a space author and the dean of print aerospace correspondents at Cape Canaveral over more than three decades.[51]

The day Sputnik I opened the space age, Benedict was the only Associated Press news features reporter at work in the New York offices and available to do a Sunday feature analysing the meaning of it all. One of the people he interviewed by telephone was von

Braun. The piece, via an accident of timing, 'made me an "expert"',
Benedict recalled. The Associated Press called on him to write any
'space story' that came up thereafter, and he found himself being
dispatched to Florida all through 1958 for the early US satellite
launchings and attempted launchings.

The world's largest news agency moved him the following year to
Cape Canaveral to cover the missile–space beat full time. Benedict
was still a 'befuddled reporter' the day he met von Braun soon after
settling in. He 'probably could surmise from my questions at a news
conference that I knew very little about missiles or rockets', the
journalist recalled. 'Von Braun [also] knew that one way to sell his
ideas, to get people thinking about them, was through the popular
press, especially those with worldwide circulation, like the Associated
Press.' Benedict continued:

> That is probably why he approached me after that first news
> conference and asked me if he could be of help in educating me
> about rocketry. I was flattered, and a couple nights later I met
> with him in his room at the Silver Sands Motel in Cocoa Beach,
> where he often stayed. For the next two hours, over a couple
> beers, this wonderful man saturated my brain with answers to
> questions like: How does a missile work? How does a satellite
> reach orbit, and how does it stay there? How will man survive
> once he rockets into space? Heady stuff for this reporter who was
> a novice in this new space era that the world was entering.[52]

Further Silver Sands tutorials followed. 'Von Braun educated me on
the history of rockets and enthralled me with tales of his early
rocket experiments in Germany,' Benedict related. To the journalist,
the engineer-scientist was 'a man with a vision, and he never
stopped pushing his dream of one day harnessing this immense
rocket power to send men into space and to the Moon and to
develop orbiting colonies'.[53]

The transfer of the von Braun group on paper from the ABMA to
NASA was assured in the autumn of 1959 when the Army
negotiated a deal to leave itself a missile R&D core capability from
which it could rebuild. The ABMA's Gen Medaris had at first objected
to losing the von Braun group. But then there was talk of giving the
team to the Air Force and arch-rival Gen Bennie Schriever; von

Braun even had a meeting with him. The Army decided there were worse things than losing the group to NASA. And so the decision was made to move most of the 4,800 people of von Braun's Development Operations Division and its facilities for building and testing rockets to the control of the new space agency. That included virtually all the Peenemünders still with the team. Most of the employees administratively making the move would stay at their same desks and workbenches in the same buildings as before, at least until new office buildings for some could be constructed.

Von Braun, who had written of man in space since boyhood, met his first real-life astronauts-to-be in 1959. He helped host the debut visit by America's original spacemen to the ABMA at Redstone. He spent hours with the seven pilots as they inspected a Mercury-Redstone booster and spoke of their coming adventures. Asked at a news conference with them if he wished he were going with them into space, he answered: 'Sure, why not? I envy them. But they just told me I'm too fat!'

During that same visit the future spacefarers inspected the Mercury spacecraft in which one of them was to ride atop a Redstone and become the first American in space. Vachel 'Val' Stapler, an engineering technician, recalled that one of the most confident of the seven, Alan Shepard, took a good look at the launch vehicle and the compact spaceship and exclaimed in mock horror:

'You mean I'm supposed to ride in that?! I quit!'

'What are you talking about, Commander?' von Braun retorted. 'You never had it so good!'[54]

In 1959, authority over the Huntsville team's Saturn rocket programme was transferred from the ARPA in the Defense Department, which envisioned no military uses for rockets that big, to NASA, which saw plenty of uses. NASA set up the Saturn Vehicle Evaluation Committee with Abe Silverstein as chairman. He was the new agency's director of Space Flight Development.

A NACA veteran, Silverstein had done years of research on liquid hydrogen at what became NASA's Lewis (later Glenn) Research Center in Cleveland, and he favoured using it as a more efficient fuel in upper stages of the future Saturn vehicles. The conservative von Braun, wary of the volatility of super-cold liquid hydrogen, preferred more conventional, safer fuels. Silverstein's view prevailed. In its December 1959 report, his committee gave its blessing to liquid hydrogen. Von Braun accepted the verdict, and later acknowledged the rightness of it. Hydrogen packs more propulsive punch per lb. Saturn ground crews in Huntsville, at contractors' facilities, and eventually at Cape Canaveral, all proved up to the task of safely handling the tricky fuel. (In the thirty-two flights of various Saturn rocket configurations from 1961–75, no mishaps occurred and all were successful, a performance record unprecedented in rocketry.)

The flap over Saturn propellants would not be the last technical wrangle between Silverstein and von Braun. The two respected scientists, even with their apparent mutual professional regard, never developed a personal rapport, according to NASA official Robert C. Seamans Jr. 'The feeling came up among von Braun's associates that Silverstein was against the von Braun group on general principles.' Stuhlinger and Fred Ordway termed that belief 'unfortunate', noting that 'von Braun and his co-workers considered Silverstein as the best rocket man in NASA Headquarters, and they would much rather have seen a brother-in-arms in him than an antagonist'. Silverstein responded that 'there is no basis for that feeling. . . . We got along very nicely. I had a great deal of respect for von Braun's capability. He was a wonderful leader.'

Years after von Braun's death, Silverstein acknowledged his awareness of the Huntsville rocket team's perception in remarks to Eberhard Rees, von Braun's successor as Marshall Center director: 'You guys in Huntsville always thought that I was against you. That simply is not so. I always respected you. In fact, ever since the von Braun team faded out of NASA, something essential has been missing in the space program.' He also cited several instances that showed his high esteem for von Braun and his team. One was the fact that he had lobbied hard to have the Army rocket-and-space team brought into NASA.[55]

In October 1959, NASA Administrator Glennan made his formal request to transfer the ABMA group to the new-born space agency. Within days Eisenhower announced his approval, subject to congressional concurrence. That action came in March 1960, and on 1 July all but a fraction of von Braun's division at Redstone transferred.

With von Braun as director, the team would staff a new NASA field centre whose main business was to be propulsion systems. It was named the George C. Marshall Space Flight Center. The name was suggested by Glennan, and swiftly endorsed by Eisenhower, in memory of his Second World War comrade – a five-star Army general, secretary of state, and the Nobel Peace Prize-winning architect of the Marshall Plan that rescued postwar democratic Europe. Marshall had died the year before.

Eisenhower presumably recognised the irony of having a centre named by him, the wartime supreme Allied commander in Europe, to honour another great US military leader in the Second World War, and having that centre's first director being a prominent enemy in that war.

It was the first time in twenty-eight roller-coaster years that von Braun was not working for the military, not developing weapons, either in Germany or America. He was 48. His most professionally fulfilling years lay ahead.

Challenge of the Moon

'We knew if von Braun was leading it, things were going to get bigger and better.' That summarised the upbeat attitude of many within the expanded von Braun rocket team facing transfer from the US Army to NASA, according to engineer-manager Robert 'Bob' Schwinghamer.[1] The mass reassignment of more than 4,500 personnel from the newly organised Army Ordnance Missile Command, successor to the Army Ballistic Missile Agency, at Redstone Arsenal to the Marshall Space Flight Center (MSFC) took place on 1 July 1960. The MSFC was given an 1,800-acre enclave on the sprawling military post and became NASA's largest centre in staff and budget. This was von Braun's first time as chief executive of his team. What did the future hold for his group in civilian space work? Would there even be a long-term future for them in Huntsville?

Those familiar with von Braun's record, like the American-born Bob Schwinghamer, shrugged off any doubts. Bigger came right away, with advances in the preliminary work on what was to become the Saturn family of super-boosters. The first was the C1 launch vehicle – originally designated Juno V. It was soon renamed Saturn I, because, von Braun explained, in the order of planets in the solar system, Saturn came after Jupiter, the name of his team's last big rocket.

President Eisenhower formally dedicated the Marshall Center on 8 September in Huntsville, a few months before he was to leave office. During a tour of the centre's facilities, he stopped at a test stand and beheld the large eight-engine cluster of a Saturn I first stage. The president turned to von Braun and confessed: 'They come into my office and say it has eight engines. I didn't know if they put one on top of the other or what!'[2]

Clearly, the earlier presidential briefer had not been von Braun. Dwight Eisenhower, the career Army man, had scarcely acknowledged the string of successes by the Army missile and space team during his eight years in the White House. His Army comrades and the news media were struck by the persistent slight, as former Army and NASA public relations official Gordon Harris noted in his memoirs.[3] Was it because the Army group was led on the technical side by von Braun and all those other former German enemies of Ike in the Second World War?

Whatever the problem, the popular President – who had German ancestry on both parents' sides – relented in dedicating the Marshall Center. 'Here under army guidance the Redstone, Jupiter, and a whole family of missiles have taken form. I share with the army its gratification in these trailblazing achievements. They have thrilled the people and won plaudits throughout the world. I freely admit sentimentally that my contemplation of these advances is stirred because so much of this dramatic achievement was pioneered by the army, which until recently was my life and my home.'[4]

As head of NASA's main new centre for launch vehicle development, von Braun knew that future projects would require major advances in large booster rockets. One day he was discussing the relative importance of factors such as aerodynamic design with Schwinghamer. To von Braun, nothing was so important in big rockets as raw lifting power, as sheer thrust. 'I can fly a beer can,' quipped the scientist, 'if you give me enough propulsion.'[5]

Things got busier fast at the new space centre in northern Alabama. In preparation for the Mercury-Redstone flights of an American one-man spaceship, the Marshall Center readied the first suborbital test of an unmanned capsule. It scored a success on 19 December 1960. After two more successful test flights – on 31 January 1961, with the chimpanzee 'Ham', and then with chimp 'Enos' on 24 March – the von Braun team stood ready to boost Mercury astronaut Alan Shepard as the first human in space.

The Russians, though, had their own plans. On 12 April, Moscow announced that Soviet Air Force Maj Yuri Gagarin had safely orbited the Earth one full revolution in a 5-ton spacecraft. Soviet Premier Nikita Khrushchev told the orbiting Gagarin, 'Let the capitalist countries catch up with our country!'

Von Braun, after offering his congratulations to the Soviet space agency, called the feat 'the shot heard around the world'. He added to reporters, knowing the White House and Congress were listening, 'We are going to have to run like hell to catch up!'

President John F. Kennedy, newly in office and already battered by the Bay of Pigs fiasco in Cuba, wearily shared his candid reaction to cosmonaut Gagarin's coup. 'No one is more tired than I am', he confessed, of America's second-place standing in the new space age. But there was no escaping the truth, JFK admitted to reporters: 'It is going to take some time. We are behind.'

Von Braun, his Marshall Center team and their partners scrambled to regain some of the lost ground. On 5 May, just three weeks after Gagarin's orbital flight, Navy Cdr Shepard climbed on top of a Redstone rocket at Cape Canaveral. His single-seat capsule, nicknamed 'Freedom 7' for the original seven Mercury spacemen, was provided by NASA's Manned Spacecraft Center (MSC) near Houston. The mini-spaceship was boosted on a curving, 15-minute ride 116 miles up and back to a splashdown in the Atlantic and recovery by the aircraft carrier USS *Lake Champlain*. It was no orbital mission, but America had put a man in space, however briefly.

In Huntsville's Courthouse Square, the site of public celebrations of US space feats ever since the first Explorer satellite launching, von Braun told the joyous crowd: 'Our opponents across the ocean, behind the Iron Curtain, thought about a month ago they had slammed the door to the universe in our face, but Shepard has let us out of our dilemma and embarrassment. . . . We will go farther and farther, eventually landing on the Moon.'[6]

Even before former Navy test pilot Shepard's baby step into space, President Kennedy had seized the initiative. On 20 April 1961, after Gagarin's stunning flight, he had asked Vice President Lyndon Johnson to recommend a specific national space project that could produce 'dramatic results' in the competition with the Soviets, one that the United States 'could win'.[7]

Johnson – like JFK, a Navy veteran of the Second World War – immediately wrote to von Braun, MSC Director Bob Gilruth, Gen Bennie Schriever, US Air Force and several others asking for their counsel on the question. Von Braun responded on 29 April with a ten-page letter to 'My dear Mr Vice President'. After noting that there was a 'sporting chance' of achieving several specific

American 'firsts' in space in the coming years, he wrote that the nation had 'an excellent chance' of scoring a manned lunar landing before 1970 – and ahead of the Russians. He spelled out the reasons and emphasised the abundance of benefits that would accrue from such an enterprise.[8]

The same day he received von Braun's response, LBJ wrote to Kennedy that the clear choice was a manned landing on the moon. The new NASA Administrator, James E. 'Jim' Webb, and Defense Secretary Robert McNamara also signed the formal report to the President. Although several others had made the same recommendation as von Braun, Johnson's letter to Kennedy closely followed the points made by the Marshall Center director.

Indications were that von Braun's argument had been the most persuasive, but he always downplayed his singular influence on the recommendation given to JFK, saying the moon was a consensus choice. He told one interviewer: 'I would not like to take credit for this myself. I participated in discussions with Vice President Johnson concerning what this country could do to assume leadership in space. The consensus was that we needed a clear, highly ambitious goal – one that was hard-hitting and long-range enough so the Russians could not do it first.'[9]

The following month, on 10 May, Kennedy made his decision. On 25 May he spoke to a joint session of Congress on the subject of space. America had 15 minutes of manned spaceflight experience and zero time in orbit. Despite that, President Kennedy challenged the nation to 'commit itself to achieving the goal, before this decade is out, of landing a man on the moon and returning him safely to Earth.'

At NASA centres and aerospace plants around the country, jubilation reigned. Von Braun and his board of directors at Marshall – the laboratory chiefs and other key players from Peenemünde[10] – had gathered in their main conference room to hear Kennedy's speech. Shouts of 'Ja!' and 'Let's go!' sounded when the specific goal and deadline were announced – even though almost one and a half years of the decade were already gone. 'For the first time, it felt like fun to be working for NASA,' recalled one lab director, Walter Haeussermann.[11]

Von Braun publicly lauded Kennedy's challenge of Project Apollo as one that 'puts the programme into focus', clearly and concisely, for the nation. 'Everyone knows what the moon is, what this decade is, what it means to get some people there – and everyone knows a live astronaut from one who isn't.'

Of the dramatic Apollo call to action, US television personality Hugh Downs recalled: 'When President Kennedy said, "We choose to go to the Moon!" and set a timetable of ten years, many of the scientists I knew were horrified. They said they were sure we'd get there, but that it would take thirty years.'[12]

Former President Eisenhower was among the vocal doubters and critics. 'Why the great hurry to get to the moon and the planets?' Ike asked news reporters, calling Project Apollo 'a mad effort to win a stunt race. To spend $40 billion to be the first to reach the moon is just nuts!'

Von Braun pointed out that some had called Charles Lindbergh's 1927 solo flight across the Atlantic to Paris a 'stunt' too, but look what it did for aviation. He insisted that Apollo would be 'the wisest investment America has ever made', stimulating advances in science, technology and the economy – and at a cost well under $40 billion. 'Even should we find out that the moon is made of green cheese', Apollo would be worth every penny spent, von Braun later argued.

He had been thinking about voyages to the moon – and beyond – decades before Kennedy issued his challenge. A magazine writer had asked von Braun in 1951 how a trip to the moon rated with him versus a journey to Mars. 'Mars is more of a challenge. It would take 260 days to get there. To the Moon it's only 100 hours. Personally, though, I'd rather go to the Moon than to Mars, even if the trip is shorter. After all, a journey to the Moon is unquestionably a possibility. The Moon's face, thanks to telescopes, is more familiar to us than even some parts of the Earth – the mountain ranges in Tibet, for example. All that's needed is adequate funds and continuity of effort.'[13]

With Project Apollo, he would get plenty of money and a high priority. Long-time close associates insisted – and he made no secret of it – that Mars was his ultimate objective, however, and would remain so throughout his life. The moon would be just a stepping-stone.

America's second manned space flight came on 21 July 1961, with Maj Virgil I. 'Gus' Grissom, US Air Force, duplicating Shepard's brief suborbital excursion. The carrier USS *Randolph* recovered the astronaut – but not his Liberty Bell 7 Mercury capsule, which sank in the Atlantic from a prematurely blown escape hatch cover after splashdown.

October 1961 saw the maiden test flight of a Saturn I. It was the first in the series of three versions of heavy-duty space-launch vehicles to be developed by von Braun's Marshall Center team for Apollo, Skylab and the joint Apollo-Soyuz manned orbital mission with the Soviets. The big rocket, with the clustered eight engines of its first stage generating an unprecedented 1.3 million lb of thrust (later uprated to 1.5 million), performed successfully.

Things might not have gone so smoothly if what was termed the 'Ostrander Affair' had ended differently, according to Walter Haeussermann, a veteran member of the von Braun team. The steering sensors that Haeussermann's Guidance and Control Laboratory planned to use for the first Saturn-class vehicle had been refined through years of development for the Army's Redstone, Jupiter and Pershing missiles. But when the team transferred to NASA, the new agency's director of Launch Vehicle Programs in Washington, Maj-Gen Donald Ostrander, US Air Force, told Haeussermann's lab at Huntsville to replace its proven sensors and associated computers with those intended for the Air Force-developed Centaur missile.

Full testing – by the manufacturer as well as Haeussermann's crew – found extensive problems with the unproven Centaur instruments, Haeussermann recalled. With von Braun's approval and a negative report in hand, the lab chief met with Ostrander in Washington. 'This is completely uncalled for!' the Air Force general exploded, summarily dismissing Haeussermann from his office.[14]

Back in Huntsville, the lab chief told von Braun what had happened and said he would have to resign. 'Walter,' von Braun replied, 'you react as you would have to in the Old Country. Here you fight for your convictions!'

'I went along,' recalled Haeussermann, 'knowing that we would fight together.' Soon Ostrander wrote to von Braun pressing for conversion to the Centaur sensors. Von Braun scheduled a meeting the next day with the general and took Haeussermann with him. The latter waited outside Ostrander's office while von Braun went in. After 15 minutes von Braun came out, smiling and happy – the general had backed off.

'How did you accomplish this?' an astonished Haeussermann asked.

'I told him he will have my resignation' over the issue, a still-smiling von Braun answered. The two rocket veterans enjoyed a good laugh. Von Braun's team used its own sensors.[15] The general later left NASA.

As was apparent from the Ostrander Affair, the otherwise halcyon Saturn–Apollo era had its share of conflicts. Friction arose among the competitive, ambitious field centres, between the centres and NASA headquarters, and at times between the centres and their contractors in industry. Usually the struggles revolved around substantive issues or differences in management approaches. But personalities clashed too, especially before time had allowed respectful, if not friendly, relationships to grow.

Von Braun sometimes aroused resentment within the space agency because of his aggressive efforts to gain added programmes, facilities and budget dollars for the Marshall Center. Jealousies also arose over his celebrity status as the spokesman for the space age at news conferences, on Capitol Hill and on the speaking circuit.

And a few officials within the NASA hierarchy found it hard, if not impossible, to get beyond his wartime service to the Third Reich. Houston's Gilruth and NASA's first flight director, Chris Kraft, for instance, had both worked on warplane design in the Second World War and had developed an acknowledged distaste then for the people behind the V2. Early in NASA's life, an 'antipathy' between Gilruth and von Braun 'was simmering just beneath the surface', Kraft wrote in his autobiography. He recalled that Gilruth had once said, 'Von Braun doesn't care what flag he fights for.'[16] Kraft did not disagree. It was the old concern about von Braun's having transferred his allegiance so readily from Germany to the United States. In the end, however, the harsh feelings gave way to a warmer relationship between von Braun and Gilruth, if not between the former and Kraft.

Perhaps one of the most unlikely personal relationships formed during Apollo was that between von Braun and Maj-Gen Samuel Phillips, US Air Force, who was brought into NASA as the Apollo programme director. As a Second World War fighter pilot based in England, Phillips had not only been on the wrong end of V2 attacks but he had also escorted bombers on raids against von Braun's Peenemünde centre. In postwar years the two men became, and remained during the push to the moon, true friends, as Phillips made clear in Ernst Stuhlinger and Fred Ordway's biographical memoir of von Braun.[17]

From the start, von Braun had a lively rapport with America's astronauts. As a mechanical-aeronautical engineer, a natural risk-taker and an avid pilot, the physicist identified closely with them. For many years he had imagined himself alongside them, travelling aboard a rocket ship through the heavens. He often expressed the hope – the expectation – of someday doing so. 'Dr von Braun always got over to see the astronauts whenever they visited,' Schwinghamer recalled. 'He liked those guys, and they knew it.'[18]

The US manned space programme at last went orbital on 20 February 1962. An Air Force Atlas ICBM boosted Lt-Col John H. Glenn Jr, US Marine Corps, into three swings around the Earth – and a dicey re-entry with his braking rockets' 'retropack' – in the first of four successful Mercury orbital missions. The destroyer USS *Noa* effected a smooth recovery, lowering Glenn's Friendship 7 spacecraft onto its deck with the astronaut still inside. President Kennedy, honouring Glenn later in the Rose Garden of the White House, took the opportunity to tell the nation: 'We have a long way to go in this space race. We started late. But this is a new ocean, and I believe the United States must sail on it and be in a position second to none.'

Von Braun praised Glenn's unflappable performance and called his flight 'a Bunyan step' for America and the rest of the non-Communist world. 'It puts us', he stated at a press conference, 'right where we belong – in space.' Von Braun's friendly personal relationship with Glenn, an alumnus of the Naval Aviation Cadet Program and a combat pilot in both the Second World War and the Korean conflict, persisted over time. Four years after the 1962 orbital flight, the rocket scientist received a postcard from the then-retired spaceman. Glenn's card had been sent from Switzerland with a message handwritten in German, the English translation of which reads:

Dear Wernher –
   Here I am in Lucerne and you are in Huntsville.
   What a switch! It is 26 years ago that I studied German and I cannot remember many words.
Cordially,
                                                                J.H. Glenn, Jr.[19]

In 1962 President Kennedy made the first of two visits to the Marshall Center to check on Project Apollo. Hosting the President's tour on 11 September of the centre's R&D, assembly and test facilities, von Braun showed JFK a model of the forthcoming Saturn V rocket. 'This is the vehicle', he dramatically assured the President, 'which is designed to fulfil your promise to put a man on the moon by the end of this decade.' He paused, glanced at the rocket model, then at Kennedy, and exclaimed, 'By God, we'll do it!'

The President impulsively asked von Braun to accompany him to his next stop, Cape Canaveral. Perhaps buoyed by von Braun's confidence, Kennedy declared in a speech there that despite the nation's late start in the space race, 'We shall be first!'

Eight months later, in May 1963, Kennedy returned to look in on progress at the MSFC. The President was treated to a raucous, window-rattling static firing of a Saturn booster stage strapped down in a test stand. After experiencing the fiery demonstration from an outdoor observation bunker, Kennedy grabbed von Braun's hand, congratulated him, and enthused within earshot of reporters: 'That's just wonderful! . . . If I could only show all this to the people in Congress!'

Within six months Kennedy was assassinated in Dallas. Von Braun heard the terrible news over the radio while returning home aboard a NASA plane after testifying before Congress. The death hit von Braun hard. A bond of 'mutual admiration' had developed between Kennedy and von Braun, according to Bonnie Holmes.[20]

On that 22 November Wernher and Maria von Braun had had an invitation in hand from the Kennedys to attend a White House reception on 25 November, which proved to be the day the President was buried, the secretary remembered. While the rest of the nation marked a day of mourning for the President's funeral, a distressed von Braun sought refuge in his work. He had asked his secretary/confidante if she would come in that day so they could attack the backlog. Only the two were in his office suite that sad day.[21] He lamented to her: 'What a waste. What a tragic loss of a friend and a great leader.'

'There was a large-screen TV in his office,' Holmes recalled. 'He would dictate a little and then we would watch the funeral on TV. I guess that was the only time I ever saw him actually cry. He was very moved.'[22]

Von Braun waited about two months – for a significant rocket launch success in the Saturn–Apollo programme John Kennedy had initiated – before sitting down to compose a handwritten, two-page letter of condolence on personal stationery to the fallen President's wife.

February 1, 1964
Dear Mrs. Kennedy:

In our elation over the successful launch of SA-5 last Wednesday – the fifth in a successful string of launchings of Saturn I rockets, but the first capable of going into orbit – I must tell you how happy and grateful we are that this test came off so well. All of us connected with this undertaking knew only too well how eagerly the late President had been looking forward to this launching, which would at last establish the long-awaited American lead in the capability of orbiting heavy payloads.

The trust he had placed in us, and his confidence that we would succeed, offered great encouragement but placed on us an even greater sense of obligation. I am enclosing a picture taken in front of the towering SA-5 rocket at Cape Kennedy on November 16th. The model at the left depicts the upper part of the rocket which is now orbiting the earth once every 94 minutes. The unit in orbit has a length of 83 feet and a weight of 37,800 lbs.

You have been overwhelmed with condolences from all over the world at the tragic death of your beloved husband. Like for so many, the sad news from Dallas was a terrible personal blow to me. We do not know a better way of honoring the late President than to do our very best to make his dream and determination come true that 'America must learn to sail on the new ocean of space, and be in a position second to none'.

With deepest sympathy – Wernher von Braun.

Within a few days he received a handwritten, two-page response from the President's widow, composed on her personal stationery. It read:

February 11, 1964
Dear Dr. von Braun

I so thank you for your letter – about the Saturn – and about my husband.

What a wonderful world it was for a few years – with men like you to help realize his dreams for this country – and you with a President who admired and understood you – so that together you changed the way the world looked at America – and made us proud again.

Please do me one favor – sometimes when you are making an announcement about some spectacular new success – say something about President Kennedy and how he helped to turn the tide – so people won't forget.

I hope I am not the only one to feel this way – it is my only consolation – that at least he was given time to do some great work on this earth, which now seems such a miserable and lonely place without him.

How much more he could have done – but I must not think about that.

I do thank you for your letter.

Sincerely,

Jacqueline Kennedy.[23]

Kennedy's death brought into the White House the tall, lanky Texan Lyndon Baines Johnson, who was an earlier and stronger space supporter than President Kennedy. When tragedy struck in Dallas, 'von Braun was working at the highest levels of the government . . . especially with LBJ as vice president', recalled close NASA associate Jay Foster. 'Von Braun had been at the LBJ ranch a couple of times working that interface, and he was scheduled to go to the ranch for dinner during what turned out to be the week after the assassination. Of course, that was cancelled.'[24]

But life – and Project Apollo – moved on. The following year, President Johnson presented von Braun with a five-gallon cowboy hat during his visit to the LBJ ranch sometime before the fifth unmanned test flight of a Saturn I rocket. The morning the rocket flew flawlessly, on 29 January 1964, Johnson telephoned his congratulations to NASA officials gathered in the launch control centre at Cape Canaveral. Then the President asked whether the hat he'd given von Braun still fitted. The space agency official on the line replied he wasn't sure, but glancing at the happy von Braun, added, 'I believe his head is beginning to swell.'[25]

Managing von Braun's part of the Apollo enterprise meant, among other things, presiding over countless meetings at Marshall. The centre director 'was superb at running a meeting', Schwinghamer recalled. 'Sometimes there would be fierce arguments – I mean, cuss fights. Von Braun would say, "Now, gentlemen, gentlemen! Let's reason this out." He had this fantastic ability to get people to stop fighting and work it out. He could be subtle, too, not heavy-handed.' He also 'had this knack: you always wanted to help him when you left' the meeting.[26]

At one special meeting, von Braun sprang an unwelcome surprise on his Marshall Center hierarchy of German-born technical leaders, all of whom had cherished direct access to him, Lee B. James remembered – von Braun's first non-German manager of a major technical programme at Marshall. 'He was going to have all these lab chiefs report to one person, who then would report to him,' James recalled. 'All of a sudden, all the lab chiefs started saying, "Wernher, this will never work because of so forth and so on!" That conversation went on for over two hours.' Finally, when the chiefs had exhausted the subject, von Braun looked around the table and asked, 'Does anybody else have any comments? No? Well, I think I'll go ahead and do what I said.' The lab chiefs, James said, 'didn't make a peep. He had let them talk themselves out. . . . That was the way he managed.'[27]

For a bold space thinker, von Braun had a reputation as an ultraconservative rocket engineer. He was inclined to rely on the safe and proven, to move ahead only in measured steps, with great caution, and to test, test and test again – down to the last component of the last subsystem. With new rockets, he strongly favoured a deliberate, step-by-step approach from development to operation. With a new two-stage rocket such as the Saturn I, for instance, his team first flew it with only a live first stage, topped by a dummy second stage. That was done repeatedly before the live second stage was added for more test flights.

As those tests proceeded, and with the Saturn IB not far behind, von Braun and his crew were counting on employing the same incremental test-flight plan with the huge, three-stage Saturn V moon rocket. Until, that is, George E. Mueller (pronounced 'Miller') became head of the NASA Office of Manned Space Flight. He soon decreed that all Saturn Vs would be launched 'all up'. That meant all three stages went live, starting with the first flight test, because of Project Apollo schedule dictates. Von Braun and his team strenuously objected to the edict – until convinced it was indeed necessary if they were going to make it to the moon, as promised, before 1970.

Meanwhile, after the remaining Mercury orbital flights by Lt-Cdr M. Scott Carpenter and Cdr Walter M. 'Wally' Schirra Jr, US Navy, and Maj L. Gordon Cooper Jr, US Air Force, were safely logged, the successful two-man Gemini flight programme in 1965–6 ran concurrently with the Project Apollo build-up. The honour and challenge of crewing the first Gemini mission, on 23 March 1965, went to Grissom and Lt-Cdr John W. Young, US Navy. With Gilruth's Houston centre in the lead role, Gemini increased the nation's man-hours in orbit, provided vital spacecraft rendezvous and docking experience, and developed spacewalking skills. Cape Kennedy became a busy place. (It had been renamed to honour the late President; however some years later the name revereted to Canaveral.)

At the Florida spaceport, US rocketry's only recorded 'de-scrubbing' of a cancelled launch attempt involved the scheduled first test flight of the Saturn IB, the mid-size Saturn. On a February morning in 1966, the crucial countdown was halted because of a problem with propulsion systems. Launch Control soon announced the mission was scrubbed for the day. CBS Newscaster Walter Cronkite reported that it would be several days before NASA could reschedule. Busloads of VIP guests and reporters left the site.

Meanwhile, Lee James and his Marshall propulsion crew continued to 'work the problem', aided by a team of experts on standby back in Huntsville. Within minutes they had a solution, and the fix was made. James quickly polled his propulsion team and got unanimity to proceed. The former Army colonel made the decision on the spot to de-scrub, and the launch countdown resumed – to the amazement of everyone still around.[28]

A perplexed George Mueller came bounding down from his seat in Launch Control, followed by von Braun, who was James's boss. Mueller asked James what was going on. James explained it all, point by point, emphasising he had full agreement from his propulsion team.

'George thought about it,' James recalled, 'and he said, "Tell me again what you did." And I said, "I did this, this, and this, and then I restarted the count." He looked up and said, "I wouldn't have done that, Lee, if I had been you." And then he turned and walked out of the room. I thought that was rather odd of George: he didn't say "Stop the count", he just tells me he's not happy with my decision.' James glanced at von Braun, 'and he winked at me – winked real big – and I let the count go. I doubt that I would have kept that decision if I hadn't gotten that wink. But I didn't need anybody else on my side at that point.'[29]

The first Saturn IB flight came off perfectly that day. James not only kept his job but was later promoted to director of the Saturn V rocket programme.

Originally designated C5, Saturn V was conceived by the von Braun team as having a first stage of four enormous F1 engines generating 1.5 million lb of thrust each. As the payload weight requirements for an anticipated manned lunar-landing project kept growing, even before Kennedy's Apollo announcement, the need for a fifth engine became clear.

Just who first cited that need became the subject of a small debate. Milton Rosen, the early NASA chief of Rocket Vehicle Development Programs, came out for the fifth engine in March 1961. Von Braun likewise saw the need early on. 'I have always pleaded for it,' he recalled. 'I said relatively early "to build the thing with four engines doesn't make sense. This great big hole in the center is crying for a fifth engine."'[30]

The addition of the fifth engine also spawned an enduring piece of rocketry folklore. Hugh Downs recalled: 'In those simpler, glorious days when the Saturn V was being developed, von Braun asked in a NASA meeting of design engineers how much safer it would be for the first stage to have five engines instead of the four . . . in the original design. When the engineers studied this and reported a much greater safety margin, Wernher said, 'Make it five.' And they did. There is no way that could be done, by anybody, now [in a modern bureaucracy].'[31]

NASA gave its blessing to the fifth engine in December 1961, upping the power of the S1C booster stage from 6 million to 7.5 million lb of thrust. It proved to be a good move.

At the start of Saturn V's development, even the more experienced, German-born members of the von Braun team found 'the tremendous size of the beast' daunting. 'We looked at drawings of the Saturn V and could not believe it would fly,' Willy Mrazek, an expert in rocket engineering and manufacturing, confessed.[32]

Standing almost 365ft tall, including the Apollo spacecraft, it towered six storeys higher than the Statue of Liberty, including pedestal. The rocket would weigh nearly 6.5 million lb at lift-off – and more than 6 million of that would be propellants. The vehicle would have more than 3 million parts, each stage containing in excess of 70 miles of electrical wiring and enough piping, tubing, valves and pumps to be called 'a plumber's nightmare'. Ex-NASA Administrator Glennan described the planned super-rocket as 'one of the most amazing combinations of engineering, plumbing, and plain hope anyone could imagine'.[33] Others would see it as a machine the size of a cathedral built to the tolerances of a microscope.

As Saturn V moved ahead, von Braun and his engineers could not resist coming up with comparisons to put into perspective the size and brutish power of what was to be the Mother of All Rockets. For starters, just one fuel pump for one of the five F1 engines in the first stage – by far the biggest power plant ever built – would exert the force of thirty locomotives. As designed, the five-engine cluster would generate the equivalent hydroelectric power of eighty-five Hoover Dams – or twice the power that would be created if all the rivers of North America were harnessed at once and channelled through turbines.

Von Braun had a strong dislike for 'moon rocket' as a description of the leviathan. He found the term much too restrictive. 'We have built Saturn V not just to go to the Moon and pick up a handful of dirt. We built it to explore all of space – to reach for the stars,' he stressed.[34] The colossus of combustion was 'the most powerful rocket in the world – at least for the time being,' von Braun noted. 'It is by no means the limit to which we can go, but it gives us the capability to do many things in space just by pushing a button.'[35] Magazine writer Gene Bylinski dubbed Saturn V 'Dr von Braun's All-Purpose Space-Machine.'[36]

Behind it all was 'the team' that von Braun continued to lead. Just how extraordinary was this group? It retained the same cohesiveness and devotion to its leader no matter how large it grew during Apollo, observers noted. Peenemünde veteran Stuhlinger described the core group as 'the most successful [technological] team in history. This was not just another team. The A-bomb, H-bomb, and Pickering [JPL rocket engine] teams all had a number of excellent individuals, but they were not a team in the same sense we were.'[37]

Long-time rocketeer Bernhard R. Tessman, whose association with von Braun dated back to 1935 and the German Army's Kummersdorf Proving Ground, could reminisce with him in 1972 about 'the fine and loyal team you led through decades – an experience which will certainly not repeat in a man's lifetime'.[38]

Von Braun always had high praise for the quality of the thousands of American-born, mostly young, engineers and scientists his organisation had attracted in the 1950s and '60s. He deflected suggestions that the team's German nucleus was 'smarter' than everybody else – especially after the group's early successes in space on the heels of others' spectacular failures. 'It's not that we're geniuses,' he often insisted. 'It's just that we old-timers have been working on these things so long, we've had twelve more years to make mistakes and learn from them.'

For the Marshall Center team, both German- and American-born, and its director, Project Apollo meant years of intense, non-stop effort. 'Von Braun didn't have a clock, as far as work went,' recalled MSFC manager Lee Belew. 'When he had a driver, he would have a light on in the car and read going home. He would do a lot of writing then and at home, too, [along with] conceptual design stuff.' During most of Apollo, the Marshall chief travelled three weeks of every month and 'was . . . gone almost every weekend', Belew remembered. 'We all spent lots of weekends working Saturdays – for years. We travelled an awful lot at night and on weekends. There just wasn't a lot of wasted time.'[39]

Von Braun set the pace, making maximum use of his time. A fast reader, he was constantly devouring books, articles, reports and technical papers. Travelling to and from work he would listen to foreign-language tapes to improve his fluency. 'He liked for people to apply themselves, maybe even reach beyond themselves,' recalled Bonnie Holmes. 'He thought you should always be doing something. From him I learned that you could relax and at the same time stimulate your mental process by maybe doing word games.'[40]

Marshall manager David Newby recalled running into von Braun one day in Florida in the mid-1960s on his way from the beach at Cape Kennedy to their motel. 'You said you were reading the book under your arm,' Newby reminded his boss years later, 'which happened to be your daughter's high school biology book. That was the beginning of the "let's sell 'Earth Resources' campaign" by NASA.'[41] (In 1966 the space agency began developing plans for a 'manned space station' using remote sensors, photography and other technical means for studying 'Earth Resources' ranging from forests, agricultural crops and mineral deposits to pollution, blight and other environmental dangers. The resulting engineering data were later applied to Skylab, the Space Shuttle and the International Space Station.)

A similar scene was painted by MSFC associate and friend Ruth von Saurma in a letter: 'I remember . . . a hot Sunday afternoon in June 1969. Almost 100 degrees and no breeze on Lake Guntersville [near Huntsville]. Everyone else cooled off in the water or dozed in the shade. But not you! Stretched out in the scorching sun on top of your houseboat, you were completely immersed in a book on Greek history. Afterwards, you came up with the most vivid account of the advanced concepts of Greek society that I ever heard.'[42]

During the Apollo build-up, von Braun used part of his office décor for making desired political points. His ninth-floor offices in the Marshall Center's newly constructed Building 4200 – dubbed 'the von Braun Hilton', to his annoyance – contained a low table near his desk. The table held a display of about a dozen scale-model rockets. Beginning with his old wartime V2, standing less than 1ft tall, the models grew progressively taller through the Redstone, Jupiter, Atlas, Saturn I, Saturn IB and, finally, the towering Saturn V – which was taller than his ceiling. Von Braun had a hole cut there and a recessed compartment built to receive the payload end – the Apollo spacecraft, managed by NASA's Texas centre.

'It was a big joke with him,' recalled colleague William R. Lucas. 'He would laugh with visitors and tell them he didn't want them to see all that expensive new Apollo spacecraft hardware up there and compare it to the old Houston capsule hardware. He'd say the Marshall stuff – the launch vehicle – was the good stuff, anyway.'[43]

Saturn–Apollo involved the work of some 375,000 people at 20,000 companies, large and small. NASA itself had a peak employment of about 34,000 government personnel during the 1960s. Roughly 90 per cent of NASA's budget went to private contractors. Critics of the Apollo effort called it a 'moondoggle' (a pun on the word 'boondoggle', meaning a waste of time and money, and denounced the spending of billions of dollars 'on the moon' instead of helping the needy and other earthly causes. Von Braun wearied of such talk and fired back: 'The NASA budget is not being spent on the Moon. It is, rather, being spent right here on Earth. It provides new jobs, new products, new processes, new companies, and whole new industries.'[44]

As if von Braun did not have enough problems during the race to the moon, even the generally supportive news media at home occasionally made waves. His Marshall public relations chief and other managers at times grew tired of the local press breaking stories before official news releases came out. On occasion, von Braun was drawn into the teapot-sized tempests. In December 1965, for instance, two senior MSFC officials had a private meeting with the publishers of the *Huntsville Times*, the afternoon daily, and its smaller morning competitor, the *Huntsville News*, to request they hold off on reporting certain new developments at Marshall until it could inform employees. Before any word was passed down to *Times* editors – if ever it was going to be – I, as the paper's aerospace reporter, got wind of the plans and wrote a couple of articles about it.

Hand-wringing over the perceived betrayal ensued in certain MSFC offices. In an executive memo that MSFC manager James T. Shepherd sent to von Braun, Shepherd proposed that the newspaper get 'the silent treatment – dry them up on news releases and stories'; that future NASA releases would come out 'after the *Huntsville Times* deadline for a specific day', thus favouring its morning competitor; and that 'the removal of Ward's badge' for MSFC access would occur. In the end, cooler heads prevailed, especially von Braun's. To the memo's reference to cutting off the paper from releases and story sources, he pencilled the notation: 'This doesn't seem to work too well. Now they are getting their stories from Sen. [John] Sparkman or NASA HQ.' Except possibly for the timing of a handful of subsequent news releases, none of the retaliatory measures was taken.[45]

Von Braun knew that his vast MSFC organisation, despite being NASA's largest centre, with nearly 8,000 of its own personnel plus tens of thousands of contract workers, was not the centre of world attention during Project Apollo. Even with its Saturn super-rockets, the Huntsville facility was no match for Cape Kennedy, with its fire-and-thunder magic at T-minus-zero, or the MSC in Houston with its right-stuff astronauts.

Still, at a dinner meeting one evening in Houston, von Braun was surprised to see a film on the Apollo effort that made no mention of the Marshall Center or its role in the coming lunar expeditions. Speaking after dinner, he pointed out the omission and then jibed: 'Compared to the astronauts, our Saturn has about as much sex appeal as Lady Godiva's horse!'[46]

Tragedy struck the moon-landing project with the deaths of Apollo 1 astronauts Gus Grissom, West Pointer Edward H. White and the rookie, Lt Roger B. Chaffee, US Navy, inside their sealed spacecraft, in a fire on 27 January 1967 during a supposedly routine check at NASA-Kennedy Space Center at the cape. Von Braun was at a dinner in Washington with hard-driving NASA administrator Jim Webb (who had replaced Glennan), MSC director Bob Gilruth and other top NASA officials when everyone got the shocking news. To the press, the Marshall Center chief lamented 'the loss of three good friends and valiant pioneers. Their deaths brought to mind the Roman saying "per aspera ad astra" – "a rough road leads to the stars".'

Back in Huntsville, von Braun wrote a personal note of condolence and encouragement to Gilruth. Webb soon appointed von Braun's long-time chief deputy, the witty but hard-nosed engineer Eberhard Rees, to head a troubleshooting team to investigate problems with the Apollo spacecraft design and manufacture by the contractor, North American Rockwell's Space Division, and to recommend corrections.

The deaths cast a pall of guilt and gloom over the NASA field centre in Texas that was to hang there for months. Von Braun helped the Houston space family come out of it. In a meeting at the Texas centre, MSC chief of Public Affairs Paul Haney suggested that it was time for a let's-get-this-show-back-on-the-road party, with drinks and dinner and laughs, 'to kind of come out of mourning, which we'd been in for months', he recalled. They picked a date in early May marking the sixth anniversary of Alan Shepard's space trip aboard Mercury-Redstone. They would roast Shepard, who had been grounded ever since by a heart murmur and made head of the Astronaut Office. They would show a spoof film, 'How to Succeed in Space without Really Flying Much', and have invited speakers roast 'Smilin' Al'.

The lone invitee from Huntsville was von Braun. He readily accepted. As Haney remembered it:

Wernher made one of the most interesting talks I think I ever heard him make in public – not that he ever made any bad talks, but he was particularly wrenching that night. He gave one of those typical von Braun, big-chin type of let's-go-back-out-there-and-get-'em-in-the-second-half talks. He would have topped [basketball coach] Bobby Knight that night. He came on strong, but he knew what we were trying to do and he did it very well.

I will never forget a syllogism [*sic*] that he used. . . . He said, in quoting one of his colleagues: 'I think we should all understand that we are not in the business of making shoes.' And he delivered it with a certain contempt. . . . He meant only that our work was on a higher plane, involving certain high risks and dangers, and he did it with extraordinary emphasis.[47]

The 6 May evening affair was 'a great success', recalled Haney. 'People started going off to parties again and being human beings again.' And a tighter-knit space agency began moving ahead to a successful first Saturn–Apollo manned flight less than eighteen months later. At last, Luna truly beckoned.

En Route to Victory

The Project Apollo push to the moon was the driving force for Wernher von Braun during the 1960s. It required leadership, management, politicking, promotion and teaching, yet it was not all-consuming. He also enjoyed relaxing with his family, scuba diving, hunting, flying, deep-sea diving in a 'magic submarine' and personal travel.

Von Braun had discovered skin diving in his youth in Germany and resumed that sport in the 1950s during trips to Florida and California. Heeding the advice of writer Arthur C. Clarke, he graduated to the scuba version and enjoyed it over the years in the Florida Keys, Mexico, the Bahamas, the Aegean, at Australia's Great Barrier Reef, and other places. He delighted in exploring caves and gullies, hunting submerged artefacts, photographing the deep-blue scenery and dazzling sea life, and fishing with a speargun. He liked the sport, he said, because it, along with piloting aeroplanes, 'seems to give me a mastery of the third dimension'.[1]

Von Braun figured – rightly – that he could go scuba diving during breaks in the work schedule on some of his frequent trips to Cape Canaveral. One of aerospace engineer-manager James S. Farrior's 'most enjoyable experiences' was a diving side trip with von Braun to the Florida Keys after a rocket launch. It was on that speargun-fishing trip, he wrote to von Braun, that he 'realized for the first time the unlimited energy you possess. Your spear went into every hole and you didn't let up until all the air tanks were empty and everybody else had long since stretched out, exhausted, on the deck.'

Preparing to head back to land at dusk, the divers discovered their boat's engine was kaput. Von Braun slid into a tethered dinghy and used 'its pitifully small outboard motor' to tow the larger boat slowly landwards. As Farrior observed to his old colleague, 'What a change from the tremendous power you had unleashed at the firing at Cape Canaveral a few days before!'

When the group at last made land at a pier, they faced an armed guard who refused to let them come ashore in the darkness. The pier belonged to an entrepreneur salvaging a nearby sunken Spanish galleon, and he didn't want any strangers snooping around his treasure. Much discussion ensued between von Braun and the hired gun before they were finally allowed to come ashore.[2]

Ernst Stuhlinger recalled one summer day in von Braun's earlier diving years, when he, von Braun, Gerhard Heller and several other ex-Peenemünders formed 'a happy gang' of skin divers aboard a small motorboat headed from Long Beach, California to the kelp beds off Catalina Island:

> Tom, owner of the boat and the diving gear, explained how to put on mask and fins, how to get seawater out of eyes and nose, how to descend and ascend, and how to use the speargun. Finally, he assumed a serious pose and said: 'Fellows, now listen. There are moray eels down there. Those beasts are vicious. They are not for you. I want you to steer clear of them! Do you hear me?' We said, 'Yes, sir', because we had heard him. And we donned the gear, sat on the rim, and plunged backward into the blue water.
>
> After a little while, you surfaced and headed for the boat, pulling heavily at the line of your gun. When you finally heaved your catch aboard, we were all stunned: it was an enormous, vicious-looking moray eel! Bursting with excitement, you told us the story at least seven times – how you first saw the head of the beast in a crevice, how you swam right toward it, and how your spear hit it through its powerful neck.
>
> Tom did not say a word. He felt that here was a man who did not need instructions and who was immune to the pitfalls of life that endanger lesser men. But the luster in his eyes was eloquent enough; it betrayed nothing but pride and admiration.[3]

Scuba and skin diving were interests von Braun also shared with his friend and famous TV news 'anchorman' Walter Cronkite. In a birthday letter to the rocketeer, Cronkite confessed to harbouring 'a secret admiration for a man whose life has been devoted to getting to the stars, and whose hobby is getting to the bottom of a shallow lagoon!'[4]

Sailing and deep-sea diving vessels were other mutual interests of the two men. Von Braun was 'terribly interested' in exploration of the sea as a 'new frontier', and he and Cronkite 'had a mutual deep interest in that and a mutual friend in Ed Link', an innovator of undersea equipment (and earlier the inventor of the Link Pilot Trainer). Cronkite said von Braun and he dived on several memorable occasions in the 1960s, although separately, with Link in his four-person submarine, 'Deep Diver', which Cronkite called a 'magic submarine'. 'Wernher and I talked more about that than we did space,' a laughing Cronkite remembered. 'Everything about space had been said, practically, between us. But exploration of the sea was so new back then. It was new to Wernher, too, so he had a fascination with it.'[5]

Hunting was another of the rocket scientist's favourite pastimes during the 1960s. With friends in the Huntsville area and elsewhere, he shot pheasant, dove, quail, duck, geese and wild turkey. A good shot, he also hunted bigger game – deer, antelope, caribou, moose, jaguar, bear – when opportunities arose. The Episcopal minister of his family's church in Huntsville, the Revd A. Emile Joffrion, invited von Braun to accompany him one weekend in the early 1960s on a big deer hunt near Greensboro in southern Alabama. The site was a plantation of several thousand acres that belonged to a friend of the priest who was the publisher of the local newspaper, descendant of the state's first Episcopal bishop, and a gentleman known for his salty language and squirrel's head soup. Joffrion recalled:

> Our host, Hamner Cobb, had an old green pickup lorry that he drove 60, 70 miles an hour through the woods and across the pastures to where we'd hunt. And God have mercy on you if you weren't sitting up front with him. Most people had to ride in the back of the lorry, of course. Well, we were getting ready to load up, and one of our host's field lieutenants who helped run these hunts said, 'Hamner, you better put Dr von Braun up front in the cab with you. He . . .' 'Hell, no!' Cobb interrupted. 'Let him ride in the back like everybody else!'
>
> Well, Wernher laughed. He loved it. I think it was the phrase 'like everybody else' that appealed to him. He was treated with

such deference everywhere he went and probably was ready to be 'one of the boys'.[6]

So von Braun climbed in the back of the battered pickup with the dogs, the black 'drivers' who helped with the hunt, and several other hunters. 'And Wernher just had a great time,' Joffrion remembered. It was the first of several deer hunts they enjoyed together, in Arkansas as well as Alabama.[7]

A frequent bird-shooting host and field companion of von Braun was Harry Moore Rhett Jr, a Huntsville community leader, cotton planter and fox-hunting master. During the Apollo build-up, von Braun was his guest on one of many bird shoots they experienced over the years. The scientist had a thoroughly frustrating off-day with his shotgun, taking numerous shots but hitting nothing all day. 'The thought occurred to me,' remembered the gentlemanly Rhett, 'but I resisted the temptation to tell him, "Wernher, I hope your aim at the moon is better than this!"'[8]

Von Braun's chance at jaguar hunting in the Yucatán came in the 1960s at the invitation of Fairchild Industries' Edward Uhl. The hunt was hastily arranged after friends of the aerospace industrialist who were cutting mahogany deep in the interior had reported sighting the big cats in the jungle. Von Braun and Uhl had flown down and then made their way by jeep to the lumber camp. Informed that the animals had last been seen at a nearby lake, the two hunters left at dusk in the old jeep and accompanied by two armed Mayan guides who spoke no English. At the lake, the rocket scientist headed on foot to one side with a guide, and Uhl took the other side.

An hour later, in darkness – the best time to see the bright shining eyes of the jaguars – loud howling and growling sounds suddenly erupted. The hunters had no idea what in blazes the beastly noises were and could not communicate with their guides. Uhl figured he had better return to the jeep. There he found an excited von Braun waiting. 'Ed, Ed, what were those awful noises?' he asked. Uhl said he hadn't a clue.

Back at camp, the hunters got their answer – and chuckles from their hosts. The animals making the ungodly raucous sounds were howler monkeys. The two men had not bagged any jaguar, but they would have stories to take home. After dinner that night in camp, Uhl returned to their experience with the vocal monkeys. 'Isn't it amazing, probably unique, that these creatures with only a loud voice can be so frightening?'

'Wernher leaned over with a twinkle in his eye and said, "Ed, haven't you ever appeared before a congressional committee?"'[9]

The decade also saw von Braun indulge his passion for piloting aeroplanes. Not everyone knew that, from his teenage years on, he had been a serious pilot. Beginning with gliders as a youth, he had flown military aircraft in the 1930s during two pre-war spells in the German Air Force reserve. In the 1960s he grew practically obsessive about it, flying anything and everything he could get his hands on: small planes, multi-engine aircraft, executive jets, fighter planes, old converted bombers, seaplanes, gliders, and every imaginable sort of aircraft and spacecraft simulator.

Although von Braun never owned an aeroplane, his job as the boss at NASA's Marshall Center gave him ample opportunity to use the government-owned aircraft assigned to his organisation, plus hired and chartered planes. But all that was just for starters. 'Sometimes on contractor plant visits, especially in St Louis at McDonnell Douglas,' associate Lee Belew recalled, von Braun 'would slip off and be out there flying with one of their pilots to see how a new jet fighter was. He was a real pilot. He loved to fly.'[10]

An aviator with an overactive sense of mischief, the rocket man delighted in giving his passengers cheap thrills – especially newcomers who were unaware of his piloting skills and already nervous about an amateur handling the controls. David Newby was aboard the centre's twin-engine Gulfstream aircraft one clear day on a 1960s trip out west when the MSFC director flew the plane 'through, not over but through, the Grand Canyon'. Flying below the canyon's rim, von Braun revelled in 'pointing out various items of interest such as, "Look up and to your right to see the Indian village."'[11]

On another occasion, flying the Gulfstream with several Marshall employees to Seattle to begin a major series of visits to West Coast aerospace contractors, von Braun headed for a Strategic Air Command (SAC) base in South Dakota to refuel. NASA's Jay Foster remembered the episode well: 'In the process of coming down to land at this SAC base, we passed Mount Rushmore. Wernher was flying. He decided to buzz Rushmore, and we went down very close. He banked that plane up on its side so that the passengers on the port side had "a good view" of it.'

'I never did see those carved faces,' Foster recalled telling von Braun afterwards. 'I don't remember what Mount Rushmore looked like. It was August, and I was too busy watching all those tourists down there staring up at this crazy airplane on its side!'[12]

It just so happened that shortly after the Marshall Center received its much-sought-after twelve-seat Gulfstream turbo-prop plane (the first of four such transports that NASA ordered), the Federal Aviation Administration (FAA) issued a regulation. It required anyone piloting a multi-engine plane above a certain weight, with passengers, to have an Airline Transport Rating (ATR) for that plane, even if he or she was cleared to fly it. Von Braun's professional contract pilots, all properly certified, broke the news to him: to stay legal at the Gulfstream's controls, he must have an ATR. Pilot Edward 'Skeets' Grubbs remembered the space scientist's determined response: 'I get me vun.' Then, 'Vot must I do to get this?'[13]

The contract pilots explained to 'the doctor', as they called him, that he must pass an extensive written test, attend flight school for a week, complete simulator tests and then pass a demanding FAA flight test. The pilots began gathering flight manuals and other literature for him to study. Pilot George Fehler recalled that von Braun pored over the material for two or three hours a day during a trip abroad, came back, took the written test and got high marks. Von Braun and Fehler then travelled to New York for the test on the Gulfstream flight simulator. The test included some complicated prop settings for different situations. 'He was good – the best one that day' on the sophisticated system, Fehler said of von Braun's performance.[14]

Finally, it was time for the flight test. They flew the Marshall Center's 'G1' to the Savannah–Charleston area.[15] The FAA inspector, impressed that the famous space scientist had progressed this far, came aboard. Fehler went along for the ride. 'The doctor took the flight check and did everything they asked,' recalled the contract pilot. 'There was only one hitch.' Flying in low-ceiling conditions, von Braun missed a landing approach and had to pull up suddenly, back into the clouds, after discovering he was headed for an inactive runway. The inspector said nothing. The situation involved a corrective action von Braun knew well. In a momentary quandary, however, he commented about being back 'in the soup' to Fehler, who responded with a thinly veiled hint: 'It's like a "missed approach" [to an active runway], isn't it? [Von Braun] picked up on that real quick,' recalled his pilot friend, and made a full recovery.[16]

On the day in 1964 when von Braun's ATR certificate arrived, he took his pilot friends out to dinner at a fine restaurant. He ordered chateaubriand and wine all around. 'It was his treat,' recalled Ed Grubbs. 'It was a thank you.'[17]

Von Braun had his share of close calls in the air.[18] One of the most harrowing came during a flight with long-time associate Donald I. Graham Jr, and others. Shortly after take-off in a NASA plane headed for Washington, von Braun had taken over the controls while the normal pilot handled the radio. Graham related the drama while reminiscing years later with the rocket scientist:

> During the course of the trip we were all talking and the subject of piloting an airplane arose. The pilot commented that it was '95 percent sheer boredom and 5 percent stark terror'.
>
> Shortly thereafter the pilot asked for clearance to land at National Airport. There were spotty clouds, but in general they said to come straight in. You lowered the nose to go down through a cloud. At that exact moment up out of the cloud came the nose of a DC-6 head on. You peeled our plane off to the right like an expert fighter pilot. If your thinking and reactions had not been so quick, none of us would now be celebrating birthdays.[19]

Moments after the near miss, recalled Graham, von Braun's co-pilot asked, 'Remember the 5 per cent?' Then he verbally tore into the air traffic controller in the tower at National Airport.

Von Braun seized the chance to pilot a very different sort of craft during a southern California work trip in the mid-1960s. On the flight west to visit several NASA contractor plants, he left the cockpit and sat beside Marshall Center engineer John C. Goodrum Sr. Von Braun asked about Goodrum's plans for that evening. The engineer, one of whose current assignments was to stay abreast of new, alternative means of transport, planned to visit the Newport Beach home of an inventor. The man had worked on several rocket programmes, Goodrum added, but his consuming hobby was designing and building personal hovercraft in his garage.

'He just lit up with interest and excitement,' recalled Goodrum. Von Braun said he was supposed to have dinner that night with a group of corporate VIPs but was weary of such occasions. 'If you don't mind, John,' he said, 'I'd like to sneak out of the hotel and go with you.'

That evening von Braun slipped out of their Santa Monica hotel alone, disguised in 'a trench coat and Dick Tracy hat pulled down over his eyes', and waited for his getaway ride in the shadows of a building across the street, a laughing Goodrum recalled. When the engineer pulled up in his hired car, the furtive von Braun slid inside, and the two men headed for Newport Beach. Goodrum could only wonder what alibi his boss had used to duck out of the VIP dinner.

Reaching their destination, the pair found their host in his garage. Von Braun, who was knowledgeable about hovercraft, became immersed in conversation with the inventor about his work and the several working models he had engineered and built. They examined the hovercraft hardware in detail. They talked until midnight. Then the men hauled one model, a two-seater about 12 or 15ft long by 6ft wide and resembling an inverted bathtub, out into the paved alley by the garage and cranked it up.

'Von Braun and the guy began flying it up and down the alley, back and forth, just inches above the pavement – back and forth!' Goodrum remembered. 'It had a kind of skirt all around the bottom, with blower fans blowing air down and sideways. They kept flying it up and down that alley, back and forth, with von Braun having a great time.' At 2 a.m. Goodrum finally dragged the boss away.[20]

Von Braun believed, as did writer André Maurois, that 'nothing is more agreeable than to travel'. He travelled widely throughout much of his life, taking delight in the adventures, new friends and opportunities for learning it brought. He journeyed from the jungles of Africa and the Yucatán to the icy Arctic and Antarctic regions, from the Greek Islands to the British West Indies, from the Himalayas to the Alps – to every continent and reputedly every state in the United States.

Some of von Braun's travels were more offbeat than others. A space-business trip to Nevada in the 1960s ended with him stage-side at a Las Vegas hotel-casino topless revue with his travelling party. The group had flown out to nearby Desert Rock to observe the test firing of a nuclear-powered rocket engine developed for a deep-space probe.

When the firing was cancelled because of wind direction – the contaminated exhaust clouds would have drifted down over Las Vegas – the group decided to head there. They managed somehow to get seats at the Tropicana for that night's edition of the 'Folies Bergère'. Marshall contract pilot Ed Grubbs remembered the evening in detail: 'They marched us right around a big line of folks waiting to get in the show, and took us – the doctor and the whole group – right down to a ringside table. I mean, it was up against the stage, the closest view possible. We were sitting there with all them bare-breasted gals right above us. The dancers were swinging their feathers out over our group. They seemed to be playing to our table, either because we were closest or they knew the doctor was there.'[21]

'Yeah,' added pilot George Fehler, 'the doctor was sitting there, real close to the stage, and those dancers were coming by with those feathers – and he was allergic to feathers.' Grubbs picked up the narrative: 'And the closer those topless gals got with their feathers, the more he scooted his chair back. They'd get closer, he'd scoot – anything to avoid red eyes and a sneezing attack!'[22]

Von Braun's work gave him opportunities to combine business travel with personal touring. He did just that in the early 1960s on his grandest journey of all. His travelling companion called it 'our round-the-world odyssey', and it lasted two and a half weeks in December 1961 and January 1962. Dr Carsbie C. Adams, a hospital administrator, space author, fellow pilot, and keen hunter and angler from Georgia, had known von Braun since 1954. Adams recalled that the plan was hatched during a weekend quail-hunting outing with von Braun at the Georgian's pre-Civil War plantation in Culloden, south of Atlanta; the whole von Braun family often joined in horse riding and other outdoor fun there.

Von Braun mentioned that he had been invited to give a university lecture series in Australia and asked his friend to accompany him. Adams needed to arrange a business meeting in London about then, and he knew that von Braun wanted to visit his father in Germany. Adams had an idea: Why not make it a round-the-world trip? Von Braun agreed. He yearned to see India, the Himalayas and Thailand, and this might be his best chance. Project Apollo could spare him for a few days around Christmas and New Year. Soon he and Adams had their globe-circling itinerary: Huntsville/Atlanta to New York to London, then Munich, brief stops in Istanbul, Beirut and Tehran, then visits to New Delhi, other Indian cities, Kathmandu in Nepal, Bangkok, Sydney, a return stop in Honolulu, then Los Angeles and back home.

Von Braun found their visits to the Taj Mahal and other magnificent structures in India humbling, Adams recalled. Guides told the two visitors of the many thousands of men who worked night and day for years and years at these sites using the most advanced building techniques and designs. 'Von Braun was very impressed and touched by the thought of such undertakings,' Adams remembered. He said von Braun commented along the lines of: we think we are doing this wonderful, enormous effort with the moon-landing programme. But these building projects in India long ago rivalled it, considering the resources expended and the technology available then.[23]

The visit to Nepal, high in the Himalayas, proved to be a highlight of the trip. The pair flew to the mountain kingdom from India aboard a DC3 airliner. It was, the plane's captain explained to them in the cockpit, the only transport then reliably capable of landing – at a mere 85mph – and taking off in the thin air at the 3,600ft-long airstrip serving Kathmandu. They touched down in a fertile valley green with tea fields. The two marvelled at the valley's sunny beauty, semi-tropical vegetation and temperatures so balmy they had to shed outer layers of clothing. It was the dead of winter – 1 January 1962 – and all around were snow-covered mountain peaks. 'We were completely stunned to find that it was extremely warm, dry and fantastically beautiful,' related Adams, who chronicled the global trip in a journal. At first, all von Braun and he could do was stare and take photographs of the astounding scenery.

Further surprises awaited them in Kathmandu and nearby towns. They visited a succession of ancient Buddhist and Hindu temples, temples with domes covered in gold, one with 2,000 carved wooden Buddhas, the magnificent temple of Swayumbhunath on top of a hill with a thousand steps, and yet another with an endless array of colourful wooden carvings depicting every human sexual act and position imaginable. In a house of worship, the visitors mused – interesting.

Leaving the Temple of One Thousand Steps in Kathmandu, they noticed a scene that aroused von Braun's 'great compassion', recalled Adams. Gathered nearby were several dozen of the most miserable, hopeless-looking people the pair had ever seen. They were refugees who had recently fled their new Chinese Communist occupiers in neighbouring Tibet. They 'had walked and crawled across the Himalayas to this valley', Adams recalled. 'We must do something!' von Braun told his friend. The two men pooled all their cash, changed it into Nepalese currency, and methodically pressed equal amounts into the hands of each member of the ragged group.[24] The money was eagerly accepted.[25]

Von Braun believed that one's religious faith must be accompanied by deeds. It was not enough merely to feel compassion for others. During a late-1960s holiday cruise he took with Maria to Greece and the Aegean Islands, a Greek mayor, excited to have the von Brauns visit his small town, gave them the red-carpet treatment. It eventually came up in conversation that the mayor was suffering from heart disease and needed multiple-bypass surgery. The operation was not readily available then in Greece, and in any event it would have been far too costly for the mayor.

According to Tom Shaner, 'Dr. von Braun took it upon himself to arrange for complimentary surgery for the mayor in Houston' by the celebrated Texas heart surgeon Dr Michael DeBakey, a friend. 'There were no doctors' bills, no hospital bills.' And it was successful.[26]

Back home during the high-flying 1960s, von Braun spent most weekends and what time his heavy business travel and high adventures would otherwise allow with Maria and their three children. Friends said it was not as much as he would have liked, nor as much as his family wanted. But along with pursuing urgent space goals, von Braun was intent on experiencing life to the fullest. His wife understood him, and by all accounts felt he did not indulge in his after-hours exploits to excess. He fitted many of these personal pursuits – scuba diving, flying, yachting and so on – into his working week. When he couldn't, then it was his time at home that suffered.

Some of his more exotic travels – to Antarctica, Nepal/India/ Thailand, Alaska, Australia's Great Barrier Reef – were taken with friends and associates. Other such journeys – to Africa, the Aegean Islands, Hawaii, India again, and back to Alaska – were with Maria and, on occasion, the children.

Summertime trips to Cape Canaveral for missile or space launches at times meant that his wife and children went too, to enjoy the ocean and beaches. But all in all, von Braun's heavy travel schedule often seemed seriously out of balance to his family. During one absence, it was Maria who had to help son Peter fashion his small 'Pinewood Derby' wooden racecar for Cub Scouts. The peripatetic papa had in the late 1950s confessed in one of his many speeches to business conventions: 'My two daughters keep telling me to quit my job, buy a drugstore, and stay at home!'

Still, there were the countless happy spring, summer and autumn weekends spent on his ChrisCraft motorboat *Orion*, water-skiing and swimming with his family on Lake Guntersville, or simply lounging aboard their small houseboat moored at the marina and helping young son Peter sharpen his fishing skills. Von Braun taught his children how to swim, dive and water-ski, 'and probably did a better job than a swim instructor could have', wrote 11-year-old Margrit in a 1963 classroom essay.

He supported Margrit in her passion for horse riding. She wrote: 'Daddy bought me a horse [Susie]. Occasionally he rides her. He takes me up to the stable often and watches me ride. . . . He goes to nearly every horse show that I am in.' He encouraged daughter Iris in learning to play the cello. And he kept his promise to take Maria on a cultural-social trip to New York at least once a year.

Von Braun's outside income helped cover such expenses. There were the sometimes lucrative speaking appearances, articles and books. Starting in January 1963 and for years afterwards, he wrote a monthly – later bi-monthly – piece in *Popular Science*. Long-time personal attorney Patrick Richardson in Huntsville advised him on several profitable investments. A Birmingham, Alabama, banker friend of Richardson's was organising a new bank in a neighbouring county to Huntsville and sought von Braun's presence on the board. On Richardson's advice he became a director of the First State Bank of Decatur in the 1960s, and served well into the next decade. 'He said in a joking way it gave him great pleasure to be a banker,' the lawyer recalled. 'He said his wife's father[27] had been a prominent banker and always looked down on him!'[28]

At work in the 1960s, von Braun often extended his workday with 'the boys' – his beer-loving, German-born senior colleagues – at a bar called the Top Hat Lounge. Proprietress Sarah Sanders Preston, who worked by day in a cafeteria out at the space centre and knew the rocketeers, opened the lounge in town in early 1964. Soon von

Braun and company began quietly dropping in at the Top Hat at 5 or 5.30 p.m.

'Dr von Braun would pull up first in our big gravel parking lot,' Preston remembered, 'and then seven or eight of the other doctors would arrive. They'd all follow him inside and head straight for our stockroom in back, where all the beer was stored. Only American beer. They would drink the beer hot in there and talk and draw all over the beer cases – rocket designs and things.' Then, after an hour or so, von Braun would open the door and invite Preston in to count the empty beer bottles he had neatly lined up so that they could pay her.

The whole routine, week after week, never varied for as long as von Braun remained in Huntsville. After he left, the Germans stopped coming by the Top Hat. And Sarah Preston, years after she sold the lounge in 1987, still wished she'd saved those beer-carton drawings.[29]

Von Braun's primary jobs during the Apollo era were, of course, as leader and manager. He led not only NASA's largest field centre but also the broader effort to help inspire public support nationally – and internationally – for mankind's new cosmic strivings. He had been 'a man with a driving force' in the mid-1950s, Col Edward D. Mohlere, US Army (Ret.), recalled. 'It was evident in everything he did. Then, of course, came the "man on the Moon in this decade". That was the schedule he had, and he was going to make it, come hell or high water.'[30]

In addition to his management duties, he made speech after speech touting the space programme, specifically Project Apollo, to citizens around the country, to politicians and to students. Representative Olin E. 'Tiger' Teague of Texas, chairman of the House Science and Astronautics Committee and one of von Braun's best pals on Capitol Hill, lauded him after a Texas speaking tour as 'really a spellbinder when it comes to making a presentation. Your ability to adapt your subject to the type [of] audience, whether scientists, high school children, or businessmen, is outstanding.'[31] Von Braun was simply 'a great salesman', recalled Lee B. James, a senior associate at the Marshall Center. 'He was so good at it, he could make a speech on any subject and be really good.'[32]

He took a special interest in speaking to college science, maths and engineering students he might attract to his rocket team in

Alabama. One such was Harry Atkins. At the only school in his coal-mining hometown of Van, West Virginia, in the early 1950s, Atkins had read a book by Willy Ley on rockets and future space travel. It persuaded young Atkins to follow a career in science and technology. He took every maths and physical science course available in high school and then in a work-study programme at a two-year college, alternating work in the coal mines with terms in the classroom. He then enrolled at Marshall University (MU) in Huntington and majored in physics.

One day in 1961 von Braun visited the MU campus. 'He came and gave this lecture and drew the V2, basically, and trajectories and such,' Atkins recalled. 'Then at the end he said, in his German accent, "I want all of you to come to Huntsville. We're going to the moon and beyond!" I never will forget it. He was like a pied piper. The students dubbed him "the space pied piper". I was in the process then of graduating and [job] interviewing. I turned down a position with IBM in Poughkeepsie, New York, and headed south to join von Braun.'[33] NASA physicist Harry Atkins made a career at the Marshall Center, with no regrets.[34]

Capitol Hill, too, saw much of von Braun's salesmanship. In hearings, he was often the star witness, keeping the members of Congress and their committee staffs informed on Saturn–Apollo progress and problems, and helping to keep those budgetary billions coming. 'Wernher was a very charismatic individual, anyway,' recalled colleague William H. Pickering of the Jet Propulsion Laboratory. 'In fact, he is the only witness at a congressional hearing I've ever heard of that, whenever he came into the room, all the congressmen came down to shake his hand.'[35]

Von Braun invested a lot of time and energy in cultivating the friendship and favour of Teague and the other Washington powerhouses who held the space purse strings and levers of programmatic authority. He went on hunting trips with them, delivered countless requested speeches in their districts and gave them the red-carpet treatment on visits he encouraged to the Marshall Center. And if they journeyed to New Orleans to inspect the vast Michoud plant for Saturn rocket booster fabrication, which Marshall oversaw, he cemented friendly relationships by making the ultimate sacrifice: he joined them on a rollicking night out in the Big Easy's French Quarter.

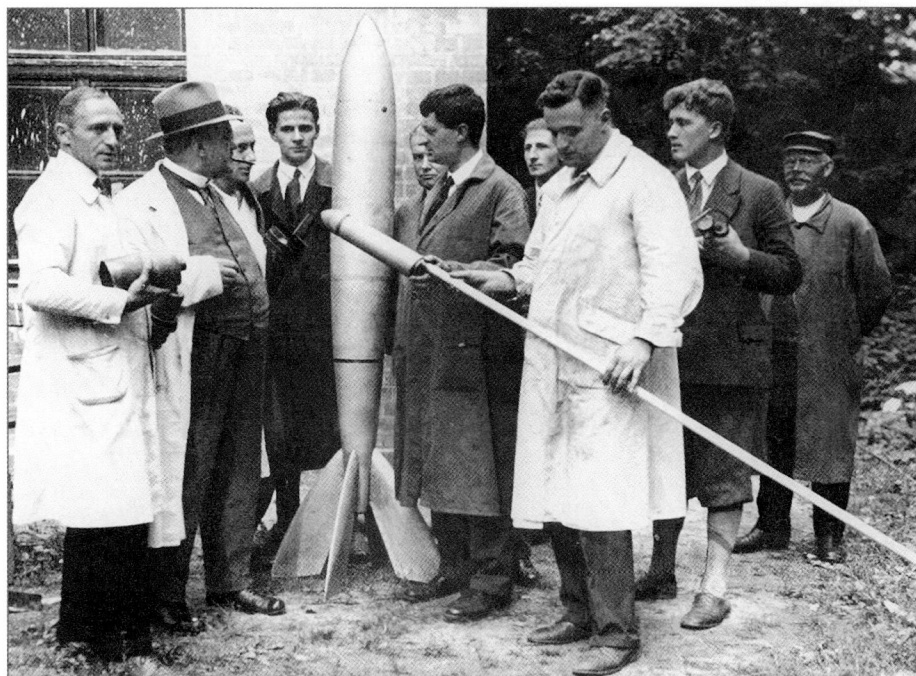

In 1930, an 18-year-old Wernher von Braun (in knickerbockers) serves as a volunteer apprentice to a group of rocket experimenters in Berlin headed by Hermann Oberth. Left to right: Rudolf Nebel, Alexander Ritter, Hans Bermüller, Kurt Heinisch, Oberth, two unidentified men behind Oberth, Klaus Riedel, von Braun, and an unidentified mechanic. *(US Space and Rocket Center)*

Rudolf Nebel (left), and von Braun, aged 18, carry their group's experimental Mirak rockets to launch sites for test flights at Raketenflugplatz, Berlin-Reinickendorf. *(US Space and Rocket Center)*

Von Braun's Kummersdorf group readies an A3 experimental rocket for a wintry test flight in the mid-1930s at its launch site beside the North Sea. This was before the German Army rocket team's move – beginning in 1937 – to the large, new Peenemünde centre on the Baltic Sea. *(US Space and Rocket Center)*

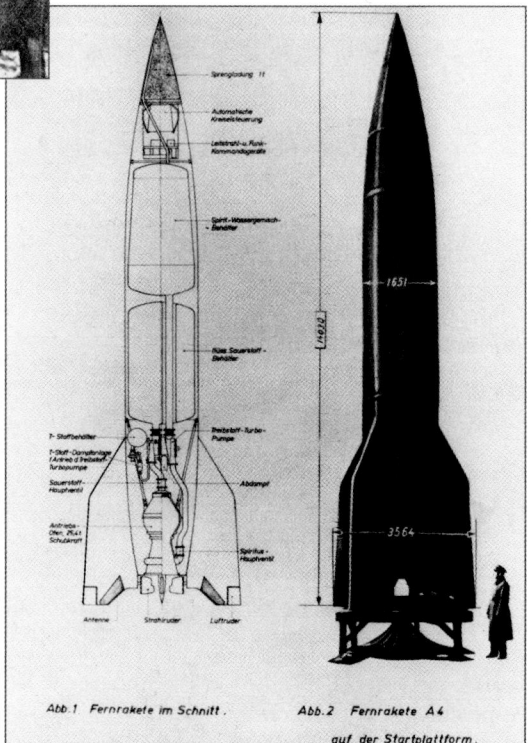

German A4 (V2) missile documents confiscated in 1945 by the US Army show (left) an engineering cutaway view and, the 46ft-tall rocket (right) on der Startplattform (launch pad), with flame deflector plate below. *(David L. Christensen)*

Adolf Hitler, third from right, visits the Kummersdorf rocket development station outside Berlin on 23 March 1939, von Braun's 27th birthday. At the German Army facility, Hitler witnessed static engine test firings and was briefed by von Braun and Col Walter Dornberger. Primary rocket development activities had shifted two years earlier to the much larger Peenemünde centre, which Hitler never visited.
(US Space and Rocket Center)

Gen Erich Fellgiebel, head of the German Army's Information Services, congratulates members of the Peenemünde team for the first successful launch of an A4 (V2) rocket on 3 October 1942. Von Braun looks on from rear centre. Left to right: Lt-Gen Richard John of the Ordnance Department; Col Walter Dornberger, commander of Army rocket work at Peenemünde; von Braun; Gen Leo Zanssen, accepting congratulations as base commander; Capt Stoelzel of the Taifun anti-aircraft rocket project; Rudolf Hermann, chief of supersonic wind tunnel operations at Peenemünde; and Gerhard Reisig, chief of field test instrumentation.
(US Space and Rocket Center)

In-development A4 (V2) ballistic missile is test fired from Launchpad 7 at the Peenemünde rocket centre in about 1943. *(Huntsville Times)*

Von Braun and Walter Dornberger, centre, host a 1943 visit to Peenemünde by Gen Walter von Axthelm (second from left), chief of German anti-aircraft weapons, for discussions of surface-to-air missile development. *(US Space and Rocket Center)*

With an A4 (V2) missile in the background, von Braun (on the right in a dark suit) accompanies a contingent of top brass during a May 1943 visit to Peenemünde. In front of von Braun, Gen Walter Dornberger turns to speak to one of the officers. In the centre in a dark naval greatcoat is Grand-Adm Karl Dönitz. On the far left is von Braun's propulsion chief, Walter Thiel (killed three months later in the first RAF bombing). *(US Space and Rocket Center)*

The 17 August 1943 Royal Air Force bombing raid on Peenemünde killed 735 people, wounded hundreds more, and inflicted extensive damage. Most of the visible A4 (V2) propellant tanks and nose cones in storage, however, survived the surprise British attack. *(US Space and Rocket Center)*

A row of finished V2 propellant tanks in a Peenemünde plant in 1944 illustrates that limited V2 production continued after the massive August 1943 RAF raid. Tidy housekeeping and strict saftey rules prevailed there; overhead, for example, is a *Rouchen verboten* (No Smoking) sign. *(US Space and Rocket Center)*

Nazi flags adorn the façade of the officers' club at Peenemünde. The SS took control of the rocket development centre away from the German Army and the Luftwaffe after the devastating RAF attack in August 1943. *(US Space and Rocket Center)*

The Project Paperclip rocket team at Fort Bliss, Texas, *c.* 1946. Wernher von Braun is in the front row, seventh from right, in dark trousers and light jacket. *(US Army Aviation and Missile Command)*

Charles Lindbergh (second from right) visits von Braun (second from left) and colleagues at White Sands Proving Ground, New Mexico, in about 1946–7, to check out the latest in ballistic missile technology. The two were together again a quarter of a century later for the Apollo 11 lift-off in July 1969. *(US Space and Rocket Center)*

In the late 1940s, a launch crew of Americans and Germans prepares a V2 for a test flight from White Sands Proving Ground (later Missile Range) in New Mexico.
(US Space and Rocket Center)

In an early 1950s moment at home, von Braun divides his time between his daughter, Iris, and a copy of a new book he had just co-authored, *Conquest of Space*.
(US Space and Rocket Center)

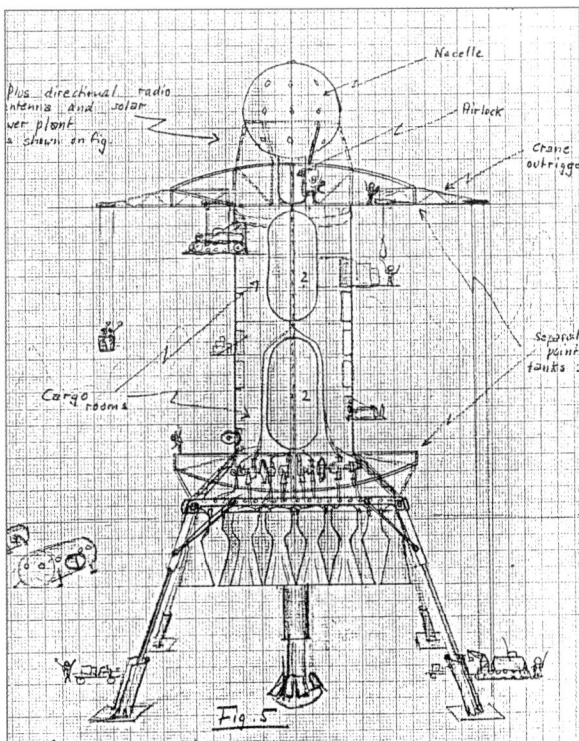

Von Braun's drawing of a lunar cargo ship described in his article 'The Journey', part of *Collier's* 18 October 1952 special issue entitled *Man on the Moon*.
(Fredrick I. Ordway III Collection)

Walt Disney (second from right) joins von Braun and colleagues in July 1954 during work on a series of educational television programmes about space at Disney Studios in Burbank, California. Left to right: von Braun, Willy Ley, Disney and Heinz Haber of the US Air Force Department of Space Medicine. *(Bonnie Holmes)*

Some forty German rocket-team members and their families – including Wernher and Maria von Braun – swear allegiance to the United States in citizenship ceremonies on 14 April 1955, in the packed auditorium of Huntsville High School. (The future Mrs Bob Ward was among the students present.) *(US Space and Rocket Center)*

Boating on the Tennessee river with their children and friends was a weekend activity the von Brauns enjoyed through most of their twenty-year stay in nearby Huntsville. Here, in a 1955 outing, Wernher and Maria, (left) with daughters Iris and Margrit, relax with Irmgard and Ernst Stuhlinger, who is holding his son Tillman (centre), and colleague Heinz Koelle (right). *(US Space and Rocket Center)*

Edward C. Uhl, Fairchild Industries chief executive (left), talks with von Braun during a visit he made to the company in Germantown, Maryland, in 1968. Four years later von Braun left NASA and joined Fairchild as vice president for engineering and development. (*US Space and Rocket Center*)

Von Braun (centre) has fun floating weightlessly inside the KC-135 'Vomit Comet' during a zero-G flight manoeuvre in 1968. Von Braun and his chief scientist, Ernst Stuhlinger, experienced several twenty-second weightless periods on the parabolic flights. In one or more, he wore a spacesuit to enhance the simulation of space flight. (*History Office, Marshall Space Flight Center*)

With wife, Maria, von Braun sports a 'five-gallon' hat given to him by President Lyndon B. Johnson in the mid-1960s at the LBJ Ranch in Texas. On the right is West German Chancellor Ludwig Erhard, wearing his own rancher's hat from LBJ. *(US Space and Rocket Center)*

A Saturn IB booster rocket lifts off from Cape Canaveral in a mid-1960s test flight. Developed by the von Braun team and its contractor partners in industry, the Saturn IB was a step in the development of the Saturn V super-rocket that launched Apollo astronauts to the moon from 1969 to 1972. *(US Space and Rocket Center)*

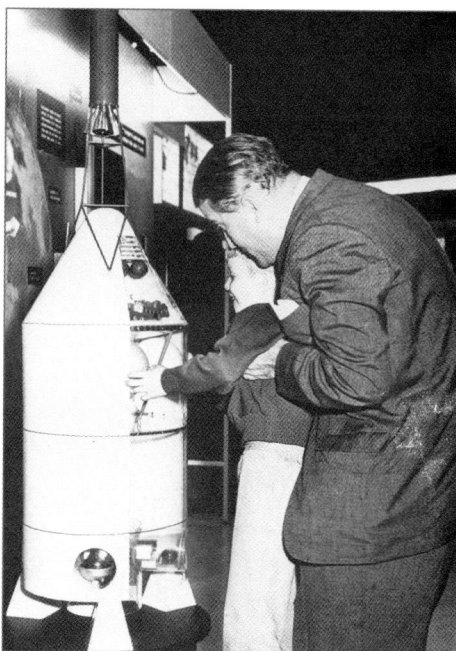

Von Braun helps a young boy get a good look at a space-theme exhibit in this undated file photograph. A promoter of education, Von Braun was delighted that so many young people were interested in space exploration. *(History Office, Marshall Space Flight Center)*

Von Braun does here what he did often, and effectively – testify before Congress. In this appearance in about 1961, he is briefing a space-related committee on Capitol Hill with Robert Seamans Jr, NASA associate administrator (right). *(History Office, Marshall Space Flight Center)*

With von Braun as host and escort, President John F. Kennedy gestures during his second visit and tour, in May 1963, of the Marshall Space Flight Center. Robert Seamans Jr is on the left. *(Bonnie Holmes)*

Von Braun points to an image of the Saturn I rocket on a large monitor in the Launch Control Center during a 1965 flight from NASA's Kennedy Space Center. *(Huntsville Times)*

Von Braun explains Saturn heavy-lift rockets to President Dwight Eisenhower during his 8 September 1960 visit to formally open the new NASA George C. Marshall Space Flight Center in Huntsville, Alabama. *(Huntsville Times)*

Von Braun (fourth from left) leads the original group of seven Mercury astronauts on a tour of Redstone Arsenal in 1960. The group inspects the bulkheads of a Redstone rocket of the type that would lift future admiral Alan Shepard (far left) and Gus Grissom (third from left) on brief suborbital space flights the next year. *(Huntsville Times)*

Seemingly nonchalant pad workers inspect a fuelled Mercury-Redstone vehicle in the pre-dawn hours at Cape Canaveral before the morning launch of Alan Shepard. The first US astronaut rocketed away on 5 May 1961, nestled in the single-seat Mercury space 'capsule' atop the modified Redstone missile developed by the von Braun team for the US Army. *(US Space and Rocket Center)*

Von Braun in 1959 outlines the trajectory of the proposed suborbital Mercury-Redstone space flights of Alan Shepard and Gus Grissom. *(Bonnie Holmes)*

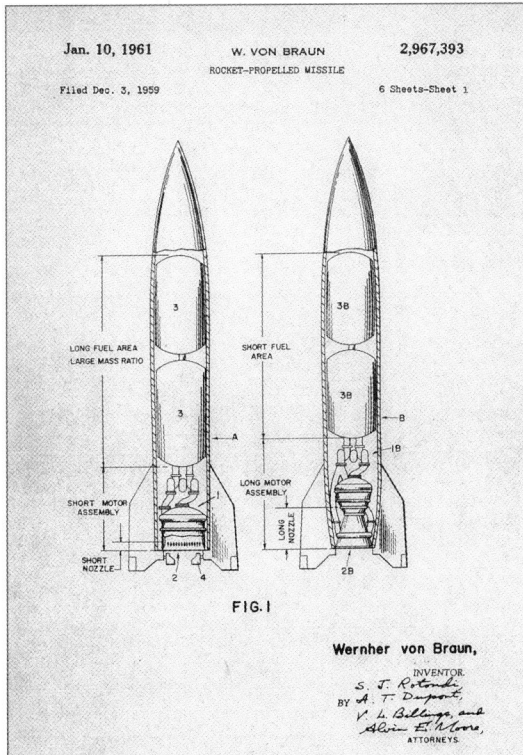

Von Braun was granted a patent in 1961 for this innovation in a 'rocket-propelled missile'. It improved 'the mass fraction of the rocket by increasing propellant volume through shortening the propulsion system,' explained another German-born rocket scientist. Exterior dimensions remained the same by lengthening the propellant tank and shortening the engine and nozzle. *(David L. Christensen)*

Early von Braun mentor Hermann Oberth (centre foreground), poses with the US Army missile hierarchy of the mid-1950s at Redstone Arsenal. Anticlockwise from Oberth: von Braun, Robert Lusser, Brig-Gen Holger N. Toftoy and Ernst Stuhlinger. This photograph appeared in 'US Races for a Supermissile' in *Life* magazine's 27 February 1956 issue. *(US Army Aviation and Missile Command)*

A Redstone missile is fielded by troops in Europe some time in 1957. Developing the 200-mile, nuclear-capable ballistic weapon was the first of two big jobs assigned to US Army missile R&D operations at Redstone Arsenal, where von Braun was technical director. The second was to develop the 1,500-mile Jupiter IRBM. *(Huntsville Times)*

Von Braun thanks his co-pilot after a 12 May 1959 flight in which he piloted the US Air Force F-100 jet fighter faster than the speed of sound at Edwards Air Force Base in California. Fanatical pilot von Braun was inducted into the Mach-Busters Club for breaking the sound barrier. *(Bonnie Holmes)*

Apollo 11 lunar lander – with astronauts Neil A. Armstrong and Edwin E. 'Buzz'
Aldrin Jr aboard – separates in lunar orbit from the command module, where astronaut
Michael Collins is at the controls – and descends for the first landing on the moon.
(US Space and Rocket Center)

Apollo 11 lunarnaut 'Buzz' Aldrin positions scientific equipment on the moon's surface
during the first manned landing mission to Earth's natural satellite.
(US Space and Rocket Center)

Von Braun experiences another sort of lift-off – an impromptu hero's ride on the shoulders of local public officials in Huntsville at the community's Apollo 11 'splashdown celebration' on 24 July 1969. *(Huntsville–Madison County Public Library)*

Saturn–Apollo 17, the final manned lunar-landing flight in the series, lifts off on 7 December 1972, in a night-time launch from Kennedy Space Center. Astronaut and Navy man Gene Cernan, the commander, became the last man to walk on the moon. *(US Space and Rocket Center)*

A simple gravestone with the reference to his favourite verse of scripture marks von Braun's burial place in the private Ivy Hill Cemetery in a church-yard in Alexandria, Virginia. *(Huntsville Times)*

Not that every politician fell under von Braun's spell in the 1960s. He sometimes came across Capitol Hill denizens who were jealously protecting their state's piece of the space pie. Senator John C. Stennis of Mississippi, for one, was very protective when it came to guarding the interests of his Mississippi Test Facility. That was the $500-million NASA site carved out of backwater acreage in the southern part of his state and placed under von Braun and the Marshall Center's control early in Project Apollo. Von Braun found out just how protective Stennis could be during a visit the senator paid him. The Mississippian mainly wanted a heart-to-heart talk with von Braun in the latter's office, recalled James T. Shepherd, Marshall facilities chief. After updating Stennis on the centre's programmes and capabilities, von Braun insisted on giving him the grand tour of the Huntsville centre's impressive facilities.

Stennis remained unimpressed through it all, Shepherd remembered. Standing on top of a huge static-test stand for Saturn booster stages, the powerful senator looked straight at his host and said: 'Dr von Braun, I really don't care what you do here in Huntsville. Just don't take it out of Mississippi.'[36] End of message. Shepherd's reaction: 'I thought, man, in Mississippi politics, they don't mince words.'[37]

Politician von Braun often played the role of teacher, too, not only to the thousands of high school and college students he spoke to around the nation but also to the junior members of his team. NASA engineer-manager James B. Odom vividly remembered a lesson on anvil clouds that von Braun gave him during a mid-1960s flight from the West Coast back to Huntsville. Aboard a hired Lear jet, flying at 41,000ft over Kansas, the Marshall Center contingent encountered just such a cloud and had to fly around the enormous, dangerous formation. Von Braun, a master on the subject of weather, left the cockpit, sat down beside young Odom and proceeded to give him an exhaustive explanation of every aspect of anvil clouds, out of a desire 'to share that knowledge he had with me . . . a young engineer', Odom said.[38]

Another time he gave an all-night astronomy lesson, NASA veteran Jay Foster remembered. On a late-1960s trip to California to visit several aerospace contractors, von Braun and a group of associates planned a last stop at the famed Mount Palomar observatory. The Marshall Center was then in the design phase of

Skylab, the nation's first manned orbital station, using leftover Saturn launch vehicles and other hardware. Skylab would be an astronomical observatory in space, among many other things, and von Braun wanted his team to have a better understanding of the closest thing to it on Earth: Palomar.

According to Foster the entourage received first-class treatment. A dinner was provided, followed by a tour; then came a briefing on Palomar's design, construction, capabilities and achievements; and after that a comprehensive lecture by a visiting university astronomer on the subject of white dwarfs, or dying stars, in their last stage. It was after midnight when the visitors headed back down the mountain for the Marine Corps base where their plane awaited them for departure the next morning. Von Braun told Foster that he had 'gotten all stirred up with the lesson on white dwarfs, and that once we got back to Camp Pendleton, he felt like giving everybody a general astronomy lesson. Of course, nobody would refuse that.'

Arriving at the base around 1.30 a.m., Foster recalled, 'we all went to von Braun's room and he lectured us on astronomy practically all night. Everyone was sitting there and he was really warmed up. He was enthusiastic about it, as he was about almost anything he got to talking about.' The men got little sleep that night. At daybreak they straggled into the camp's mess hall for breakfast, made their way onto the NASA plane and took off towards home. Then they slept.[39]

As the director of the Marshall Center all through the 1960s, von Braun devoted vast amounts of his time to directing: managing, chairing meetings, fostering internal communications, making decisions and getting daunting jobs done within deadlines. Although something of a hands-off manager where pure administration was involved, von Braun was the ultimate hands-on manager in engineering matters. He haunted his centre's technical labs and shops and drafting rooms, as well as the innards of contractors' plants. He preached individual responsibility and absolute perfection of product, having learned the hard way in the rocket and satellite business that near perfection 'is the equivalent of disaster'. Earlier, he had said of his team's satellite launcher: 'We were almost cocky about our equipment. We had to be, because we knew that there is no "98 percent successful" satellite launching.'[40] It either reached orbit – the proper orbit – or it did not.

While he had his strengths as a manager, von Braun was less than perfect. He had an aversion to personally reprimanding or disciplining anyone on his team, usually leaving that to his deputy. He would overspend if not watched closely. At one point in the 1960s, he allowed the number of managers with direct access and reporting responsibility to him – including all his German-born lab directors and other technical chiefs – to reach thirty-eight, according to senior Marshall manager Lee B. James.[41] Eventually, von Braun was persuaded to correct the situation. Against resistance, he ordered the lab bosses to closet themselves and choose their own overlord of Marshall's R&D operations, through whom they would report to von Braun between regular staff meetings. They were simply to inform him of their choice after his return from a holiday trip with his wife. 'I won't come back', he added, 'until I see "the white smoke" of agreement', as James T. Shepherd remembered it.[42] The lab directors acquiesced, selecting one of their number, Hermann Weidner, for the superchief's role. 'Thereafter,' team member Charles Lundquist recalled, 'von Braun used to introduce him as "my pope"' chosen by his cardinals, the lab bosses.[43]

Von Braun was adept at choosing and keeping good people, however. He held regular and frequent meetings with his senior staff and others, and allowed all to be heard, interminably if need be. He had a knack for motivating people and keeping morale high. In any reorganisation he was good at anticipating and personally handling difficult individual cases. He fostered the concept of 'automatic responsibility', empowering personnel at all levels to take part in the decision-making process.

And yet von Braun was also a decisive, action-oriented manager. He abhorred the indecision that too often besets bureaucracy. Jim Shepherd recalled the time when Hurricane Camille struck the Marshall-managed Mississippi Test Facility (MTF). The devastating 1969 storm took several lives among the rocket test centre's employees and their families. It destroyed homes and schools, and damaged the facility itself. The distraught MTF on-site manager, Jackson 'Jack' Balch, quickly contacted von Braun's executive staff up in Huntsville.

'Jack wanted help,' Shepherd remembered. 'He told one of our managers, "We have to do something down here!" Our man's reaction was, "We'll get a team down there and see what they need." But von Braun said, "They don't need a survey. They know what they need. We are going to get a convoy together in the morning, and we are going down there and help."' The disaster-relief convoy rushed construction equipment, radio gear, medicine and other emergency aid to the storm-ravaged Test Facility and vicinity before the Mississippi National Guard showed up – and stayed for weeks. 'I thought that was a lesson in action to meet a situation,' Shepherd observed. 'It said: "Let's [not] study about these things. . . . Get it fixed." And it was.'[44]

Occasionally, von Braun practised the management philosophy of 'Better to ask forgiveness than permission.' Von Braun and company conceived plans for a 'neutral buoyancy simulator' at Marshall in the 1960s. The idea was to build a large water tank in which the microgravity conditions of space could be simulated. Engineers in diving spacesuits could use it to test flight-hardware designs for placing, say, foot restraints and handgrips for spacewalking astronauts. But this R&D tool would require a new building. Extra funding for it would be hard to come by and, if secured, would probably bring rigid congressional restrictions. So Marshall improvised. A 'temporary building' was unilaterally erected with funds scrounged from existing accounts. The centre used its own expert welders from in-house shops to fabricate the spacious – 35ft in diameter by 33ft deep – tank within.

Operations at the new facility got under way, and all was going well. NASA's Houston centre, which had belittled the scheme after learning of it, later used the tank extensively for astronaut training. When news of all this reached Congress, the Space Subcommittee sent word it wanted to inspect this new structure. 'They came down, and we took them over to see it,' facilities boss Jim Shepherd recalled. 'The big doors opened up, and these guys from Washington saw this massive, 1.3-million-gallon tank that von Braun had built on the cheap. He said it was just a piece of "equipment", not a "facility". They didn't say anything then – not one word – but they later put a lot of restrictions on it.' Shepherd remembered von Braun saying to his inner circle: 'You build the facility first, then take the slap on the wrist. But you have the facility. They are not going to burn it.'[45]

The tank proved a success as an engineering and training tool, Shepherd and others said. Rigged for lowering large structures inside, it was operated by scuba-qualified engineers at Marshall. Fifteen or more astronauts trained in it, as did several Soviet cosmonauts later. Navy SEALS used it, and aquanaut Jacques Piccard tried it out. Scuba enthusiast von Braun made frequent dives – as soon as a helmet ring large enough to accept his imposing head was found, recalled James Splawn, head of the tank staff.[46] The tank offered a bonus not lost on the PR-minded von Braun: with viewing portholes on the sides, it became a popular stop on the centre's tour for visitors, especially when in use by astronauts or others.[47]

One of von Braun's severest Apollo-era managerial challenges involved dealing with North American Rockwell, prime contractor for the S2 second stage of Saturn V. The Marshall Center's resident engineer at the company's Downey, California, plant, where the S2 stage was being developed and built, found himself shut out of the information flow on technical issues. Visiting Marshall groups were made to feel 'very unwelcome' at Downey, recalled MSFC official Mohlere.

The problem stemmed in part from a clash of inbred management cultures – and of two titanic personalities. The von Braun team had a history, first with Army missile agencies and then with NASA, of working in intimate contact with its contractors. North American Rockwell, the former North American Aviation, with its proud aircraft-building heritage, was accustomed to the Air Force system: agencies choose a contractor, specify what they want, write it into the contract, and then stand back and let the company perform. Von Braun and company were especially uneasy over the S2 stage, whose power plant would burn higher-energy – and highly volatile – liquid hydrogen as fuel.

And as the Marshall team saw it, there was a significant further problem: the president of North American Rockwell's Space Division, Harrison B. Storms Jr, 'was known as "Stormy", and he was well-named,' remembered Mohlere. 'He was a well-known, highly successful aeronautics engineer. . . . He also had some personal characteristics which tended to keep outside influence outside his plant.' Von Braun and his deputy director and unofficial chief engineer, Eberhard Rees, informed NASA higher-ups that 'this system was compromising the programme's schedule, and that we wouldn't make it if something wasn't done,' related Mohlere, who worked directly for Rees. A resultant visit to Downey by a special contingent headed by a NASA headquarters official and including Rees, Mohlere and others got a frosty welcome and found little to allay their concerns. But no major corrective actions came about.

A substantive flash-point between the Marshall Center leadership – especially Rees – and Storms had not made for a loving relationship. Marshall strongly favoured an S2 stage design with separate bulkheads between the liquid hydrogen and liquid oxygen tanks, for safety's sake. North American and Storms argued for a common bulkhead for the two adjoining tanks, with a weight savings of about 2 tons. Storms had won that dispute.

Sometime around 1966, shortly after the group from NASA headquarters made their unsatisfactory visit to the Downey plant, von Braun attended a meeting there with briefings on the S2 programme given by Storms and others representing the Space Division. Also in attendance were North American executive George Moore and other corporate honchos. When the briefings ended, 'von Braun just got up and . . . said that the continued interference of the president of North American's Space Division [Storms] was impossible if this programme was to continue,' recalled Mohlere, who was present. Von Braun turned to face Storms. 'He right then and there called for the resignation of Storms. There was dead silence [except for] a lot of gasping. That was a monumental thing . . . the most courageous thing he ever did' as a space leader-manager.[48]

Storms, however, remained in place. On a follow-up visit to the Downey plant, an unrelenting von Braun – the customer – tried another tack during a private walk between buildings with J. Leland Atwood, the head of North American. Only Jim Shepherd of the Marshall Center accompanied them. Shepherd recalled von Braun's blunt statement: 'Lee, Stormy has got to go', and Atwood's equally blunt reply: 'We can't replace him.' 'And it ended there,' said Shepherd.[49]

Actually, it ended for Storms the following year. The trigger was the Apollo 1 spacecraft fire on 27 January 1967, and the deaths of astronauts Gus Grissom, Ed White and Roger Chaffee during a dry-run countdown at Cape Kennedy for the first Apollo manned mission. The legendary North American executive's Space Division was also the prime contractor – to NASA's Manned Spacecraft Center (MSC) – for the command and service modules of the Apollo spaceship. Design flaws and evidence of shoddy workmanship later came to light. Schedule and budgetary shortcomings had been highlighted in a damning internal report made by NASA Project Apollo Director Sam Phillips in December 1965, but it did not surface until after the fatal fire. The principal heads to roll in 1967 were those of Joseph Shea, Apollo spacecraft manager at the MSC, who was summarily transferred to NASA headquarters, and the president of North American's Space Division. In addition, George M. Low, Shea's immediate boss at the MSC, had to step down a notch at the Texas centre.

Three months after the fire, on 1 May, Lee Atwood called a meeting and informed Storms he was being replaced and transferred to a staff job at company headquarters.[50] Some said he got a raw deal; von Braun and Rees were not among them.[51]

Lunar Triumph

Reaching the moon and being the first to do so was of great importance, von Braun made clear as Project Apollo rolled on, but he also wanted people to understand that the nation stood to gain much in the process. Scientific advances and technological spin-offs would come, along with a boon to the national spirit, the opening of the gateway to the planets and stars, and the confirmation that mankind can do anything it sets its mind to.

During the Apollo push von Braun further observed: 'The Moon is as much or as little a goal as the city of Paris was to Lindbergh on his immortal flight. After all, if all Lindbergh had wanted to do was to reach Paris, he could have taken a boat.'[1]

The race to beat the front-running Soviets in putting men on the moon was deadly serious. Space-aviation author Martin Caidin had captured the mood earlier in the 1960s when he wrote: 'There emerges a new note, an undercurrent in our nation's lunar exploration program. That undercurrent is one of urgency. It may be that we will not realize our goal to be the first on the surface of the Moon, that the Soviet head start will be too much to overcome.'[2]

Building on the first Saturn I test flight in October 1961, the series of heavy-lift vehicle missions continued successfully in the apparently unstoppable march to the moon. The ten unmanned flights of the Saturn I used both NASA-produced stages fabricated in the 'Fab Lab' of the MSFC (Marshall Space Flight Center) and contractor-built hardware. Several of the sixteen-storey-tall rockets carried scientific payloads such as the large, winged, Pegasus meteoroid-detection satellites. The Saturn I flights were followed by the first launch of the project's even larger test-bed rocket, the Saturn IB, in February 1966. It had a redesigned booster stage and a new second stage – the S4BC – that would serve as the third stage of the goliath Saturn V.

The time arrived for the critical first 'countdown demonstration test' of the Saturn V in mid-October 1967; it would lead to the giant vehicle's first flight. In charge of the demo test was Peenemünde and Mittelwerk veteran Arthur Rudolph, director of the Saturn V Program Office at the Marshall Center. He was positioned in Launch Control at Florida's Kennedy Space Center. As several abortive attempts at the run-through were made, other Marshall officials monitored the situation in real time from the operations support facility in Huntsville. At long last, after repeated false starts and glitches, the full test was accomplished. Congratulatory calls flowed in to Rudolph on an intercom hotline. Listeners soon heard another caller come on the line from Huntsville.

'Arthur, congratulations on a successful test!'

'Who is this?' asked a weary Rudolph.

'This is Wernher.'

'Werner who?' asked Rudolph, still not recognising the caller's voice and knowing several 'Werners' on the team back at Marshall (there were several Werners but only one Wernher, with an 'h', the classic German spelling).

'You [expletive unknown]! I am the one that goes to Washington and gets all the money for you to play your funny games!' the caller fired back.

Brief pause. 'Oh, Wernher von Braun!' an embarrassed Rudolph replied at last.

Rudolph is said to have enjoyed a good laugh later with von Braun and others over the whole thing. At his retirement party, however, he became upset when some of his associates played a recording of the exchange.[3]

After Peenemünde and Mittelwerk, Rudolph had worked for the US Army at Fort Bliss and Redstone from 1945 to 1960, and then spent almost a decade with NASA's Marshall Center. His cherubic appearance and easy charm at social events belied a managerial toughness. Boeing-Huntsville space engineer-manager Gene Cowart recalled the day Rudolph lost his temper with a member of his Saturn V staff who was unprepared at a meeting: 'You do better, or I send you to Dr Mueller and you'll have blood on your face!'[4] The man got up and left the meeting.[5]

Health problems – reported heart trouble, along with a palsy that caused his head to bob up and down, then drop and turn to one side – led Rudolph to retire in 1968. That was the year after the first Saturn V flight and after high honours for contributions to his adopted country's defence and space efforts had been bestowed on him.

In retirement he and his wife moved to California to be near their daughter. There, separated from colleagues and friends, he was contacted in 1984 by investigators from the US Justice Department's Office of Special Investigations (OSI) assigned to hunt down Nazi war criminals. (Wernher von Braun had been dead seven years at that point, and many speculate that this action against Rudolph would never had occurred had his friend still been alive in 1984.) After two voluntary interviews with OSI interrogators, Rudolph was informed that his own words about Mittelwerk and his role in the use of forced labour for V2 production had incriminated him. He was threatened by OSI with prosecution, loss of pension and more if he refused to cooperate, yet Rudolph did not consult anyone back in Huntsville for advice or help. His only counsel was an immigration lawyer whose name he had found in the yellow pages.[6] Rudolph quietly signed papers agreeing to surrender his US citizenship and return to Germany, while keeping his full US government pension and all awards given him. He acknowledged no persecution of forced labourers at Mittelwerk.[7]

The action came to light only when Rudolph failed to show up for the Fort Bliss old-timers reunion in Huntsville that spring. Later that year the OSI issued a news release on Rudolph's deportation that characterised him as a Nazi war criminal, setting off a rush of interest from both detractors and supporters, who complained that he had not been given due process of law. Alabama's Republican junior US Senator Jeremiah Denton applauded the action. Coming to Rudolph's support were Alabama's other senator, Democrat Howell Heflin, Huntsville's City Council and its mayor, the local American Legion post, Rudolph's Huntsville Lutheran pastor, the Revd Curtis E. Derrick, as well as Maj-Gen John Bruce Medaris, US Army (Ret.), Rudolph's ex-commander at Redstone Arsenal and then an Anglican bishop in Florida. Ex-Peenemünder Walter Haeussermann and others began collecting documentation to present to the Department of Justice for reconsideration of the case.[8]

In Germany, the Bonn government investigated Rudolph's wartime past, absolved him of any war crimes and restored his German citizenship. In Hamburg he grew increasingly embittered over his treatment by the OSI and the US justice system, especially because Washington had known of his Mittelwerk role when he was invited to work in the States in 1945. Encouraged by old rocket-team colleagues and others, he attempted to return via Canada in 1990 to overturn the agreement, regain his US citizenship and resume residency. After detention by Canadian Immigration, eight hours of questioning and an extensive court hearing, Rudolph was found not guilty of the OSI allegations but was denied entrance to the United States on points of immigration law. Arthur Rudolph died in Hamburg on 1 January 1996, aged 89.[9]

During the tense push for the moon, a number of honours were conferred on von Braun. One of the happiest occasions came on 6 June 1967, when he and his family appeared at the Smithsonian's National Air and Space Museum, where he was awarded the gold Langley Medal for his 'creative vision' in advancing rocketry, for leading the way to America's first Earth satellite and for 'technical leadership in development of the Saturn class of large launch vehicles'. In its fifty-nine years, the medal had been awarded to just twelve aviation and space pioneers, among them Orville and Wilbur Wright, Charles Lindbergh, Robert H. Goddard and astronaut Alan Shepard. Von Braun was all smiles as he joined that short list.

Although Apollo's objective was well on the way towards being achieved in the latter half of the 1960s, von Braun was gravely troubled. As early as 1966, well before the Apollo 11 attempt, Washington had begun to cut back the space programme. This prompted von Braun to vent publicly, 'Our main effort today is busily destroying the very capability that we have built up to put a man on the Moon.' He told reporters that if the trend was not reversed, the nation might as well hang a sign on the moon saying 'Kilroy was here' and declare 'the show is over'. In September 1967 he continued in the same worried vein: 'To make a one-night stand on the Moon and go there no more would be as senseless as building a railroad and then making only one trip from New York to Los Angeles.'[10]

By early October 1968, the once-stricken Apollo spacecraft programme was back on track and the Saturn V was up and flying. Von Braun, whose nature was to accentuate the positive, remained uncharacteristically gloomy over space budget cuts, and he continued to sound the alarm. In a high-profile magazine interview that month, he emphasised that, in contrast to the growing Soviet space programme, '[NASA funding had] gone down and down and down for the last three years. . . . It may surprise you to hear this, but for the last two years my main effort at the Marshall Center has been following orders to scrub the industrial structure that we had built up at great expense to the taxpayer, to tear it down again. The sole purpose seems to be to make certain that in 1972 nothing of our capability is left. That's my main job at the moment. And we haven't even put a man on the Moon yet.'[11]

He blamed the slippage in space support on several factors: the draining war in Vietnam, other troubles abroad, 'riots in the cities', civil rights issues, the assassinations of Martin Luther King Jr and Robert Kennedy, and the long hiatus in both US and Russian manned space flights. That lapse had caused some loss of American public interest. The sense of a quite real 'moon race' had disappeared, he said, replaced by the feeling that America was far ahead. 'We aren't ahead,' he emphasised.

By the time of that October 1968 interview, von Braun noted one especially troubling result of the slide in US space priorities. 'It is an awfully difficult thing for anybody participating in our lunar program to try to run as fast as he can if the environment is one of building down rather than building up. It is like being ordered to disarm while the war is still on.' What was really needed, von Braun appealed, 'is sustained support of this space program over a great number of years. We must stop blowing hot and cold. This thing is for the long haul.'

And it does matter whether America or Russia is first to reach the moon, he stressed. That feat 'will not go unnoticed on earth', and the 'scientific status and technological quality of the two countries will be compared . . . for many years, perhaps for generations to come'. After all, von Braun observed, 'Who remembers the second man to fly the Atlantic Ocean?'[12]

Von Braun did more than simply vent his anguish over the downturn in space funding in the latter half of the 1960s. While 'intensely focused on the success of the planned lunar missions', space scientist Pat R. Odom recalled, 'he was concerned that . . . the US taxpayers could see longer-term returns on their enormous investment in the Apollo program'. That meant 'exploring ways the Apollo hardware – spacecraft and launch vehicle – and facilities might be adapted to other important space missions, manned or unmanned'.[13] Maximising Apollo's returns would also strengthen von Braun's hand politically in his pitches to Congress and the public.

For example, von Braun conceived of exploring Jupiter with a large unmanned craft built from elements of the Apollo lunar lander and launched by a Saturn V, Odom later pointed out. That concept was passed for feasibility study through Marshall's laboratory system to a contractor, Northrop Services in Huntsville, where Odom headed the advanced studies team. The resulting study led to a favourable report, which led years later to NASA's amazingly successful Galileo mission to Jupiter.[14]

Amid von Braun's angst over the impact of budget reductions, the spectacular first launch and flight of an unmanned Saturn V on 20 November 1967 had gone well. So had the last of five Saturn IB flights, almost a year later, in October 1968, which was also the first in that series to carry an astronaut crew. In their Apollo 7 warm-up for the first manned Saturn V flight, ex-Navy test pilot Wally Schirra, Marine Corps flyer Walter Cunningham and US Naval Academy graduate-turned-US Air Force pilot Donn Eisele had flown aboard the new spacecraft in Earth orbit. And during the memorable Christmas 1968 flight, Saturn–Apollo 8 made history in carrying Air Force pilot Frank Borman, US Naval Academy alumnus Jim Lovell and fellow Annapolis graduate Bill Anders on ten swings around the moon and back home; they took turns reading passages on creation from the Book of Genesis.

Next came the Apollo 9 flight of the all-Air Force crew of James McDivitt, David Scott and (by then, civilian) Russell 'Rusty' Schweickart in March 1969 in an Earth-orbit performance test of the spacecraft's command-service and lunar modules. Then, in a May dress rehearsal for the big show, on stage strode the Apollo 10 astronauts: Thomas Stafford, former US Air Force test pilot leader; Cdr John Young, US Navy, two-time Gemini veteran; and Cdr Eugene Cernan, US Navy, Gemini 9 pilot and the future 'last man on the Moon'. They were boosted by a Saturn V into lunar orbit for a fly-by that dipped teasingly close to the moon's pockmarked surface.

And, suddenly, it was almost time. Just two weeks before the scheduled July lift-off of Apollo 11 for the first landing mission, von

Braun somehow slipped away with Maria for a holiday trip to Greece. He visited the ancient Temple of Apollo on the island of Delos, birthplace of the god. 'Since our space project is named Apollo,' he explained to Greek reporters, 'I thought it only proper that I come to Greece and pay my respects to the god Apollo before we finally made this big moon-landing attempt.' As friend and co-worker Ruth von Saurma wrote to him a few years afterwards: 'And who but you would squeeze into the weeks of check-out procedures at Kennedy Space Center preceding the first manned lunar landing a quick trip to the Greek isles and return full of enthusiasm and new impressions!'[15]

In Washington, NASA Administrator Thomas O. 'Tom' Paine, who had succeeded Jim Webb in early 1969, was asked what he thought of von Braun's leaving the country at such a crucial time. 'Well,' Paine told reporters, 'Wernher is beseeching the Greek gods, and we just sent astronaut Frank Borman to see the Pope – so I guess we have all bets covered for Apollo 11.'

Returning to the Marshall Center, von Braun gave assembled employees a trip report plus a history lesson on ancient Greece. He threw in a memorable mini-lecture on Prince Henry the Navigator, recalled Marshall's Jay Foster. He drew parallels between the Portuguese mariner's African voyages, their unforeseen impact on European maritime exploration, the resultant expansion of sixteenth-century Europe's holdings, and the approaching Apollo flights of lunar exploration, discovery and adventure. Foster found the talk a typical example of von Braun's intellectual breadth and penchant for synthesising.[16]

At a news conference in Huntsville just one week before Apollo 11's appointment with history, von Braun struck a balance between confidence and nervousness. He predicted success in man's first attempt to set foot on the moon and stressed that extreme caution remained the norm: 'This is no wild-blue-yonder project – no come-hell-or-high-water attitude.' But he also admitted at a press conference on 10 July to anxious pangs 'just like you and all of us have. Keep in mind we are pioneering . . . and the element of chance is always present. There are pitfalls, and we have to take chances to get from here to there.' He added, 'The public should not be complacent. Something may happen and the public should be prepared for the shock.'[17]

(Similarly, a couple of years earlier, Jim Webb, then the boss at NASA, had pointed out the great dangers when he testified to Congress after the Apollo 1 fatal fire. He had likened a fuelled and loaded Saturn V to 'an atomic bomb' out there on the pad.)

At the press conference, von Braun also noted that NASA had helped the odds for success by choosing a landing site – the Sea of Tranquility – that was easiest for the astronauts, if not the most promising for scientific discovery. 'The maria [seas, or dark spots] on the Moon are about as much value to science as a gravel pit,' he acknowledged.[18]

As the Apollo 11 lift-off approached, the space scientist expressed two regrets. One, the demands of his administrative workload had removed him increasingly from the hands-on engineering pursuits he had so enjoyed during his German and postwar years. Two, it saddened him that he must remain earthbound when the Saturn V roared away on 16 July.

'I would love to go to the Moon first,' he told enquiring reporters, 'but we must be realistic. As this gray hair will show you, I'm fifty-seven years old. And astronauts must be trained test pilots who can react to emergencies with precision, speed, and accuracy. But if things go right, within ten years from now I could very well be going to the Moon as a passenger – an opportunity many ordinary people in many walks of life can also experience.'[19]

The scientist also said he doubted man's propensity for warfare would spread to the heavens. That's mainly because 'we have enough capability to destroy ourselves right here on Earth, without taking it into space'. Emphasising NASA's 'peaceful' role, he said he was thankful for that, adding, 'I'm happy to be out of weaponry.'[20]

With a week to go, von Braun acknowledged to reporters that the Russians would not beat America to the moon. The United States was on the verge of winning the race, he said, 'because we decided to do it'. He did not address the disastrous problems the Soviets, who had also decided to do it, had suffered in their Saturn-class rocket development programme.[21] He said only that they remained ahead of America in certain other areas of space exploration, such as successful soft landings on Venus.

In a heavy touch of irony, the USSR launched its Luna 15 probe to the moon on 13 July in an effort to steal some of Apollo 11's thunder. The unmanned probe's mission was to land a lunar rover, collect soil samples and return to Earth. Luna 15 reached lunar orbit just two days before Apollo 11's scheduled arrival, but the Soviet ploy failed. The probe circled the moon fifty-two times. Then, instead of making a soft landing, it crashed into the lunar surface on 21 July.[22]

As for the cost to US taxpayers of the anticipated victory on the moon, von Braun asserted to reporters: 'I am absolutely convinced that the $23 billion spent on the manned space program has not made the United States $23 billion poorer but many billions richer in new knowledge acquired.' Not only had the nation seen dramatic advances across the board in science and technology, he added, but the space programme was acting as the 'cutting edge for progress in many other areas', from industrial management to quality control.[23] Von Braun himself, commanding NASA's largest centre, with the biggest hunk of the agency's multi-billion-dollar budgetary pie, did not add much to Project Apollo's total tab. His yearly pay then was less than $35,000.

At T-minus two days, von Braun was one of five NASA officials holding a packed news conference at the Kennedy Space Center. As a member of the press corps on hand, I saw two revealing incidents. First, the rocketeer drew the most admiring reaction from the press corps with his unequivocal response to the question of how he would rank the impending Apollo 11 lunar landing with other historic events. 'About with the importance of aquatic life first crawling on land.' Then, at the end of the briefing, the reporters swarmed around von Braun. The other space agency leaders were allowed to leave the hall virtually unmolested, while eager questioners surrounded the always-quotable von Braun.

The tradition of 'little get-togethers' – big, lively parties hosted by aerospace and news media companies, as well as smaller affairs – all over the Cape Kennedy area the night before major launches was in full bloom on the eve of Apollo 11's scheduled take-off. For these pre-launch soirées von Braun was known to helicopter from party to party with a few associates in tow. All the big Saturn–Apollo contractors threw parties before each launch, and all wanted von Braun to attend theirs. A chopper was the only practical way he could make the rounds.

On the night of 15 July 1969, the hottest ticket was to a party hosted by Time-Life at a country club in Titusville. The place was jammed with entertainment stars and other celebrities – James

Stewart, Charles Lindbergh, Norman Mailer, among others – as well as US and foreign government officials, industry bigwigs and key NASA leaders. At one point in the festivities von Braun took the stage to join aviation pioneer Lindbergh and his earliest mentor in rocketry, Professor Hermann Oberth, who, at the age of 74 had been flown to Florida from Europe for the epochal flight. The Marshall Center chief also invited the programme director of the all-important Saturn V, Col Lee James, US Army (Ret.), onstage for special recognition. Von Braun, with Maria looking on, spoke to the crowd and took questions. He gave his confident promise for the next morning's event: 'Of course it will be a success!'[24]

The morning of 16 July dawned bright and blue-skied on Florida's east coast. Lee James had left the big pre-launch bash about 11 p.m., dropped off his wife, Kathleen, at their motel and headed straight for the countdown in progress at Kennedy Space Center's (KSC's) Launch Control Center. He got no sleep that night. Von Braun fared little better. He went from the party to his Cocoa Beach motel room to turn in – but instead spent an hour reviewing the launch plans. He telephoned his old Peenemünde comrade, KSC Director Kurt Debus, to wish him luck and to double-check several details. At last he crawled into bed and said a prayer. He did not sleep well. He woke before dawn, showered, shaved and dressed in a suit and tie for the day's business.[25]

An estimated 1 million spectators were taking up positions at viewing sites a few miles from the launch pad. Hundreds of invited VIPs, the list topped by former President Lyndon Johnson and his wife, Ladybird Johnson, began to assemble at the grandstand provided. A horde of 3,000 news media representatives, the most ever recorded for a space story, gathered at the press site in the darkness before sunrise. Three miles away, the Saturn V–Apollo 11 stack, nearly as tall as a forty-storey skyscraper, stood glowing a bright white under brilliant spotlight beams in a scene reminiscent of an old-fashioned Hollywood film premiere.

Von Braun arrived at 4 a.m. at Launch Control, where a team of more than fifty were working in the main room. He checked in with Debus and learned the count was going well. He took his assigned place, put on his headset, adjusted the earphones and tuned in.[26]

Later that morning, after the sun had risen, von Braun left Launch Control for the press site to do live pre-launch interviews inside the network television trailers. Afterwards, he was walking alone in the sunlit press area when he encountered a familiar newspaper writer-editor from Huntsville. I stuck out my hand and said the only thing there seemed left to say: 'Good luck, Doctor.' We shook hands, and I was reminded how large his hands were. He smiled and said, 'Thank you.' He then turned and headed with his thoughts back to Launch Control.

Von Braun took his seat again in the main control room as the countdown advanced. The hour he had worked towards for so long was at hand. What were his final thoughts as the Apollo 11 mission's first moment of truth neared, a reporter had later asked. 'I quietly said the Lord's Prayer,' he replied.[27] His emphasis was on the passage 'Thy will be done.'[28]

Ignition of the S1C booster stage's ground-shaking power plant came at 9.32 a.m. For nearly nine interminable seconds the von Braun team's Saturn V went nowhere, remaining clamped to the pad while the five engines built up to full mega-thrust. Hundreds of thousands of gallons of fuel were consumed in an inferno before the leviathan at last lumbered upwards with agonising slowness and pounding thunder that was felt as sound waves slapping reporters' chests back at the press site. The rocket gradually gained speed and, dropping spent stages along the way, headed for a parking orbit around Earth. From there, its third stage would propel the spacecraft carrying former naval flyer Neil Armstrong, Buzz Aldrin, whose PhD was in astrophysics, and gifted writer/handball champ Michael Collins to the moon, three days and about a ¼ of a million miles away.

Minutes after the launch, I spied von Braun's mentor, Hermann Oberth, walking (evidently unrecognised by others at the press site) under the escort of MSFC chief scientist Ernst Stuhlinger. I rushed across the grass to intercept the pair for a brief interview, with Stuhlinger serving as translator. 'It was marvellous,' Oberth said of the Saturn–Apollo 11 take-off. 'When I began thinking of this flight I was a boy of 11. It was just as I imagined – only more marvellous.' He speculated that the first manned mission to Mars might come 'in 1982 at the earliest, or if we go first-class [taking time for a non-crash-programme], about 1986'.

Just as the Pilgrims did not envisage the nation that would evolve when the *Mayflower* landed on American shores, neither von Braun nor anyone else could accurately predict 'what will eventually spring from the bosom of Apollo 11', the elated scientist commented after the launch and before the lunar landing. 'But I do hope that history will record that we were aware . . . of the enormous implications of the lunar journey, that . . . we are reaching out in the name of peace, and that, while we take pride in this American achievement, we share it in genuine brotherhood with all nations and with all people.'[29]

Later during that launch day, the von Brauns and their 9-year-old son, Peter, enjoyed a relaxing private dinner in the area. Their companions were an old friend, Capt Bill Fortune, US Navy (Ret.), his wife and another couple of American-born friends, Austin and Margaret Stanton.

Before the meal, Peenemünde veteran Gerhard Heller and his wife came by to join in a round a celebratory drinks. Peter von Braun was fishing in the waters behind the place where they were eating. Soon the boy excitedly brought in his catch. His proud father praised him and showed him how to clean the fish. 'We were not only greatly honored' to have the von Brauns as dinner guests on that epochal day, Fortune later wrote, 'but had a most happy time, with Peter providing the fish course.'[30]

Two days later, von Braun headed for Houston aboard his NASA Gulfstream with a coterie of Marshall Center team members in order to be at Mission Control for Apollo 11's final approach to the moon and the landing attempt. He invited George Fehler and Ed Grubbs, who were among his regular contract pilots out of Huntsville, to join him in the control centre as soon as they had the plane serviced and secured at the airport.

'What a treat to be at Mission Control for that [lunar] landing!' Fehler recalled. 'But when we finally got there, we didn't have a badge or anything to get in. They didn't recognise us and weren't going to let us in. . . . About that time Dr von Braun walked nearby and saw us there. He said, "Get them a badge." They got us a badge.'[31]

Von Braun had been in his seat in Mission Control for several hours on Saturday afternoon, 19 July, when the Apollo 11 spacecraft reached lunar orbit. He was also there the following afternoon when Armstrong and Aldrin separated their lunar landing craft from the orbiting mothership carrying Collins and headed down towards the grey, grim surface. He was there for the white-knuckle, running-on-empty touchdown by cool customer Armstrong. And he was there for Armstrong's welcome words: 'Tranquility Base here. The Eagle has landed.'

Some hours later, von Braun was back in his seat watching a man preparing to become the first man to walk on the moon. 'When that leg appeared on television as Armstrong was about to get down to the lunar surface', the rocket scientist acknowledged afterwards, it was 'a pretty emotional moment'.[32] Armstrong then took the plunge, saying, 'That's one small step for a man, a giant leap for mankind.' An estimated 500 million people around the world saw it live on television, and an even larger number heard it on radio. To reporters, von Braun said it was 'a quantum leap' for science and technology, and a long-burning dream realised for a naturalised American citizen.

NASA administrator Tom Paine described the Apollo 11 drama to the press as 'a magnificent triumph – and the entire world was applauding'. Arthur C. Clarke told a reporter: 'I haven't cried or prayed for twenty years, but I did both today. It was the perfect last day for the Old World.' The next day, Lord Duncan Sandys sent a telegram to his old enemy, von Braun: 'Warmest congratulations on your great contribution to this historic achievement. I am thankful that your illustrious career was not cut short in the bombing raid at Peenemünde 26 years ago.'[33]

Navy veteran Richard M. Nixon, in his first year as President, made a point of joining the returned Apollo 11 crew aboard the aircraft carrier USS *Hornet* after their splashdown and safe recovery from the Pacific. The excited chief executive gushed, 'This is the greatest week in the history of the world since the Creation!'

Not surprisingly, von Braun, too, had several further comments on the subject. 'What do we have now that Neil Armstrong and Edwin Aldrin have stepped on the moon's surface? A new element of thinking is sweeping across the face of this good earth. For the first time in history, life has left its planetary cradle, and the ultimate destiny of man is no longer confined to the Earth.'[34]

After von Braun returned home to Rocket City, the community staged a celebration and conquering hero's welcome for him in the Courthouse Square. Four members of the city council carried him on their shoulders to the rostrum. 'To be borne by a rocket to the moon is impressive,' an overjoyed von Braun told the ecstatic crowd that day, 'but to be borne up these steps by admirers is almost as impressive!' As the cheering continued, the space leader emphasised what the team had accomplished. 'The space flights must continue,' he exhorted. 'The ultimate destiny of man is no longer confined to Earth.' He referred to the remaining lunar-landing missions and wistfully added, 'Maybe one of these days we'll even have a man on Mars.'[35]

An immediate post-Apollo 11 visitor to his office was Associated Press correspondent Howard Benedict, who had been a junior reporter at Cape Canaveral in 1958 and 1959, when von Braun began tutoring him on the fundamentals of rocketry and space exploration. In the 1960s, after von Braun joined NASA, Benedict often travelled to Huntsville to interview him and to follow, and write about, the development of the Saturn V rocket. And then, shortly after the moon landing, the ace space chronicler remembered, 'I walked into von Braun's office for an interview. His first words: "I told you we could, and we did it."'[36]

Fully one-half of the more than 100 German rocket engineers and scientists who had accompanied von Braun to America in 1945–6 were still with the team when Saturn–Apollo 11 rocketed into history. In mid-August 1969, the von Brauns and key members of his Marshall group were among the guests of honour at a star-studded state dinner hosted in Los Angeles by President and Mrs Nixon to celebrate the lunar-landing triumph. Whole galaxies of film actors and other celebrities[37] attended the formal affair, Huntsville's Lee and Kathleen James vividly remembered.[38]

The Apollo 11 flight crew, now national heroes, presented von Braun that autumn with a leather-bound copy of their book, *First on the Moon*. It bore an inscription signed by the three: 'To Wernher, who postulated, predicted, advertised, conned, pulled and finally pushed to make us first on the moon.'[39] Later, after several more successful missions to the moon, von Braun expressed his awe over the exploits of the Apollo flight crews: 'Those men took craft they had never flown before. They went where men have never been before, and came back. And they did it time after time, time after time.'[40]

Fate, or luck, or divine blessing had been with them, and with von Braun. He could look objectively, with admiration and even a sense of wonder, at what his Marshall Center rocket team and its network of contractors in industry had accomplished in Project Apollo.

'Sometimes I marvel and sometimes I shudder at the things we did in building Saturn V, and at our naïveté,' he confessed, adding:

> Our biggest problem was always the totality of the problem – getting all the parts working together. There were other difficulties, too – controlling violent combustion in the rocket engines, serious reliability problems with new computers, an inordinate amount of manufacturing difficulties, particularly as concerns the structure of the fuel tanks, and the fact that welding is more an art than a science. . . . A mountain of things went wrong, but this is precisely what we expected. At no time have I had any doubt that it would be completely successful.[41]

Some observers rated the scope of the development of the Saturn V alone on a par with the atomic-bomb-creating Manhattan Project during the Second World War. In its entirety, Saturn–Apollo was later widely ranked as mankind's largest engineering effort of all time. Jim Webb lauded Marshall's management scheme as a model for large government–industry enterprises everywhere.[42]

Despite the technological mountains that had to be climbed, Apollo 11 and its successor missions had become, as the Associated Press later wrote, 'a promise kept'. It is for meeting the enormous engineering challenge of Saturn V that von Braun and 'the Peenemünde boys' will be the most remembered of all the Apollo participants 500 years from now, Walter Cronkite predicted in 1999.[43] Furthermore, he said, history will regard von Braun as 'the New Columbus', as Cronkite himself already did. Columbus's voyages of exploration and discovery stood out in 1999 as the greatest events of 500 years before. 'Five centuries from now, I believe that the one date that will be remembered from our century is 1969 – the year that the human race first journeyed from the Earth to the stars.'[44]

In an interview with me in 1999, Cronkite envisaged that 'as people later look back for those heroes that made [the moon landings] possible, they will fixate undoubtedly on the astronauts who did it – Armstrong and the rest. But they will also look at the engineers. They will recognise it as an engineering feat. And when they do, they will fix on von Braun as certainly one of the greatest space engineering pioneers.'[45]

Could America have made it to the moon without the German-born rocket man and his team? The consensus at the time was: not so soon, but eventually, yes. One-time Project Apollo director Sam Phillips, after years of reflection, changed his view. He said, in essence: no, developing the impossibly daunting Saturn V probably took the unique experience and skills and ingenuity of von Braun and his team.[46]

Von Braun's own answer to the question usually boiled down to this: man was destined to reach the moon, with or without his help. He tended to give credit to the rocket pioneers of the past as well as his enlarged team with its German-born nucleus. He often deflected personal plaudits by saying, 'Space exploration is not a one-man show.'

Rocket City Legacy

The achievement by 1969 of Project Apollo's objectives of depositing earthlings on the moon and returning them safely home presented an occasion for much reflection. For the people of northern Alabama, this was a time to contemplate not only that momentous event but also to reflect on the profound changes wrought in that lower Appalachian locale during the two decades since the rocket people had come, and even since the decade of Apollo had begun.

Socially, demographically, culturally, economically – the community of Huntsville had been transformed from a small cotton town to a relatively progressive, racially desegregated, urbanised, high-tech mecca. Residents could trace the metropolitan make-over directly to the arrival and continued presence of Wernher von Braun and company.

The 1950 influx of the Germans from Fort Bliss, Texas, to Redstone Arsenal had found a city still struggling to recover from a post-Second World War depression. Huntsville then had maintained its position as the largest cotton market east of Memphis; surrounding it was a fertile sweep of cotton country that was ranked the most productive in the state. But as older townspeople also remembered twenty years later, a quarter of the city's population back then had fewer than five years of formal schooling; less than a fifth had finished high school; and college graduates were almost as scarce as hen's teeth.[1]

By 1950, the community's historic high of a dozen textile mills had dwindled to just three. Its once-thriving wartime ordnance and chemical arsenals, where employment had soared from zero to more than 14,000 workers at their peak, were back near zero again.

With the advent of Project Apollo just eleven years after the arrival of von Braun and his rocketeers, however, the town's roller-coaster economy had turned skywards once again. The Tennessee river valley town at the top of Alabama had literally boomed: thunderous static-firing tests out at 'the base' made cash registers jingle in the city's new shopping centres. The cotton town now embraced nicknames like 'Rocket City, USA' and, its favourite exercise in hyperbole, 'Space Capital of the Universe'.

As the lunar triumphs of the 1960s neared an end, two decades of state- and community-oriented actions and influences by von Braun also drew to a close. All across Huntsville and Redstone, now the home of NASA's Marshall Space Flight Center, was an array of buildings, institutions and other entities whose existence, or growth, bore the space leader's imprint.

Because von Braun took 'such an intense personal interest' in it, as Army missile official John L. McDaniel put it, the Redstone Scientific Information Center grew into the world's largest library for rocketry, missiles, astronautics and related sciences, and the US Army's largest technical library of any sort.[2]

The von Brauns and several German-born colleagues were influential in forming the Huntsville Symphony Orchestra. More than a dozen of the original rocket team members, spouses and offspring played in the orchestra. A laboratory director, Werner Kuers, who was also a gifted musician, was the first to be appointed first violinist. Eberhard Rees, von Braun's top deputy, served on the orchestra's first board of trustees. Wernher and Maria often attended the orchestra's Saturday night classical concerts and the shorter series of pops concerts. (In 2005, the Huntsville Symphony Orchestra would mark its 50th anniversary. It had long since become a well-regarded, all-professional orchestra, drawing big-name guest artists Beverly Sills, Yitzhak Perlman, Kathleen Battle, Van Cliburn, Yo-Yo Ma and others.)

Long-time residents remembered that at the weekends in the 1950s von Braun had physically helped build the Rocket City Astronomical Observatory and Planetarium. The mountaintop facility had opened with the second-most-powerful telescope in the south-east, and was later renamed after von Braun.

By the close of the 1960s, von Braun had taken a hand in one significant initiative after another in the Huntsville area. They ranged from promoting peaceful desegregation to lobbying for slum clearance, from pushing for equal job opportunities to attracting new aerospace industries, from encouraging economic diversity to aggressively criticising Alabama's 'Jim Crow' (segregationist) practices and narrow-minded politicians.

Redstone by 1969 was home to multiple Army missile-related agencies, other defence organisations, and several major on-base rocket propulsion R&D contractors – all in addition to NASA's Marshall Center. All could trace their ancestry directly to the coming of the band of German and American missile experts two decades earlier. It had become arguably the world's pre-eminent rocket development centre – civil and military.

The 4-square-mile town of 1950 had become a sprawling Huntsville of more than 100 square miles by 1969. The number of its residents mushroomed from 15,000 to almost 135,000. It was a southern 'Silicon Hollow' in the making, with expanses of computer hardware, systems software and other technology-based industries, an education-rich population, and the busiest public library system in the state. Modern Huntsville had become a city of brainpower – the city that Wernher von Braun made possible. One remembers the epitaph of Sir Christopher Wren (1632–1723), the great architect-designer of St Paul's Cathedral, London. It is carved at his simple burial place within the cathedral: 'If you seek a monument, look about you.'[3]

Von Braun had somehow made time to pursue high-priority community interests through two busy decades. The city's business leadership in 1958 had named him as the first recipient of the annual Distinguished Service Award. He was just getting started. In the early 1960s, he and a committee of associates joined forces with city leaders to enhance graduate education in science and technology at the struggling University of Alabama in Huntsville (UAH). Several of the rocket experts – German- and American-born – taught there for years as adjunct professors in their hours off.

The city's aerospace organisations had found it difficult to attract and retain enough bright young engineers and scientists, because the community had no university where they could advance their capabilities and careers by working on master's degrees and doctorates. The Alabama Agricultural & Mechanical College (later university), high on a hill just outside the city, was a historically black state institution in that era of *de jure* racial segregation in the South. It did not offer the necessary technical courses and advanced-degree programmes, and its leadership showed little interest in moving in those directions. The same held true for the community's private Oakwood College, a four-year, largely liberal arts, international Seventh-Day Adventist institution for blacks. Both institutions' position on the matter had changed beneficially within a few years.

After the 1960 start of construction of a permanent UAH campus, and hard on the heels of President John F. Kennedy's lunar challenge, von Braun had spoken in June 1961 to a joint session of the Alabama Legislature. He explained Huntsville's need for advanced-degree programmes and asked for approval of a $3-million bond issue for a UAH-based research centre. He noted that space was already big business in Alabama and stressed how this action would benefit the entire state.

'Let's be honest with ourselves,' he had told the legislators. 'It's not water, or real estate, or labour, or cheap taxes that brings industry to a state or city. It's brainpower.' It was the educational triangle of Boston University, Harvard and MIT (the Massachussetts Institute of Technology), 'not beans', that had brought the great electronics boom to the Boston area. 'For a $3-million investment now, I promise you that you'll reap billions, easily billions.'

Before the speech, von Braun and his assistants had been anxious about the reception he would find at the state Capitol. 'This was just sixteen years after the big war was over, and there were some members of the legislature who had fought in Germany and may not have had a great deal of fondness for him,' remembered Foster A. Haley, a senior MSFC public affairs officer.[4]

Using slides, humour, and the basic speech that Haley had written for him, von Braun turned on his charm. He received a standing ovation. 'He went down there,' Haley recalled, 'and he just wowed them.' The legislature passed the measure unanimously. In the Alabama statehouse, it was tough to get unanimous consent on Mother's Day proclamations, let alone a piece of funding legislation. The state's voters later that year approved the bond issue by a wide margin. A year or so after that, Haley was with von Braun as they drove past the construction site. 'Don't you wish you had asked for $10 million?' Haley asked. 'You'd have gotten it.' Von Braun could only agree on both counts.[5]

The $3-million University of Alabama Research Institute opened in 1962, strengthening the university's graduate teaching and research programmes. The institute's first director was Rudolf Hermann, von Braun's chief aerodynamicist from the Peene-münde days. UAH became a rarity among universities in having master's and doctoral programmes in several technical disciplines before it offered undergraduate degrees in those fields. By 1969–70, UAH had matured from a fledgling 1950 extension centre of the mother campus at Tuscaloosa to a flourishing, 5,000-student, autonomous campus that offered advanced degrees in science and technology as well as the humanities. The impact of von Braun and his native-German-led rocket team on the college prompted one of the enrolled students to propose in fun that UAH's name be changed to 'the University of Bavaria'. After all, the student reasoned, 'If it had not been for the Germans, there would be no college here.'[6]

In 1963 von Braun did an encore performance before the Alabama Legislature. This time he asked the lawmakers for almost $2 million to create the Alabama Space and Rocket Center in Huntsville as an educational facility, visitor attraction and showcase for the space programme and Army missile R&D activities. 'When von Braun spoke, you could have heard a pin drop,' then-State Representative Harry Pennington of Huntsville and Madison County recalled. 'Everyone was hanging on his every word. . . . The bond issue was approved without a single dissenting vote. We should have asked for $20 million!'[7] Voters ratified the measure by more than three to one.

The resultant 'space museum', as it was called, would open in 1970 on acreage the Army sliced off Redstone Arsenal, with Pentagon and congressional approval. Von Braun helped obtain many of the needed aerospace artefacts. He finessed acquisition of the star exhibit for Rocket Park – a leftover full-size Saturn V.[8] He arranged an official 'transportation test' of the segmented super-rocket via flatbed carriers from the Marshall Center to the site. And there it simply stayed, reclining its full length of more than a sports field. (The busy, renamed US Space and Rocket Center, which houses the Wernher von Braun Library and Archives among its many other facilities, would continue to draw some 400,000 paying visitors a year into the twenty-first century.)

Von Braun was also behind the original Space Camp. He was strolling through Rocket Park with Executive Director Edward O. Buckbee of the Huntsville aerospace visitor centre and museum when he noticed a teacher with a group of children taking notes on what they were learning. A lightbulb came on. 'You know,' von Braun mused, 'we have all these camps for youngsters in this country – band camps and cheerleader camps and football camps. Why don't we have a science camp? It would give boys and girls a more involved experience with science, technology, and the principles of rocketry and space flight.'[9] US Space Camp would materialise in 1982. The series of week-long camps was an instant smash hit.[10]

Also in the 1960s, von Braun let such major space contractors as Lockheed, Boeing, IBM, Douglas Aircraft, McDonnell and Rocketdyne know that it would be to their advantage to locate branches of their engineering operations in Huntsville. Most took the hint. This step put the big companies close to their customer and the purse of billions. This allowed face-to-face dealings between von Braun's people and their Project Apollo contractors without blowing the Marshall Center's travel budget on constant trips to the West Coast and elsewhere.

Von Braun had done what he could to help local aerospace companies too. One was Brown Engineering, headed by former cotton trader Milton K. Cummings. Brown's wealthy CEO was a master at building friendly relationships with von Braun and key members of his team – and at making sure that his friends in Congress and Democratic administrations helped the Marshall Center at budget time. At a meeting of MSFC officials one day, the agenda included awarding a modest $1.5 million contract for a piece of work that any one of the centre's technical contractors could handle.

'Von Braun brought this up,' recalled Haley. 'He said, "Look, as long as we've got this work here and anybody can do it, let's keep the money locally."' In what several associates said reflected von Braun's philosophy of 'take care of the little guy', he noted that Cummings needed contracts for Brown, '"so let's give this million and a half to him",' Haley remembered. Almost everybody at the table nodded their heads in agreement.

But Marshall's deputy [director] for administration spoke up and said, 'Well, wouldn't it be easier if we tack this onto a contract we already have with North American?' And von Braun said: 'Yes, but most of our money goes to California. Let's keep what we can locally.' Then he went through the whole rigmarole again and everybody nodded, 'Let's do it.' But then a guy from contracting spoke up: 'Well, I don't know. It's just too much trouble to set up a separate contract for a million-and-a-half dollars.' Von Braun went through the whole thing again and got everybody to nod their heads, yeah, yeah, yeah [for awarding it to Brown].

At last, remembered Haley, an out-of-patience von Braun settled the matter: 'Well, goddammit, let's do it!' And that was that.[11]

Heeding a cue from von Braun, Cummings and his Brown Engineering president, Joseph C. Moquin, had taken the first steps in 1961 to create a research park on vacant land bordering the new UA-Huntsville campus. The company bought 150 acres, in part for a new plant complex for itself and the rest for resale at cost to other contractors. The ever-expanding site became the location favoured by most of the big aerospace companies that von Braun gently persuaded to build facilities in the Rocket City, and by local industries that he and Army missile agencies at Redstone helped sustain with contracts. (By 2005, Cummings Research Park would rank as the second largest such development in the United States and fourth largest in the world; it now sprawls over more than 5,000 acres, its scores of plants, labs and offices employing in excess of 20,000 people.)

Also in the mid-1960s, von Braun had joined the civil rights fray in Alabama and the South. The rocket scientist spoke out in support of school desegregation, voter rights and equal employment opportunities. Much of the motivation clearly stemmed from NASA headquarters' wishes and the need to improve Alabama's poor race-relations reputation, so that the Marshall Center and its contractors in the state could attract and keep top technical and management talent.

Von Braun had supported the campaign to abolish Alabama's literacy tests and poll taxes, which discouraged many blacks, along with poor whites, from becoming voters. 'All these regulatory barriers form a Berlin Wall around the ballot box,' he declared. 'They discourage the qualified as well as the unqualified from registering to vote. It's a little ridiculous to give a literacy test to a person with a college degree. Our schools aren't that bad!'[12] Years later, Fred S. Schultz of General Electric's Space Division reminded von Braun, 'Your brave address at the Huntsville Chamber of Commerce annual meeting in . . . 1964 gave the Association of Huntsville Area Companies [Contractors] the backing it needed to launch a successful drive for equal opportunity for all our citizens.'[13]

In another speech during the civil rights movement, von Braun had taken a swipe at feisty Alabama Governor George C. Wallace and other white-supremacist officials' defiant attitudes towards federal authority. The space leader jabbed, 'In the vernacular, it is okay to disagree, but I wonder if we have to be so disagreeable?'[14]

Segregation, racial discrimination, denial of civil rights – all were wrongs that von Braun had both spoken out against and taken action to stop in Alabama in the 1960s, 'way before there was anything like an Equal Employment Opportunity Office', recalled Guy B. Jackson, a von Braun speechwriter, MSFC public affairs specialist and Baptist lay leader. Von Braun pressed for hiring more blacks and other minorities at the Marshall Center and at contractors' plants in the area, Jackson said. He added: 'Dr. von Braun was genuinely concerned about minorities. He tried to find jobs for them, and when [a minority] quit the program or dropped out without explanation, von Braun would find out and want to know why, what had happened.'[15]

Was the former 'Nazi rocket scientist's' outspokenness and proactive steps against segregation motivated solely by marching orders from his NASA boss and by von Braun's self-interest as the Marshall Center's director, as commonly assumed? The record shows that he took certain anti-segregation steps before NASA chief Jim Webb's directives came down from Washington. But how did von Braun, the man, feel towards apartheid, southern style? Did he act on moral grounds as well? James Shepherd, the facilities manager at MSFC in those days, recalled that von Braun 'was quite concerned about segregation, and he was pretty outspoken'. Shepherd remembered a private moment when von Braun shared his thoughts on his special relationship to the subject of racial injustice.

Harry Gorman [MSFC deputy director for administration] and von Braun and myself were in Washington, and we had an hour or so to kill. . . . We were in a lounge talking. It was during the height of the civil rights unrest. Harry was concerned that von Braun was going to have a [Ku Klux Klan] cross burned in his yard. He was cautioned against that. It was a serious thing . . . and there were the problems with his different nationality and background.

Nevertheless, von Braun said that when he landed here [in America] people asked him, 'Where were you? What did you do [during] the persecutions? What did you do when people were disappearing?' He said to us, 'That's not going to happen again. . . . I am not going to sit quiet on a major issue like segregation.'

And he didn't shut up. . . . You sensed he felt like 'the heck with the crosses burning', that it wouldn't bother him. I tell you this from the standpoint of his sensitivity [to the issue]. . . . He had been through a lot, and he knew what silence meant and the deep depression when they were aware of something [going on] and never spoke up.[16]

In the mid-1960s von Braun had figured in an entirely different contribution to his adopted country, remembered Randy Clinton, the defence intelligence official based at Redstone. As the Vietnam War escalated, the US military prepared to bomb Hanoi and needed someone on the scene to provide information on 'the functions of certain facilities and the reaction' there to the raids, Clinton recalled. A Japanese businessman who travelled to North Vietnam and knew his way around Hanoi agreed reluctantly to fill that role, on one condition. An admirer of Wernher von Braun, he said he would do it for the rocket scientist's autograph dedicated to him, aside from other compensation.

US military intelligence agents speculated that the man believed he had set an impossible condition and might thus safely dodge the covert assignment. Officials turned to Clinton's foreign missile intelligence arm at Redstone, just down the street from the MSFC where von Braun worked. Clinton paid a call on his friend and explained the situation. He left with one Wernher von Braun autograph and inscription. He later learned that in due time the recruited Japanese businessman produced the desired intelligence reports.[17]

In the end, for all von Braun had done for his adopted hometown, state and the nation he had chosen to serve, the community at large did everything but canonise him.

After Apollo, What?

'Where do we go from the moon?' was the question, literally and figuratively, that the von Braun team and others in aerospace had begun wrestling with several years before the mid-1969 triumph of Apollo 11. The NASA-run Marshall Center's Saturn V workload and staffing levels had peaked as early as 1966–7 and then started to decline. That reflected the way of life for a centre whose primary purpose was R&D – working well ahead of launch day – versus an operations-oriented organisation such as the Kennedy Space Center.

One Apollo add-on project the Marshall team had taken on – with the blessing of its sister field centre in Houston – was managing the design, development and production of the Lunar Roving Vehicle. With Boeing as prime contractor, this 'moon buggy' was intended to further the exploratory range of the 'lunarnauts' during later Apollo missions. Von Braun regarded it as 'a full-fledged spacecraft on wheels'. The 'lunar jeep' would prove a scientific and public relations boon.

According to senior Marshall official James 'Jim' Shepherd, however, it was a diversionary bone tossed to Marshall by the Manned Spacecraft Center to keep the Huntsville centre out of lunar soil analysis and other science programmes Houston wanted all for itself. Shepherd was present for the apparently inadvertent revelation by Houston's Chief Engineer Max Faget in a meeting in the late 1960s attended by von Braun and Texas centre leaders Bob Gilruth and George Low. The meeting, on roles and missions and 'the division of labour' between the two rival centres regarding Skylab and other post-Apollo projects, 'sort of came unglued' after Faget's slip, recalled Shepherd.[1] Ernst Stuhlinger, who was later briefed by von Braun about the meeting, called it 'typical of the strained atmosphere between Houston and Huntsville'.[2]

'I don't remember that' episode out of the 'many meetings' between the two centres, Faget commented from the Houston area more than three decades later. He did observe that although 'most of the time they [Huntsville and Houston] got along pretty good', some at the Manned Spacecraft Center thought that many at Marshall 'were jealous of our role [as] the centre of attention because of the astronauts'. More than a few at Houston thought that 'Marshall's role was just to help us', Faget said.[3] Boeing-Huntsville's chief engineer for the Lunar Roving Vehicle, Gene Cowart, remembered how a team leader of a Marshall Center engineering contingent that was leaving for a meeting at NASA-Houston cautioned the group, 'Remember now, we're heading into enemy territory.'[4]

All of that aside, much more than a moon buggy was needed to keep the Huntsville centre alive and productive as Apollo wound down. In looking ahead, von Braun and others conceived a follow-on project that came to be known as the Apollo Applications Program. It would apply the flight hardware and expertise developed in the moon race to different, future purposes – and would preserve their precious Saturn–Apollo production lines.

One major piece of forward planning was conducted during a private weekend retreat arranged by von Braun and his staff at a hideaway lodge on Lake Guntersville in northern Alabama that included George Mueller, head of NASA's Office of Manned Space Flight. (Two of the successful projects to have emerged from such planning sessions were Skylab, the first-generation US orbital space station, and the joint United States–USSR Apollo–Soyuz manned flight in Earth orbit.) This post-Apollo brainstorming also produced some of the early planning of the Space Transportation System, more commonly known as the Space Shuttle programme.[5]

In September 1969, with one manned lunar landing accomplished and another due soon, there was already talk of dropping several of the nine planned additional flights. The thinking of some in Washington boiled down to this: the science aside, how many times must we prove that we beat the Russians to the moon? The Vietnam War and domestic problems were creating severe pressures on Congress to cut back on space spending.

That month the Space Task Group, headed by Vice President Spiro T. Agnew, came out with its positive report, *The Post-Apollo Space Program: Directions for the Future*. The report called for a bold programme capitalising on Saturn–Apollo hardware, experience and momentum. It included a manned landing on Mars. Von Braun was quick to speak up in support of the report, as were NASA administrator Tom Paine and Mueller.

What else could these men do to make the proposal a reality? That same month, Paine approached von Braun with a confidential, qualified offer: leave the Marshall Center, transfer to Washington and become head of planning for NASA. He would develop, with Paine's support, solid and detailed plans for near- and far-term programmes, including the manned Mars mission. He would also lead a crusade to sell this twenty-year package to the White House, Congress and the nation, because only he, Wernher von Braun, could do it. (Mueller had also encouraged von Braun to accept the new challenge – and may have been the first to broach the idea with him.)[6]

Von Braun later recalled that he was prepared to accept the post then and there. But he said the NASA boss needed to check around and get back to him. That is not exactly how Paine remembered it years later. He recalled that von Braun at first 'was somewhat on the fence'. His initial reaction was that while it was nice to be asked, he loved his job at Marshall and he would have to think long and hard about moving to Washington. Meanwhile, Paine – who had reasoned that he should first check on von Braun's level of interest before proceeding – would test the water in the necessary high political circles to gauge reactions; von Braun intended to say nothing.[7]

The 57-year-old von Braun weighed many factors in considering his decision, friends and colleagues remembered years later. 'Both Tom Paine and George Low [NASA deputy administrator] wanted Wernher up there,' recalled former von Braun aide Frank Williams. 'He was going to be the visionary, the future activities planner, the promoter. They even used the term "Mr Outside" [for him], because he could mesmerise Congress.'[8]

But von Braun had conflicting feelings about going to Washington, said Konrad Dannenberg, a long-time member of the rocket group. On the one hand, he hated to leave his old team. On the other, he knew Paine would push as hard as he could for the Mars mission. 'Wernher always liked to talk about "future projects",' Dannenberg remembered. 'That was his baby.'[9]

There were suggestions that von Braun felt forced to accept. Werner Dahm disagreed: 'It was an invitation. [Paine] couldn't command him.' Dahm said he wanted to believe that mutual trust and respect existed between von Braun and Paine. And yet, he added, 'I also heard that this whole thing had been a political arrangement. As center director, he was unassailable. In Washington, he [would take] a step back. There were people who were tired of being in his shadow, and they had arranged this. . . . The story . . . was that an American colleague here [at the Marshall Center] was talking to a colleague from Houston, who explained, "It was arranged for [von Braun], to get rid of him – an agreement",'[10] Later events would give the conspiracy theory a ring of truth. Jay Foster, an inner-circle figure at the Marshall Center and later in Washington, remembered von Braun thinking mainly in these terms:

From his perspective, one of the big reasons he wanted to leave Marshall was to have time to think and dream about the future. It's really hard to find time to do that when you're running a multi-thousand-man organisation with a 2- or 3-billion-dollar-a-year budget, and with all these people wanting to see you and spend time with you and get on your calendar. So his driving influence, other than the fact that Tom Paine really wanted him to go up there, was to be able to have more free time to dream and think about the future.[11]

There was another, perhaps crucial influence on von Braun's acceptance of Paine's offer: his 40-ish, aristocratic wife, Maria. Von Braun gave great weight to 'her desire to get out of this little town [Huntsville] and into [Washington] society and the arts', said Lee B. James, adding, 'She never did like Huntsville. He was a big man in a little town, and she wanted him to be a big man in a big town. Maria once said to me, "So-and-so is going to Washington. Maybe we will someday." . . . I think that bothered von Braun, because he wanted to make her happy, of course. . . . I think he was quite happy in Huntsville. With von Braun, happiness was in his work. He was doing here what he wanted to do – and what he couldn't have gotten done most anyplace else.'[12]

Dorette Kersten Schlidt, whose ties with von Braun dated back to her days as his secretary at Peenemünde, agreed about Maria's influence: 'Mrs von Braun was very happy to leave for Washington. You know, she is a big-city girl. She grew up on that family estate in Pomerania, but then her parents left for Berlin and she lived there. Huntsville was a small city. She was so well known here [in Huntsville], and she was so shy.' Averse to any public attention, 'she wanted her privacy, and in the big city she could have that'. In Schlidt's judgement, Maria 'pushed him to go'.[13]

Some believed that Maria von Braun, by all accounts a devoted mother, thought that the educational and cultural advantages of the Washington area would be far superior for their children, then aged 9, 17 and 20. She also believed that her husband's long workdays – and workweeks – might lessen in a new role in the nation's capital. 'She was excited about it because of the availability of the arts – the museums and the concerts and all that,' remembered von Braun assistant Tom Shaner.[14]

Paine's recollections leave no doubt where von Braun's wife stood on the matter. 'I had a great ally: Maria,' Paine told an interviewer years later. 'She wanted to come up to Washington. She had been down there [in Huntsville] long enough.'[15] And yet no one should make the mistake of thinking that she did anything that she felt was not in her husband's best interests. According to family friend and attorney Patrick Richardson, 'Maria always was devoted to Wernher. She would walk through fire for him'.[16]

Along with everything else, von Braun considered the questions of historical and political timing. Apollo had been almost magical. Could the magic be recaptured for Project Mars? The space pioneer thought it over. Ultimately, his loyalty to Paine – and his wife's strong feelings – won the day. Von Braun felt a natural rapport with the NASA chief. Scientist Paine and the rocket leader shared, among other interests, an affinity for the sea. The son of a retired US Navy commodore, Paine had served as a Navy submarine officer and was once a deep-sea diver. Von Braun let the NASA administrator know his answer would be yes. 'I indicated readiness [for the move] to Dr. Paine then if he felt it necessary,' von Braun was later to explain.[17]

Then, in November 1969, came Apollo 12, commanded by Navy flyer and live-wire Ivy Leaguer Charles 'Pete' Conrad Jr. With the second moon-landing mission safely in the books, von Braun left with his family in early December for an eight-week holiday in the

US Virgin Islands and the Bahamas. There he planned to teach his daughters to scuba dive. It would be by far his longest break from work since his student days. Any announcement of a move to Washington would await his return from the Caribbean and another talk with Paine, or so von Braun thought.

In late January 1970 the confidential plan fell apart. With the von Braun family still away on holiday, word of the agreement was leaked to the press – evidently from within NASA headquarters, according to several observers. Unconfirmed news reports revealed that the rocket scientist was leaving his old team and heading for Washington. 'Not only no, but hell no!' responded von Braun's feisty Public Affairs chief at the Marshall Center, Bart Slattery, to enquiring reporters – before he learned the reports were indeed true.

'It was not supposed to have been announced until he got back,' confirmed Shaner. The plan was for von Braun to finish his holiday and then go to NASA headquarters, where Paine would introduce him at a news conference and they would jointly announce he was moving to headquarters with White House approval. 'But you can't keep secrets at headquarters,' observed Shaner. 'It got out from some of the underlings, and it messed up the plan.'[18]

Some von Braun allies suspected the leak – by parties who were less than friendly towards him – was intended to make sure the transfer came about. It precluded any possible change of heart by von Braun. His long-time secretary Bonnie Holmes – who reflected, 'I don't think he was really all that happy about moving' – and others deplored what they called the 'ignominious' chain of events that ensued.[19]

With von Braun believed to be somewhere in Florida and unreachable, Paine decided on immediate action for damage control. He flew to Huntsville on Tuesday 27 January, and confirmed to a hasty assembly of stunned Marshall Center employees that their leader would be departing soon for duties in the capital as NASA deputy associate administrator for planning, a new position. 'There was a sense of disbelief at Marshall that he was leaving,' recalled MSFC engineer-manager Bob Schwinghamer.[20] Paine announced that the move would be effective as of 1 March, praised von Braun, noted that President Nixon had been informed and proudly pointed out that their departing director would get a pay rise with this promotion – from $33,490 to $36,000 a year. Paine also announced deputy director Eberhard Rees's appointment as von Braun's successor at the Marshall Center, and added, 'We face a period of austere budgets.'

'There was nothing sinister or suspicious about it,' Shaner contended, 'because it had been talked about for a long time. . . . [Von Braun] told me that his only reluctance to go was because of his German colleagues.'[21] Other more senior MSFC officials felt he was simply leaving sooner than he had anticipated. Von Braun had earlier let Paine know that he wanted Rees to succeed him at Marshall. That would provide a measure of job security for the ageing ex-Peenemünders, at least for a while.

A great awkwardness surrounded Paine's pronouncement. The 6,000 government employees of the Marshall Space Flight Center – their number already down from a Saturn–Apollo peak of nearly 8,000 – had heard the news from someone other than their leader. Von Braun was nowhere in sight. He had made no public statement. Paine said the plan had been for von Braun to be present for the announcement, but we've had the Florida State Police out looking for him', the NASA administrator said to nervous laughter from the auditorium audience. 'I might add that we were unsuccessful.' The next day, Paine reported von Braun had been located in Palm Beach Shores later that Tuesday afternoon, 'when he was good and ready to be found'.

Von Braun, declining to take telephone calls at his Florida hotel from reporters (even from hometown newspapers), quietly flew back to Huntsville with his family on the evening of 28 January. He stayed in seclusion, ostensibly to enjoy the remaining few days of his long holiday. In fact, insiders later confided, he was not yet prepared to face the members of his rocket team, including his loyal German-born colleagues of decades.

Next, a press statement was issued in his name. 'Leaving my Marshall home and teammates of many years was no easy matter,' von Braun emphasised. But, he added, 'now is the time to con-solidate and utilize the gains that our past successes in the space program have won us'. He expressed his gratitude and 'tremendous enthusiasm' for the 'immense opportunity' to work with Paine and other leaders at NASA headquarters 'to develop a space program worthy of the 'Seventies and Eighties'. Von Braun also stressed in the release: 'While it would be most difficult to deliberately terminate relationships that go back, in some cases, nearly four decades, I am not really being called on to do that. I will continue to work with my colleagues here, and I'll be back in Huntsville time and time and time again.'

At last he stood in the flesh before his Marshall Center co-workers on his first day back at work in 1970, 2 February. In one month he would be gone, separated for the first time from the team, both the old and the new incarnations. He felt like 'a navy skipper leaving his ship – but it's not a sinking ship!' he quickly stressed to the gathered employees. 'I'm leaving Marshall with nostalgia. I have my heart in Marshall. I love this place. I helped build it up. I feel I'm a part of it,' von Braun said, his voice filled with emotion. And yet it was now time for him to move on, he added, to do some essential planning for the decades ahead, at the urging of his boss, the administrator. 'I've spent ten years here doing what was urgent, and regrettably not doing what was essential,' he explained.[22]

With his audience hanging on his every word, von Braun gave his account of how recent events had unfolded. Not long after Apollo 11, Paine had begun discussing the Washington job with him. 'I indicated readiness [to accept] to Dr Paine then, if he felt it necessary,' he said. But Paine had said he wanted time to check out the idea politically. So von Braun left for his long family holiday, he said, confident nothing would happen before his return.

But toward the end of his holiday he began getting calls alerting him to the fact that the word was out. 'I did not know if the way was greased. I did not know if it meant a green light or a wave-off. I regret the announcement that I was going to Washington came from outside sources,' he apologised to the hushed space workers. 'It wasn't planned that way.' Appearing to struggle with his emotions, von Braun told the Marshall employees he wanted 'to remove any doubts that the transfer is for real'. Although it was a difficult decision to reach, he said, it was made easier by knowing 'that a man like Eberhard [Rees] will be at the reins'.

Von Braun went on to address rumours that he was being groomed to become NASA administrator and that he had even been offered that job in years past. Not true, he said. As the auditorium grew even quieter, he explained that as a 'former enemy' of the United States in war, he considered himself a political liability for such a high position and would not have accepted had it been offered. 'There are far more effective men for that job,' von Braun said softly. 'Far better men . . . Dr Paine, for one.'

Von Braun suggested that he might have seven or eight more years of useful government service left. He said he would not mind spending all that time in NASA planning activities, while keeping alive 'the spirit of Apollo'.

He had emerged from his long holiday in the tropics wearing a full beard. Why? 'For once in my life,' he joked to reporters, 'I had the urge to look like a real scientist. But a couple of friends . . . said the beard wasn't enough. They said what I really needed was a pipe, tennis shoes, and a pair of blue jeans, preferably dirty.' He noted an odd thing about beards: 'It's a sign of youth these days, isn't it? It once was a sign of age!'

With his lush salt-and-pepper beard, deep tan, and robust looks, the scientist appeared the picture of perfect health. Very few knew that a while back he had undergone a routine medical exam at a Texas oncologist friend's clinic in which polyps in his colon were detected. The friend, Dr James R. Maxfield Jr, wanted to operate right away to remove them. Dealing with the initial lunar flights, von Braun faced an exceptionally busy schedule. For that and other reasons he decided the surgery could wait. After all, he had never suffered any serious health problems. He would just have the growths excised later.[23]

A week before his move to Washington, the people of the Huntsville area observed 'Wernher von Braun Day' on 24 February. They honoured the man who had been their neighbour and benefactor for what he described as 'twenty pleasant, happy years'. The focal point of the occasion was the city's historic centre, where Alabama had entered the Union in 1819. The rainy-day events included a parade, presentation of gifts, the unveiling of a permanent stone monument to him on the Courthouse Square and speeches by both of Alabama's US senators. Everyone put the best possible face on what was for most a sad, worried day of saying *Auf Wiedersehen* to the leader who had put the now-fortunate city on the map.

'This is a happy occasion, because a "local boy" has made good,' a smiling Governor Albert P. Brewer told the crowd. 'The people of Alabama truly have a great pride in this man. He represents the success of the United States in space exploration – in all of our space efforts.'

Von Braun's former commander at Redstone Arsenal, Maj-Gen John Bruce Medaris, US Army (Ret.), had come from Florida to take part in the farewell tribute. 'I'm glad that somebody had the good sense to put Wernher someplace where he can do some selling!' cheered Medaris. 'Space needs selling, and I think he's the one to do it.'

The then-chief of the US Army Missile Command, Maj-Gen Edwin Donley, gave von Braun a portrait of one of his earlier Redstone commanders, the man who had brought the 'paperclip' rocket team to America, and later to Huntsville, the late Maj-Gen Ludy Toftoy.

The Huntsville City Council president was chemist Kenneth Johnson, who had a big surprise gift for von Braun. He presented the rocket scientist with a copy of a council resolution approved earlier that day that named the city's forthcoming $15-million cultural, entertainment and convention complex the Von Braun Civic Center, in recognition of 'your interest in and impact on the total community'. Von Braun thanked the City Council and, noting the centre's planned spacious concert hall, quipped that he was a 'frustrated pianist'. In an aside to the council president, he joked, 'Someday I'll come back and play in it!'[24] A large sign stretched behind the speakers' platform said it all for the city, including the plaintive hope that he would some day return.

Dr. WERNHER von BRAUN: HUNTSVILLE'S FIRST CITIZEN . . . ON LOAN to Washington, D.C.

The inscription on the city-centre monument dedicated that day closes with the words: 'Dr. von Braun, whose vision and knowledge made possible the landing of the first man on the Moon by the United States, contributed significantly to the life of this community. He will forever be respected and admired by his local fellow citizens.' In truth, not all his fellow townspeople agreed with those sentiments or with the naming of the civic centre after him. More than a few members of the city's prominent Jewish community, for example, had not forgotten, or quite forgiven, his past in Nazi Germany. Three decades later, that remained the case.

An emotional farewell speech by a 'deeply touched' von Braun was the climax to the public ceremonies. With his family and some fifty members of his original German missile team seated nearby, he paid tribute to the people of the area. 'All this wonderful success here on Earth, and our flights to the moon as well, would not have been possible without your steadfast and heartfelt support. . . . I spent here the finest days of my life.' The space leader noted that the Apollo 11 triumph would be 'a hard act to follow'.

He went on to recall earlier scenes at this same Courthouse Square, and then offered a dose of realism tinged with characteristic optimism. 'My friends, there was dancing in the streets of Huntsville when our first satellite orbited the Earth. There was dancing again when the first Americans landed on the moon. There is only one moon. I'm afraid we can't offer any such spectaculars like that for some years to come. But I'd like to ask you, don't hang up your dancing slippers.'[25]

After the Rocket City's send-off, events culminated with a big going-away party out at Redstone. A reception and banquet that evening drew about a thousand of his Marshall Center associates, friends and their family members. The mood was less than upbeat, especially among his German-born colleagues. They could see the end of the line for the old team.

'It was presumed then that he was being groomed for NASA administrator', his denials notwithstanding, Schwinghamer recalled. In the receiving line von Braun shook his hand and said he hoped to see him in the future. 'Tears were running down his cheeks,' the senior MSFC engineer-manager remembered. 'They didn't look like tears of joy. He didn't look like he wanted to go.'[26]

Scientist John D. Hilchey had joined the Marshall Center in the early 1960s filled with what he termed a 'hero worship' that drove him 'to be a part of the "von Braun team"'. That feeling had turned into 'profound respect', he told the rocket pioneer in 1972. Furthermore, he reflected that since his leader's departure from the Marshall Center, 'it's like trying to "carry on" at Camelot without King Arthur'.[27]

DC and the Gods

The story of Wernher von Braun 'reminds me of a Greek drama that deals with the fights between the gods', mused Peter Petroff in June 1999. The Bulgarian-born master engineer and multi-millionaire entrepreneur, then in his 80s, added: 'Zeus, the big-shot god, gives selected people just enormous capabilities. Then the lesser gods are jealous and throw up all these hindrances.' Petroff, whom von Braun recruited to work at NASA's Marshall Space Flight Center in the mid-1960s, continued with his analogy: 'Zeus created a guy, von Braun, who was way, way above everybody else. The gods gave him too much talent. He was not only a top-notch engineer and physicist, but he was a visionary. On top of that, he could convince anybody of anything.'[1]

Then, on 1 March 1970, von Braun, at the apogee of the Apollo 11 and 12 moon-landing triumphs, left his impregnable home base of twenty years and moved north to take on a challenging position at NASA headquarters in Washington.

'This was a tragedy,' Petroff observed years afterwards. 'Little guys, from jealousies or for fun, messed up what [von Braun] was doing, a typical Greek drama. This was tragic.'[2]

The downfall was yet to come. Full of hope and purpose, von Braun moved into his new offices – two doors from administrator Tom Paine's – at the space agency's central command at 400 Maryland Avenue. As deputy associate administrator for planning, he nominally ranked fourth in the agency's hierarchy, yet he was outside the chain of command. His task was to develop the grand plan for US space activities for the next twenty years or more, map a master campaign for selling it to the White House, Congress, the scientific community and the American people – and then go out and implement it.

His first step in gathering a planning staff of twenty to thirty was to line up two trusted lieutenants in Huntsville. Frank Williams, NASA engineer-manager and former Air Force officer, had been assistant to the director at the Marshall Center in the early 1960s; Jay Foster was a technical management specialist at the MSFC. Another valuable ally assigned to him was James L. 'Jim' Daniels Jr, a former MSFC management associate who had been with the NASA headquarters Executive Secretariat for two years when von Braun arrived.

Also helping with the transition were young engineer Tom Shaner, his last assistant at the Marshall Center, who drove von Braun's cherished, late-model, deluxe Mercedes saloon to Washington,[3] and Bonnie Holmes, von Braun's secretary of nearly two decades at Huntsville. Holmes temporarily moved to the capital to help her departing boss get settled in his new job, screen secretarial candidates, and select the best. That proved to be Julia E. 'Julie' Kertes, a professional career woman experienced in working with high-level aerospace and defence executives in Washington and elsewhere.

Certain workaday realities at NASA headquarters paralleled those at the Marshall Center. That meant tons of fan mail, piles of requests for speeches, much work over long hours – and much laughter, at least for a time. Nina Scrivener, secretary to George Low, the deputy administrator, had an office next to von Braun's in the top-floor executive suite. 'She could hear the loud laughter next door,' recalled Kertes. 'She often remarked how wonderful it must be to work for someone with Dr. von Braun's sense of humor. We were the envy of the suite.'[4]

For his first few months in Washington, the flood of post was so heavy and the telephone calls so incessant that Kertes could hardly get any work done, she remembered. 'Our mailroom said he received more mail than the other four [top NASA] executives combined!'[5] Well after the first few Apollo lunar landings, von Braun continued to be 'inundated with requests for his autograph', Kertes recalled. 'If he had signed everything sent to him, he would never have accomplished anything else.' So she had to sign his name with a stylus on half or more of the stuff flooding in.[6] All of this attention on von Braun did not delight everyone at headquarters.

Before being offered the post at NASA's headquarters in Washington, von Braun had not inspired affection among some working there – and certainly not from his NASA colleagues in Texas – with one of his stronger displays of bravado. He had lobbied the agency for practically an exclusive role for the Marshall Center in managing the development of the proposed Space Shuttle. Each centre director was to present a proposal to headquarters on how the shuttle programme ought to be run and which responsibilities should go where. The main office had given guidelines and directions to each of the field centres.

'They were telling Marshall, "Now, you ought to propose this, this, and that",' recalled Shaner. 'Of course, von Braun listened to all that and took it all in and had his meetings with his top guys at Marshall, and they discussed it all. He was going to be the man to pitch it . . . and everybody thought they knew what he was going to do.' Shaner continued, 'He went up to headquarters, and I mean he proposed that Marshall . . . manage the total programme – everything, lock, stock, and barrel, even the [programme] management, [rather than] headquarters. [Marshall was] going to develop the orbiter. Marshall was going to train the astronauts . . . going to do it all. And politically, that didn't go over well at all. But he would always go for the whole enchilada on everything.'

'That didn't endear him to Chris Kraft out at Houston, and . . . others. There was a lot of animosity towards Dr von Braun – from headquarters, from other field centres. They were extremely jealous of him, number one, because of his tremendous intellect, and, number two, because the man was famous, and most people in the country thought that he was the head of NASA.'[7]

But things might be all right now. After all, von Braun was in Washington at the invitation of the real administrator, his friend. 'Tom Paine and Wernher von Braun', Shaner emphasised, 'had a close personal relationship and tremendous respect for each other.'[8] No one doubted that.

Getting resettled in Washington also meant finding a new home. The von Brauns had sold their late-1950s, split-level ranch house on a Huntsville hillside to the State of Alabama, which envisaged it as a kind of annexe to the new Alabama Space and Rocket Center a few miles across town. When those plans fell through, the house was sold at public auction to a married couple, George and Pam Philyaw, coincidentally both aerospace professionals and licensed pilots. Moving in, they found a note from Maria von Braun on an interior wall of the garage. The note was tacked beside a couple of sizeable holes punched in the plasterboard at front-bumper level. It read: 'This is where the family practiced takeoffs and landings.'[9]

In Virginia, the von Brauns had found a spacious, comfortable home in a neighbourhood of good-sized wooded lots in Alexandria. Only Peter Constantine – 'Pete' to his sisters – was now living year-round with his parents. The Alabama-born boy was well aware of his background. One day at work, Wernher recounted to his staff: 'Well, Peter's in trouble. They got into the Civil War in his studies at school, and Peter said he's a Confederate! And so he held forth for the Confederate cause!' Von Braun laughed as he told the story, adding that, after all, Peter was an Alabamian.[10]

Iris Careen was away at Oberlin College in Ohio, and Margrit Cecile was at a girls' school in Atlanta. As their daughters grew to adolescence and young adulthood, Wernher and Maria shared the usual parental worries over peer pressure and drug use in the turbulent 1960s and '70s. At times their concerns rose above the norm: Iris was present at Kent State University in Ohio when the pivotal 1970 student protests over the Vietnam War's widening erupted on campus and ended with deadly bloodshed. She escaped injury.[11]

The von Brauns' Alexandria property had room to add a heated pool and small domed observatory to accommodate the scientist's prized new 8in Celestron reflector telescope. It was a farewell gift from Huntsville friends and colleagues so that he could indulge his passion for star-gazing on clear nights.[12] The couple liked their new home's secluded location on a cul de sac.

The von Brauns soon began enjoying an active social life in Washington. One evening they attended an elegant dinner party at which the space scientist was seated next to popular Washington newspaper columnist Betty Beale. Wernher apparently turned on the charm. She soon informed her readers, 'One of the most fascinating men in the world has just moved to town. From now on, Wernher von Braun will be so sought-after as a dinner partner that he may wish he had stayed in Huntsville, Alabama. The rocket genius is a brilliant conversationalist, extremely handsome and socially charming. His lucid conversation covers everything from the atom to God, in whom he believes deeply, and he can make cosmic science seem perfectly clear even to a society columnist.'[13] The smitten society writer added that, while von Braun said he would delight in travelling to the moon, he admitted, 'It's better to have a live astronaut than a dead scientist.'

After settling into their new home in Alexandria, Maria von Braun got to know the neighbours, including the children. 'She loved kids,' recalled Frank Williams. 'She loved her kids, and she loved kids in general. In fact, she used to baby-sit quite a few of the neighbours' kids there, and she loved doing that.' Maria had toys for them, was 'a good storyteller', and even came up with an invention to rig her kitchen 'in Rube Goldberg [Heath-Robinson] fashion so the kids could not get into any of the kitchen cabinets', Williams remembered. 'She looked forward to keeping those kids (free of charge, of course). It was just having them and helping.'[14]

Several early events did not augur well for the success of von Braun's mission in Washington, however. George Mueller, the NASA associate administrator for Manned Space Flight, was an admirer who had encouraged him to make the move. Mueller left the space agency in December 1969. That was after the rocket scientist had committed himself to the job, but before he had made the transfer. At the same time, Low, the Apollo spacecraft programme manager at the Manned Spacecraft Center (later named the Lyndon B. Johnson Space Center) in Texas and no fan of von Braun's Huntsville operations, was promoted to NASA deputy administrator and shifted to the capital.

Also in late 1969, after the positive space report by the Agnew Commission had been circulated, Paine had talked with President Nixon about publicly calling for a manned Mars mission – von Braun's ultimate personal dream – as a future national goal. Nixon responded by announcing that landing men on Mars by 1982 would be a worthwhile objective for America. It was hardly a stirring, Kennedyesque challenge. Then, in early 1970, Paine briefed the news media on the severe impact of the 'drastic reductions' newly made by the Nixon administration in NASA's proposed budget and programmes for the next fiscal year. Nixon appeared to have forgotten all about Mars.

And so, by the time von Braun headed for Washington to draft plans for an aggressive, long-term space programme, support had been undercut at the White House and on Capitol Hill. With the continuing Vietnam War, growing budget deficits, urban unrest, campus protests and other headaches, bringing about any turn-around in space efforts would be a steep uphill struggle. Even supreme optimist von Braun knew it. But he had given his word, and maybe there was a chance that he and his friend Tom Paine could somehow pull it off. He honoured his decision to enter what he would soon call 'the Washington jungle'.[15]

Paine was indeed a bright spot as von Braun began his Washington sojourn. A close, friendly rapport between the two quickly deepened, Kertes remembered. They had almost daily, one-on-one discussions in the administrator's nearby office. Their collaboration fortified von Braun in his planning task.[16]

Among von Braun's early steps as he set out to chart America's future in space was to ask Ernst Stuhlinger, his chief scientist back at the Marshall Center, to give him a list of the top ten US scientists with interests in space experimentation. Because of his reputation as someone much more attuned to engineering than science, von Braun sought through these contacts to get input from, and stir up further interest within, the scientific community in what he saw as enormous future opportunities in space science.[17]

He also took a coast-to-coast, fact-finding tour of NASA centres with the added purpose of building support for his mission. Flying into Los Angeles one afternoon for several days of visits to area space facilities, von Braun had dinner at a restaurant with assistant Jim Daniels and two other NASA men. He then suggested the four of them go out on the town for some entertainment. One of them mentioned a nightspot that might do. Arriving there, the group entered a club featuring nude dancers. The men tried to make their way inconspicuously to a dark, secluded booth as the famous rocketeer softly cautioned, 'Now, let's not make too big a deal of this.'

About that time a loud voice rang out: 'Hey, Wernher! Wernher von Braun!' 'My God,' murmured the object of the attention. The big mouth turned out to be a colleague from headquarters. He stood up and led his party over to join von Braun and friends. The men had a few drinks, watched the sexy performers, joked and chatted.

As the hour grew late, they began leaving. Night owl von Braun said he wasn't quite ready to go and asked Daniels to stay with him. The two found stools near the stage where the dancers undulated to the music. The appreciative men carried on light talk, and some of the dancers flirted with von Braun. 'Jim, you need to watch their eyes,' he advised his younger friend. 'It's in their eyes you can see their true dedication to their art.'[18]

At one point Daniels excused himself for a visit to the men's toilets. He returned to find von Braun had taken a new seat – at a piano next to the stage. 'He was playing some jazzy stuff for the girls to dance to!' Daniels remembered. 'There were even some bar hangers-on gathered around him. Probably none of them knew who he was. He certainly wasn't drunk. He was just enjoying himself playing that piano while those gals danced.' At closing time, around 2.30 a.m., the two men left and returned to their hotel. 'Nothing else happened,' the associate recalled. 'He just wanted an entertaining night out with the boys, to let his hair down. That was "Wernher von Braun the human being", I guess you could say.'[19]

Life back in his Washington office, too, had its lighter moments. For one thing, his secretary remembered, 'he kept falling out of his chair!' Von Braun's new, executive-style chair was top-heavy and sat on a thick plastic mat on top of the carpet. He would lean back in the chair and put his feet up on his desk. Before he could catch himself, the chair would roll, topple and send him crashing to the floor.

'The first time it happened,' recalled Kertes, 'there was a loud crash in his office and I ran in. He was on the floor, thankfully unhurt, ensconced between his desk and credenza. The third or fourth time I heard this crash, I thought, "Oh, no! Not again!" It struck my funny bone and I started to laugh, but not out loud.' She could not go into his office right away for fear she would burst out laughing. Soon she heard a loud 'Julie!' She rushed in, trying not to look. All she could say was, 'I'm going to do something about that chair – today!' A new, more stable, chair arrived before the day ended.[20]

When the boss had meetings in his office with his male associates he would close the door to her office, Kertes recalled, 'because they would sometimes get carried away with their language'. One day a male member of staff brought her a 'cuss [swear] box' for the top of her desk. Thereafter, when a spirited meeting in von Braun's office broke up, the men would file out and usually drop coins in Miss Julie's box. One day, even with the door closed, 'they were unusually loud and I could hear almost every word they said. So I went into his office and placed the cuss box in the middle of the table.' Von Braun protested that the cussing didn't count if the door was closed. 'If I can hear it,' Kertes replied, 'it counts.'[21]

A different side of von Braun showed in the case of Kevin Steen, a boy from Carefree, Arizona, who had written in the late 1960s to von Braun in Huntsville about his enthusiasm for space flight. Kevin developed cancer at an early age. Von Braun responded with a letter of encouragement, and soon the two became pen pals. Not long after the scientist had transferred to Washington, his secretary received a call from Kevin's father. He wanted her to tell von Braun that the boy had undergone surgery at the Mayo Clinic but that 'his body was filled with cancer and the surgeons could do nothing', Kertes recalled.

'Dr. von Braun said [to his staff], "We must all pray for Kevin", and we did,' the secretary said. 'To everyone's amazement, a miracle happened and Kevin began to recuperate. A few months later we learned that Kevin had been back to Mayo and there was no sign of cancer in his body!' The secretary remembered well her boss's comment to the staff: 'Now we can see what the power of prayer can do.'[22]

A steady stream of interesting and famous visitors flowed through von Braun's Washington office. Two in particular caused considerable stirs. One afternoon Jacques Cousteau, the undersea explorer, paid a lengthy visit, and the fellow skin and scuba divers strengthened their friendship. The meeting ran on so long that Kertes had to run downstairs and hold a taxicab so Cousteau could get to the airport during rush hour. Departing employees of the Department of Health, Education and Welfare, which occupied the lower four floors of the NASA building, were astounded to see Cousteau there and asked why. They seemed equally surprised to learn that Wernher von Braun worked in the building.[23]

Another day, actor Hugh O'Brien came calling. The space enthusiast was there with Mike Ross of the Kennedy Space Center to work with von Braun on a documentary for television. At the time, the handsome, square-jawed star played the title role in the hit TV Western series *Wyatt Earp*. Word spread that he was waiting in the reception area, and soon NASA employees swarmed in to ogle at and meet him, take his picture and get his autograph. Julie Kertes entered her boss's office and cheerily announced, '"Dr. von Braun, Wyatt Earp is here to see you." He said, "Who?" He had no idea who Wyatt Earp was.'[24]

Not every visitor to von Braun's office was welcome. He received a stream of calls from a former US Army associate who had succumbed to the bottle and was trading on his acquaintance with the space leader for business purposes. 'Dr von Braun finally told Julie not to put the man's calls through to him anymore,' recalled Daniels. 'He told her not to lie, but just don't put him through.' Eventually, when the man simply showed up at the office one day, 'Dr von Braun went out to lunch with him – strictly out of compassion for an old associate who had become a sot.'[25]

Equally memorable was the visiting sculptress from Finland who had admired von Braun from afar. She had gone through NASA Public Affairs chief Julian Scheer and arrived at von Braun's office lugging one of her large abstract stone sculptures as a gift. 'It was this convoluted, unrecognisable piece that was too big for his desk,' remembered Daniels. 'The lady personally presented this modern masterpiece to Dr von Braun with praises for his accomplishments.' After graciously accepting the purported work of art, thanking its creator, chatting briefly and walking her to the door, he asked Daniels, 'What the hell should I do with it?' Daniels suggested it would scare away the crows from his garden at home. Von Braun decided the piece should grace the NASA building's garden, for whatever use the birds there chose to make of it.[26]

Kertes soon discovered that one of the 'finest qualities' of her dynamic boss was 'his treatment of other people. He treated everyone the same, whether it was the garage attendants or a high government official. In other words, he treated everybody like human beings.' Such qualities led her to develop a 'great admiration' for him – although admiration and respect were not always his preferred sentiments. 'I recall a letter from a young woman to him in which she said "I respect you very much." He wrote a note to me on her letter saying, "The fact that she says 'respect' instead of 'love' makes me feel old."'[27]

An easy rapport blossomed between secretary and boss. She also became close to his family, as had Bonnie Holmes in Huntsville. 'I love Mrs. von Braun; what a lady,' Kertes wrote many years later.[28] Unmarried, she occasionally stayed with Peter von Braun when his parents went out. When the couple was about to leave on a trip to Africa, the secretary asked her boss to bring her back an elephant. 'Okay,' he had agreed, 'but you'll have to pay the shipping!' He later sent her a postcard from Kenya with news of having sighted more than a hundred elephants that day. It bore a postage stamp depicting one elephant with huge tusks. With von Braun's interest in elephants, and his upper-class background, 'I was convinced . . . he must be a wealthy Republican,' Kertes said, 'but he informed me he was not!'[29]

Frequent little notes moved between von Braun and his secretary. One day she passed along a commemorative foreign stamp picturing von Braun, President Kennedy and the Saturn V super-rocket, with a request for an autograph on its face. He complied, and then returned her memo with the notation: 'Could you get me such a stamp for myself? Be glad to pay up to $1 for it.' Kertes checked around and later sent her boss a note stating that the stamp was sold only as part of a souvenir sheet – priced at $6. 'Forget it,' von Braun wrote back. His secretary consoled him: 'It's not a very good likeness, anyway.'[30]

Another time, she sent him a clipping from a press release that was available during an international space conference in Bremen, Germany. In it, von Braun was quoted as saying he anticipated walking on the moon within ten years. It further reported that he envisaged a stay of 'at least eight or ten days' at a fifty-person US lunar research station. Von Braun returned the clipping – on it he had written 'Baloney'.[31]

In addition to the heavy flow of post and phone calls to von Braun's office, his secretary recalled that he received a preposterous number of invitations to speak – from members of Congress, trade associations, colleges, primary school classes and so on. Even so, he never saw some of the speaking invitations addressed to him. Washington staff member Jay Foster said Julian Scheer, who 'actively disliked' von Braun because of his past in Nazi Germany, intercepted these and quietly substituted himself without von Braun's knowledge.

It reached the point, Foster recalled, that Paine stepped in and issued an edict: the only people henceforth allowed to sub for von Braun at speaking engagements were (1) any astronaut and (2) Robert Jastrow, director of the Institute for Space Studies at NASA's Goddard Space Flight Center in Maryland, in whom Paine and von Braun had implicit confidence.[32]

Von Braun eventually baulked at the flood of routine speech requests from Capitol Hill. 'His treatment of members of Congress was unheard of,' Kertes remembered. 'In Washington, congressmen cater to their public, but they expect government employees to cater to them. But Dr. von Braun wouldn't even return their telephone calls!

. . . His problem was that he couldn't say no!' The lawmakers always wanted him to appear at functions they – or their big contributors or important constituents – were hosting, because 'he was a very big draw', the secretary said. Still, they couldn't be ignored. And yet Kertes couldn't force her boss to return the calls. The solution was to steer the more important ones to Arnold Frutkin, the space agency's congressional liaison officer. 'He would talk to Dr. von Braun,' the secretary remembered, 'and the matter was taken care of.'[33]

Von Braun also received more than a few German and other foreign visitors to his Washington office – too many to suit Frutkin, who handled international, along with congressional, affairs for NASA. He confronted von Braun one day and insisted that all these foreign visitors first come through his office as a matter of protocol. 'I had never seen Dr von Braun lose his temper,' Kertes said, 'but he lost it with Dr Frutkin. He said, "Arnold, the last thing I want is your job. These people were not here on NASA business. If and when they come to talk about the space programme, then I'll send them to you! Look, these are friends of mine, and they will remain my friends."' Kertes added, 'That cleared the air.'[34]

Von Braun assistant Daniels helped with management of the planning operation during the space scientist's first months at headquarters. Like other staff-level members of the Executive Secretariat at NASA headquarters, Daniels had dual assignments – a primary job with the Secretariat and a secondary role serving an individual executive there. The two travelled together, worked together, and became close.

During the Second World War Daniels had been a teenage aviator on B24 bombing raids in Europe, and sometimes the two discussed the war. After von Braun had been in Washington six months, the day came for Daniels to leave Washington for a year-long stint at Syracuse University to advance the teaching of management within government. Before a going-away party, von Braun asked Daniels to step into his office.

'We sat down – and he brought me to tears,' Daniels remembered. For 20 or 30 minutes von Braun talked about his gratitude for the departing associate's help, about the Syracuse assignment, about his own plans and hopes, 'and about things like friendship and loyalty'. He then gave Daniels something he would always treasure – a signed photograph inscribed 'To Jim Daniels, with regards – who helped me find my way through the Washington jungle.'[35]

As part of a new offensive to win White House support for a reinvigorated space effort, and to gain more input, Paine, von Braun and Jastrow organised a major brainstorming conference, Space Programs in the Year 2000. It covered five days, including a weekend, in June 1970 at the NASA facility on Wallops Island, Virginia. It was attended by two dozen of the leading thinkers inside NASA and the aerospace community at large, among them astronaut-hero Neil Armstrong. Von Braun and Jastrow presented their new papers on advanced propulsion concepts. The keynote speaker was von Braun's long-time friend Arthur C. Clarke.[36]

Material presented and discussed at the conference incorporated much of the new input that von Braun had considered as he worked on his comprehensive space plan for the nation. The resultant twenty-year plan included several main elements. There were to be a series of three separate Skylab space stations in Earth orbit; a smaller, simpler Space Shuttle than the mainstream concept then being pushed; and continued production and use of Saturns and other launch vehicles for scientific, unmanned missions to Mars, the sun, and on multi-planet tours. Later would come a larger shuttle, a bigger, permanent space station, a permanent lunar base and ultimately a manned mission to Mars.

'Tom Paine was excited about it,' Shaner recalled von Braun saying. The administrator thought it was 'the greatest integrated space plan he'd ever seen'. It called for a gradual increase in the space budget to a peak of $12 billion to $14 billion, and then a slow decrease over a period of years thereafter.[37]

After months of preparation of their draft proposal of the aggressive, long-term national space plan, von Braun and Paine were ready to test Richard Nixon's level of excitement over it.[38] Paine took it to the White House in the summer of 1970. The response was negative: the timing for another grand space initiative was all wrong. The word was that Congress would not buy it, the taxpayers would not buy it and the Nixon administration did not buy it. The White House said, do only what you can within a total budget of $5.5 billion a year and maintain the programme on an even keel.[39]

Paine and von Braun were crushed. The NASA administrator soon told von Braun of his decision to leave the agency and rejoin his old company, General Electric (GE). Paine said he had no interest in 'carrying this agency', if that was all they were going to do. He suggested that von Braun also leave, the latter confided to Tom Shaner.[40]

Von Braun's friend and sponsor had been heading the space agency on borrowed time. Paine was a Democrat appointed by President Johnson in early 1968 to be NASA's number two leader under administrator Jim Webb, a Kennedy appointee. When Webb resigned in the autumn of 1968 after Nixon's election, Paine became acting administrator. When Nixon took office in early 1969, he had tried to replace him with a good Republican but failed to get the person he wanted. And so he had made Paine permanent administrator.

With the Nixon administration's rejection of a bigger and brighter space future, Paine was now ready to leave. He submitted his resignation on 28 July 1970 and declared his intention to return to GE, effective as of 15 September. The announcement came just five months after he had uprooted von Braun and brought him to Washington. Paine told reporters that recent cutbacks in the space budget had nothing to do with his departure. 'When he left, Dr von Braun was just devastated,' recalled secretary Kertes.[41] The rocket scientist and master planner could not imagine the changes of fortune the next two years would hold.

Perigee in Washington

With NASA administrator Tom Paine's mid-1970 resignation, speculation immediately focused on von Braun and George Low, the deputy administrator, as possible successors. Von Braun didn't have a chance. Charles Sheldon was the White House senior staff member of the National Aeronautics and Space Council from 1961 to 1966 and later chief of Science Policy Research for the Library of Congress. He observed, 'There was always a lingering resentment at the Washington end toward von Braun and his team. There were always rumors that von Braun would someday be the next head of NASA. But there is a great sensitivity in Washington about racial and ethnic interests in government. People said, "Don't pay any attention to the rumors. Von Braun would never be given any political position. No one who had worked with Hitler and the Nazi government could be trusted."'[1]

Low, a superb engineer, experienced technocrat and Austrian-born Jew, was soon appointed acting administrator by Nixon until a permanent choice could be made. That would take more than six months. With Paine gone, von Braun was now on his own at NASA headquarters, caught in the Washington morass. The climate there soon turned arctic.

Low had been the number two leader at NASA–Houston, arch-rival of the Huntsville operations headed by the hard-driving von Braun. More than a few of those in charge at the Manned Spacecraft Center in Texas viewed the German-born Marshall Center director as a grasping, scheming glory hound. Low's personal background also stood in contrast to that of von Braun, the aristocratic former Nazi Party member.

'The first time I met von Braun I was prepared not to like him, but we became the best of friends,' Low told Lee B. James, who was then the manager of the Saturn V Program Office at the Marshall Center.[2] Also, as NASA deputy administrator, Low had assured the MSFC's Frank Williams in early 1970 that he was 'pleased' von Braun was coming to Washington to take charge of space planning. He encouraged Williams to accept von Braun's invitation to join his team there.[3]

Whatever friendly feelings and healthy working relationship had existed between the two, a drastic change occurred nevertheless in the treatment accorded von Braun at headquarters – for whatever reasons – after Paine left and Low took over. When the agency's top officials testified on Capitol Hill, NASA's erstwhile perennial star witness was not among them. He was required to submit all his other speeches for review and approval, and advised to stick to the script and forgo his trademark ad-libs. Few headquarters officials ever asked his advice or input on decisions. There were no more chatty sessions in the administrator's office, von Braun's secretary recalled; he had difficulty even getting an appointment to see Low, whose office was two doors away.[4] Clearly, he had become *persona non grata*.

'Wernher and Low had problems,' recalled Jay Foster of von Braun's planning team. 'I don't know that it was necessarily personal, but I don't think Low ever forgave Wernher for the Nazi connection.'[5] Others, however, doubted that this was the major cause of the friction between the two.

Now essentially out of the loop, von Braun was not invited to most meetings of the NASA hierarchy. He may have been nominally the fourth-highest official in the agency, but he was not considered part of the command chain. One example was a high-level meeting von Braun learned of informally. As Lee James recounted, based on a private conversation he had afterwards with a pained von Braun: 'He walked in a little late, as he always did, and the room got kind of quiet. So, Wernher said, "I looked around at so-and-so [James could not remember who von Braun said was presiding] and said, 'I am welcome here, aren't I?' And they said, 'No, you're not.'" Can you imagine that? It was just that they had decided they didn't want von Braun up there.'[6]

'From that meeting on,' said James, 'von Braun was rather crushed. He said, "Lee, how do I handle something like this?" . . . Obviously . . . he had gotten into something where he was not going to be happy.'

Insiders attributed the antagonism shown von Braun to a range of political and personal circumstances. Among the more benign political factors cited was that Low considered himself a caretaker interim administrator in a time of deflated White House and congressional support and tight space budgets. He was not about to buck the tide and allow von Braun and his 'tiger team' of long-range planners to run loose with their grand schemes. Von Braun staff members also suspected that darker internal NASA politics included payback pressure on Low from Houston partisans to give a now-vulnerable von Braun his comeuppance for past intra-agency sins.

Several personal factors were also at play. Homer E. Newell, NASA associate administrator, chief scientist and then the official directly above von Braun in the pecking order, had been in charge of planning. A cautious, deliberate man, Newell was not pleased with surrendering that role to the full-throttle von Braun, and he let it show, several of von Braun's staff said. Another headquarters official who tended to be 'anti-von Braun', remembered Jay Foster, was William E. Lilly, chief of plans and analysis in NASA's Office of Launch Vehicles, and later the agency's first comptroller general. In government service since 1950, 'Bill Lilly thought Wernher was not enough of a bureaucrat,' said Foster. 'He felt that Wernher spoke too candidly in public and rocked the boat by saying what he thought.'[7] Some von Braun staff believed that the so-called Jewish Mafia within NASA and the larger space community had applied pressure on Low to drive von Braun out while he had the opportunity. Tom Shaner thought such talk was 'overdrawn and overly dramatic'.[8]

More often than not, simple human animosities flared over von Braun's having been a wartime enemy, along with pure jealousy over his celebrity status, high accomplishments and charisma. One senior German-born member of the old rocket team, who kept a close eye on the Washington machinations from Huntsville, said he believed Low's ill-will towards von Braun sprang much more from 'outright jealousy' than from any 'ethnic factors'. 'For instance, von Braun in the past would always testify at a congressional committee hearing in a full chamber, with only standing room,' the retired rocket expert continued, on condition of anonymity. 'When George Low testified, there would be only a few people in the audience, and some of them would fall asleep. It was just jealousy.'

The Peenemünde veteran took pains to note that Low 'had acquired worldwide admiration for his work and his influence at the future Johnson Space Center'. And yet, he added, Low 'became a different person when he transferred to NASA headquarters. He allowed his actions and his decisions to be controlled much more by personal feelings – and by pressures from others – than by a desire to provide strong leadership to the national space programme.'

No one believed that von Braun had entered the high-powered world of Washington bureaucratic politics as an innocent, however. Before leaving for his new post, he had told his closest associates at the Marshall Center that he was headed for a place where a great deal of horse manure is shovelled. The trick, he had said, is making sure you stayed at the right end of the shovel and did not live 'the life of a mushroom'. 'What's that?' someone asked. The departing director had explained: 'You are kept totally in the dark; every now and then someone opens the door a bit and throws in a shovelful of horse manure on you. And if you dare try to poke your head above the surface, it's in danger of being chopped off!'[9]

Hermann Weidner, von Braun's chief of R&D operations at the Marshall Center, may or may not have been aware of the irony in what he wrote two years later to his ex-boss. 'Dear Wernher . . . You also may be able to tell by now which end of the shovel you finally ended up with in this deal: whether you have become mushroom-number-four in NASA or rather the handler of the stuff at the other end, as you predicted you would.'[10]

There had been notions before 1969–70 of persuading von Braun to leave Huntsville and move up to headquarters. However, he 'had been advised by Mr [James] Webb that it would be very unwise for him to go to Washington', said Edward D. Mohlere, a Marshall Center manager under Eberhard Rees. '[Webb] had said, "They will eat you alive."'[11] Webb, a bulldog-tough Marine Corps veteran of the Second World War, had specific notions about who 'they' were. An in-depth 1987 NASA history of the Marshall Center expanded on the subject. 'Von Braun's relationship with Webb had always been proper but distant, and was tinged with the Nazi question. Paine claimed that Webb wanted to keep von Braun out of Washington: "I think Jim had the feeling that, well, the Jewish lobby would shoot him down or something."'[12]

Paine believed that Webb's fears were baseless regarding the fate awaiting von Braun. Perhaps Webb had not wanted to compete with von Braun in Washington, DC. Paine elaborated in a 1986 interview with author Joseph Trento:

I think most people felt that he had a damned unfortunate past and nobody liked a Nazi . . . but he had kind of paid his dues and that he really helped us get to the Moon in developing the Saturn V and showed himself to be a worthy citizen of the country; and while we won't exactly forgive and forget, politeness dictates, at least, we won't get into a disgraceful knock-down and drag-out. So it was sort of a neutral thing. He was neither the terribly charismatic or popular figure Jim feared, nor was he the great target of the anti-Nazis who very properly would object to having a prominent member of the Hitler regime ensconced in Washington in a policy area.[13]

But von Braun had moved to Washington, and it had not taken long for things to start coming apart for him. John Goodrum Sr, a veteran engineer-manager at the Marshall Center, was in the capital on business in the Nixon–Henry Kissinger era of the early 1970s and spied his ex-boss one evening in a bar on K Street. Goodrum joined him for a drink. Von Braun looked unhappy. Goodrum asked how his work was going. 'Well, John,' replied the space leader, 'I've found out up here I'm just another guy with a funny accent.'[14]

In von Braun's beleaguered tenure at headquarters, his wife's sympathetic ear was apparently his salvation. Maria related those grim times years afterwards to family friend Ernst Stuhlinger. 'On many evenings we walked around the block for hours and hours while he talked and poured out his soul . . . all I could do was just listen . . . he was so deeply depressed; for him, a world was falling apart. . . . I felt that the only relief I could give him was just encourage him to talk.'[15]

Despite the frustrating downturn of events for the space programme and for him personally in the capital, von Braun tended to keep up a positive front with his former compatriots during visits to the Marshall Center, according to Stuhlinger and others. He was able at last to do some serious, big-picture thinking, he said. And he was busy with the proposed Space Shuttle and other challenging projects, he emphasised, reassuring his old team members that better times for space activity would come again.[16]

Still, he felt bitter disappointment over decisions by the national leadership to abandon the Saturn–Apollo capabilities created at dear cost and to sound a retreat in the space arena. Those feelings burst through in a never-published, August 1970 interview in his offices with then-NASA contract writer Robert Sherrod:

We are saying, 'Well, okay, we now have landed on the Moon, so let's do something else. Let's clean up the rivers, let's do something with the air, do something with the polluted cities . . . do something for the poor', and so forth. But instead of now cleverly utilizing that new capability, we just kill [it]. We can't replace it overnight for something on the other side.

So, this is again a case where we don't rule by rational thinking but by emotion. People say somehow instinctively, 'These goddamn Apollo guys had their day in court, they had all the fun. But now that we have landed on the Moon, let's quit – you know, walk away from it and do something entirely different.' By quitting in this area prematurely you don't get one alternate problem solved overnight. . . .

We surely learned one thing in Apollo as we were looking back at this spaceship Earth . . . coasting in the universe with limited resources. . . . What can we do with that new space capability to help solve the problems of that planet Earth? Let's turn that capability gently around and use it in these new areas.[17]

It made no sense, he lamented in the interview, to 'get the fire axe out' and whack a successful space programme to pieces. Good sense or not, however, it happened.

Life in Washington went on for a demoralised von Braun, with occasional sparkling social and cultural events to brighten his otherwise largely unhappy existence. Maria and he were among the guests of President and Mrs Nixon at a White House dinner early in 1971 honouring Apollo 14 astronauts Alan Shepard (promoted to admiral that year), Edgar D. Mitchell, also a Navy flyer, and Stuart A. Roosa, of the Air Force, plus other key people behind the successful mission. Included were several of von Braun's old colleagues invited up from Huntsville and Cape Canaveral.[18]

In April 1971, von Braun travelled to northern Alabama to take part in the dedication of the collected papers of his US writing collaborator, friend and colleague from 1930s Germany, the late Willy Ley. With encouragement and support from von Braun and others, the library at the University of Alabama in Huntsville had purchased the Ley collection from his widow after the death in 1969 of the early rocket enthusiast and space author.

The working climate for von Braun at headquarters took a turn for the better in April 1971. James C. 'Jim' Fletcher became President Nixon's appointee to head the space agency. Low resumed the post of deputy administrator. Fletcher, an Air Force missile scientist in the 1950s, had been favourably impressed by von Braun's participation in the eye-opening series of space articles in *Collier's* magazine and had later met him during a visit to Redstone Arsenal. 'They had a cordial relationship' at headquarters, remembered Frank Williams of the planning team. 'They joked a lot.'[19]

One day Fletcher, von Braun, Williams and Daniels, among others, were flying aboard a NASA plane. The steward announced that the bar was open and called for drinks orders. Von Braun and Daniels ordered cocktails. Fletcher, a Mormon who had come straight from the University of Utah presidency, said, 'I will have milk.' When the drinks arrived, Daniels recalled, von Braun called Fletcher's attention to 'these fine mixed drinks' he and others held while the boss had a glass of plain milk. 'Jim,' von Braun teased, 'it's too bad your religion doesn't allow you to have more than a milk drink. Cheers!' With hardly a pause Fletcher responded, 'Wernher, it's too bad your religion doesn't allow you to have more than one wife. Cheers!'[20]

Under Jim Fletcher, von Braun's sense of usefulness was less than satisfying, although not nearly so gloomy as in the months during Low's interim reign. Fletcher and von Braun were 'never impolite' to each other, Williams remembered, but von Braun was left somewhat adrift; 'Fletcher asked him to do very little.' Williams added:

Von Braun was much better known [than Fletcher], much better liked, and had a much broader clientele of contacts and friends, and so I think there was a lot of jealousy. He [Fletcher] didn't know what to tell von Braun to do. We were just about to lose our thrust [in the space programme]. There was a window of opportunity, and if Jim had let von Braun go do his thing, I think we could have aroused enough interest and enthusiasm on [Capitol Hill] to put pressure on the White House to let us do more.[21]

But Fletcher did not seize the moment, if indeed there was one. Von Braun did put considerable thought and effort into the proposed Space Shuttle. The operative design concept called for a large, two-stage, all-reusable system. Von Braun argued for a transitional smaller, simpler, cheaper vehicle that was not fully reusable, saying the scaled-up model could come later. Development of the lower-cost version gained Nixon's eventual approval.

Otherwise, recalled Williams, the von Braun planning team 'wrote reports, made surveys, did all kinds of tasks'. Von Braun's own discomfort intensified. Of what use was he now in Washington? 'He was being wasted, and it was eating him alive,' Williams said. 'He shared [that frustration] with quite a few people. He talked with Eberhard [Rees, his successor at the Marshall Center] about it. He didn't run to a bar and talk, or go to any therapist, but he did discuss it with some of us. He said, "I'm just wasting my time."'[22]

Wernher von Braun the natural optimist was treading water in a sea of pessimism. He was a strong leader with hardly anyone to lead any more and no goal to lead them towards. To some he seemed like a great conductor without an orchestra. Word of his predicament evidently reached Walter Dornberger, his old military boss, patron and friend all through the V2 era at Peenemünde. The former German general, who had retired to Mexico from Bell Aerospace early in 1972, urged von Braun: 'Always remember our old battle cry: "Never give up!" And let nobody ever break your spirit. Much to do is left.'[23]

Also early that year a plan emerged for what proved to be 'Wernher's Excellent Sailing Adventure'. He loved sailing, and although he never owned a boat, he nevertheless enjoyed sailing borrowed ones and occasionally helping to crew aboard friends' fine vessels in America.

The idea for a getaway sail in the Caribbean to mark his sixtieth birthday was his wife's. Maria von Braun knew that her husband would take little pleasure in the big Washington birthday bash that inevitably awaited him unless he escaped town. She knew his growing disaffection with his work would have only added to his dread of such a party. So she broached the idea, and her husband jumped at it. Maria excluded herself from the event; this was to be Wernher's birthday escape. Besides, she wasn't much of a sailor, and hardly relished the idea of being part of the boat's working crew.

An enthusiastic von Braun recruited staff member Frank Williams, who was an experienced small-boat sailor and fellow pilot adept at navigation – in the air, anyway. The two hatched a plan for a ten-day, island-hopping cruise aboard a chartered, 41ft sloop that von Braun had already identified in the British West Indies. The plan came together when von Braun signed up expert big-boat sailor, NASA colleague and good friend Hans Mark and his wife to round off the crew.

Then seas got choppy: yachtsman Mark had to cancel shortly before the departure date. Williams hastily recruited a willing couple as substitutes: James Ian Dodds, a young employee in von Braun's office with rowing-crew, but no sailing, experience, and his wife, Ruth, a nurse and good cook. They had a crew again – but one inexperienced at sailing in open seas. That meant the venture would be a 'bare-boat' sail, or one without a single professional-level sailor aboard. But the four had a boat, charts and a general course mapped by von Braun in the West Indies from Grenada north as far as St Vincent and back again. They also had a book on sailing.

At St Georges, Grenada, after stocking *Josephine III*, their graceful vessel, with a full case of Caribbean rum and other basic provisions, skipper Wernher and crew were surprised by word of a required test of their knowledge of ocean sailing. The novices crammed for the exam and the next day managed to convince the local authorities that they were seasoned sailors. They set rules for the cruise, prepared a logbook to be kept and established a required daily radio code-phrase that involved reporting the number of bottles of rum 'in bond' remaining on board. This was to show all were safe and not taken captive by pirates, drug smugglers, Communist agents or aliens.

The four set sail aboard *Josephine III* on 23 March, von Braun's 60th birthday. It was to be a voyage of exotic islands and idyllic harbours, of snorkel dives among tropical fish above coral reefs, champagne birthday toasts, skies sunny and stormy, seas calm and rough, happy encounters with the locals and succulent seafood feasts. There would also be ample refreshments of beer, rum, gin, fine wine and brandy, as warranted. It all came about, island after island, somehow with boat and sun-tanned crew relatively intact at the end.[24]

Towards the finish of the adventure, over drinks ashore with his crew and a few hearty islander acquaintances, von Braun conspired to have some fun with his secretary back in Washington. He would send her a note in a bottle. With the heading '*Josephine III*, March 30, 1972, AD', in his bold script he wrote:

Dear Julie:
    Had a rough trip. Down to one torn sail and two bottles of rum in bond. Whales to port, sharks to starboard and reefs ahead. Sun merciless. Twenty minute turns on the pumps. One crew member dying of tetanus. Scurvy rampant. Skipper delirious. Finishing our last days with the two remaining bottles of rum – and drifting into the sunset.
    Ruthie Wernher
    Frank Ian
    P.S. Please answer by return bottle: Is Nixon still running?[25]

He wrote a second note addressed 'To Finder', advising that both it and the letter to his secretary should be dispatched to 'Miss Julie Kertes' at the NASA headquarters address. At the bottom of that second note he stated, 'Finder will be rewarded with 10 bottles of Tang' (the astronauts' official space drink). The bottle was then capped and supposedly tossed overboard into the sea. Von Braun arranged for the bottle to be 'found' by an equally fun-loving island acquaintance. Two weeks later the acquaintance wrote to Miss Kertes:

Bottle enclosed was found on deserted island while I was in search of turtles. Found no turtles, but am big on Tang, as you Americans call it. No sign of boat or crew but local legend has it rum-soaked scurvy crew still haunting Caribbean. Send Tang in bond to H. Richardson, Petit St. Vincent, Petit Martinique, Grenada, West Indies.
    P.S. Who is Nixon and where is he running?[26]

After his return to the office von Braun asked his secretary every day for several days if a package had come for her from the islands. When at last it did, she recalled, 'He was so excited! He was like a

kid!' Kertes then shipped finder H. Richardson one bottle of Tang –
ten would have been too expensive. She kept the note and bottle.
And she always suspected that Dr von Braun had been drinking a
bit, perhaps, when he wrote that letter in a bottle.[27]

After his homecoming from sailing the seas, von Braun was also
surprised with a slightly delayed 60th-birthday gift from friends,
colleagues and associates all over the country, and the world. In a
variation on a German tradition, it was a bound album of hundreds
of one-page personal letters recounting shared memories with von
Braun over the decades.[28]

Later that spring, von Braun decided to end his free fall at NASA.
As Frank Williams later put it, Wernher decided to 'bail out, cash in
and just say sayonara' to the agency.[29] The frustrated space planner
had done all he could at headquarters. 'Wernher . . . came to the
conclusion that the long-range plan was . . . in being,' recalled
NASA veteran Jay Foster, 'and the government and the country and
the world could attack that [space challenge] at whatever pace the
politicians decided they wanted.'[30] Time would tell when his twenty-
year plan might be activated – perhaps forty or fifty years hence.

Von Braun quietly advised his staff to seek other positions and
acknowledged their frustrations along with his own. Daniels had left
in late August 1970. Foster returned to the Marshall Center for fresh
assignments and Williams soon departed to build a new career
elsewhere within NASA and later with Martin Marietta.

At NASA headquarters on the morning of Friday 26 May 1972,
von Braun announced his retirement from the agency, effective as of
30 June. He had worked the last twenty-seven years for the US
government. He would now join Fairchild Industries in nearby
Germantown, Maryland, as vice president for engineering and
development and would pursue 'some space projects I feel are of
particular importance'. Administrator Fletcher said everyone at
NASA 'will miss the daily stimulation of his presence'. He praised the
departing rocket pioneer as the individual most responsible for
leading America into the space age.[31] Von Braun's staff later agreed
that neither Fletcher nor anyone in power at NASA had lifted a
finger to try to keep him there.

Von Braun expressed deep gratitude for his twelve years with
NASA, but he was less than effusive about the last two and a half
years at headquarters. Daniels, back in Washington for the occasion,

put the brightest possible face on the situation: 'Wernher is quite happy with his accomplishments up here. But in the current climate, he felt his job was done. And, as he told me, he wanted to try a new approach to making a contribution, or, as he put it, "I'd like to try it on the other side of the fence now."'[32]

The immediate reaction of one of his former rocket team members in Huntsville was to remark, 'What really surprises us most is that he stuck it out up there that long.'[33] But the announcement of his retirement 'broke my heart', said Kertes. 'With Dr. von Braun, there was never a dull moment. They were the busiest two and a half years of my life.' He was 'probably the best boss I ever had, and I had many fine ones'.[34]

One afternoon at the end of June there was a farewell gathering for von Braun at NASA headquarters that included a contingent of his long-time associates from Huntsville and the Marshall Center, a few close friends in Congress and his Washington staff members. Administrator Fletcher spoke. So did George Low, whom, perhaps surprisingly, Maria von Braun was said to have liked especially well. Her husband responded briefly to the two men's remarks.

'It was a sad occasion,' Kertes remembered.[35] For von Braun, though, his 'happiest moment' in Washington occurred during that unfestive party. Low had pulled him aside at the going-away affair and thanked him for lobbying for the smaller and simpler Space Shuttle over the 'grandiose concept' then favoured by most within NASA. Low, one of the pillars of Project Apollo, quietly told von Braun he had done a great service to the agency and the nation by saving the shuttle with his input and advocacy.[36]

Thereafter, with von Braun gone, Fletcher and Low abolished his planning position, office and staff. Fletcher invited a member of the von Braun group to join his staff. 'Before I took [the new job] I called Dr von Braun and asked him what he thought about it – and whether Dr Fletcher had been responsible for his leaving,' the staff member recalled in an interview many years later. The former staff member said von Braun replied he had no problem with him joining Fletcher and that no, another official, whom he named, had been responsible. After a while, the new member of Fletcher's staff grew confident enough to ask Fletcher why he had not tried to keep von Braun on board. Fletcher said he realised he was blamed, but that he had not insisted that the rocket pioneer should go. The administrator named the official responsible, although granting he could have 'overruled' that individual, as Fletcher put it.

'It was George Low' whom Fletcher identified, the retired Washington staff member revealed a quarter of a century later, on condition of anonymity. 'He was very jealous of Dr von Braun.' It was also Low whom von Braun had privately identified to his former staff member.[37]

Peenemünde veteran Adolf K. Thiel ran into von Braun after his departure from NASA, when both were visiting the space agency's Ames Research Center in California. They had dinner together in what turned out to be their last meeting. 'It was clear he was very disappointed over the way things had gone in Washington,' Thiel recalled. 'George Low . . . resented von Braun.'[38]

But von Braun's undoing at NASA headquarters resulted from far more than just one man's resentment or jealousy or other form of ill-will. A fateful confluence of individuals and circumstances cut short his hoped-for seven or eight 'good years' of government service in Washington. He managed to put in little more than two, and they were hardly good ones, for the most part.

On the Private Side

Word that Wernher von Braun was joining Fairchild Industries on 1 July 1972 had caused an excited buzz to run through corporate headquarters at Germantown, Maryland. The general feeling was that in bringing him aboard from NASA, Chairman Ed Uhl had scored a coup for the aerospace company.

The two men had known each other since 1946. The American-born Uhl was a young Army engineering officer and small-rockets specialist before and during the Second World War. He had been the lead inventor of the anti-tank bazooka. After the war, he went to White Sands, New Mexico, to help build a rocket test stand in conjunction with the V2 firings there by the imported von Braun missile team and its new American partners. Their paths crossed occasionally until the Pegasus satellite project in the early 1960s brought them together in a more direct way.

As the Saturn–Apollo enterprise got under way, NASA's Marshall Space Flight Center decided that more research on meteoroid activity outside the Earth's atmosphere was needed to gauge the threat posed to manned space vehicles with their thin skins. Von Braun and his team had proposed that the burly new Saturn I rockets launch a series of three large satellites whose wing-like surfaces would unfold in orbit to measure possibly deadly micro-meteorite hits. Critics at the time had said it was just 'busywork' (of little value) to justify the Saturn I.

Uhl had determined that Fairchild would seek the job, which was named Pegasus after the winged horse of Greek mythology. When he had become Fairchild's president in 1961 at the age of 43, it was an aeroplane company with no missile or space capabilities. One of his first steps was to help his company diversify by luring away two dozen aerospace engineers from other companies. For the Pegasus competition he decided not to join forces with another contractor but to use only this small technical team.[1]

On the day of the review of Fairchild's bid to design and build the satellites, Uhl sat near von Braun. 'We went down the agenda item by item,' he remembered. 'An engineer [from Fairchild] might come out and brief on structures, and later the same engineer came back to brief on weights. Our people being innovative, they changed their clothes in between their appearances – a sweater at one point, then jacket, then shirtsleeves. Part way through the presentation, Wernher leaned over and said, "Ed, didn't I see that chap before?"'

'Yes, Wernher,' Uhl replied, thinking fast. 'We have a small but very capable team, and each person covers more than one discipline. Everyone you see today will be 100 per cent on the Pegasus, and our manufacturing division is very capable in bonding [the solar energy panels and detection sensors]. We can do the job.' Von Braun said nothing more.

Fairchild went on to win the contract and build the satellites, which were launched successfully in February, May and July 1965. 'All three performed without a glitch,' recalled Uhl. 'The schedule was made – no overruns. We made an excellent profit, which made me happy. Fairchild went into the satellite business, and space was declared safe for manned flight, the micro-meteorite danger having been determined to be minimal.'[2]

As time went by, the relationship between von Braun and Uhl grew into friendship, partly through the hunting trips they took together. Along the way, Uhl had mentioned more than once that if von Braun ever wanted to leave NASA for private industry, to please consider joining Fairchild. That day arrived with the approach of summer 1972. The space scientist telephoned Uhl and asked whether the offer still stood, and if so, would Uhl mind coming into Washington for lunch?[3]

From that meeting a plan emerged. After forty years in the employ of government – thirteen in Germany, twenty-seven in the United States – von Braun at the age of 60 would enter the world of private enterprise as Fairchild's vice president for engineering and development. That would make him the top-ranking technical executive at corporate headquarters. It also meant a flashy increase in annual salary – from under $40,000 to somewhere between $200,000 and $250,000, associates later said. For a man who'd never before cared a pfennig for making a lot of money, he and his still-youngish family would now do more than all right.

In signing on with Fairchild, von Braun was able to keep his residence in nearby Alexandria, Virginia; this also pleased Maria, who had embraced the cultural advantages of the Washington area. His new employer provided him with a car and driver for commuting and for area appointments, enabling him to think, read and make notes during the drives. It also did not escape the notice of passionate pilot von Braun that Fairchild had its own airstrip right there at its headquarters.

Von Braun's happiness at joining Fairchild in 1972 was mixed with sorrow over word of his father's death that year at the age of 94 at his home in Oberaudorf, Bavaria. The space scientist made the difficult decision not to attend the funeral. His attendance at his mother's funeral in Bavaria in 1959 had drawn a disruptive horde of reporters, camera crews and curiosity-seekers to the services. Von Braun, knowing that his two brothers would be at their father's funeral, was determined not to cause a repeat fiasco.

At Germantown, Ed Uhl and Fairchild were fully aware that they had a celebrity on the payroll, although the rocket pioneer never acted like one, Uhl said. Both present and potential customers fell over themselves in order to have lunch with von Braun. 'He could open doors anyplace in the world' for the company, Uhl recalled. Moreover, 'wherever he went, he was the guest of the head of government'. Both Uhl and von Braun resisted brazenly exploiting the latter's celebrity, they maintained. But if, in planning a business trip to, say, Iran, the space scientist was invited by the shah to stay at the palace, how could he refuse?[4]

The shah personally invited von Braun to come to Iran and give briefings to his government and university people on the benefits of the Applied Technology Satellite (ATS) network of large communications orbiters that Fairchild was touting. As Uhl remembered it:

> We always sent a projectionist with Wernher. But since he was staying at the palace, the shah insisted on providing the royal projectionist. The first address was at the University of Tehran.

The auditorium was packed wall to wall. After the introduction Wernher said, 'First slide, please.' The slide came on and it was upside down and backward. Well, the projectionist was really upset, the hosts were upset. They were more than just worried – they were fearing for their lives.

So Wernher took over, and he said, 'Okay, while the slides are being corrected let me talk about space.' And he ad-libbed about twenty minutes, got the ready signal, and said, 'First slide, please.' It wasn't backward this time – it was [only] upside down. Wernher said, 'Okay, maybe if you turn the projector and hold it upside down, we can go on with this.' Well, when they did, all the slides fell out! The Iranians were devastated.

Wernher said, 'Look at that. I helped put a man on the moon but I can't even help a man project his slides!' Well, everybody laughed and chuckled. The projectionist recovered, Wernher ad-libbed and then [he continued], 'First slide, please.'[5]

The unstoppable von Braun proceeded to make his full presentation, said Uhl – and nobody left the hall. It was not an isolated case. Tom Turner, Fairchild's vice president for marketing at that time, had also been with von Braun once in Tehran for a space lecture before a large audience in the US Embassy's outdoor auditorium. 'As Wernher was about to begin, the power in the city crashed, and all was in darkness,' Turner recalled. 'But with the help of a bullhorn [loud-hailer], Wernher ad-libbed; his presentation was long, and as it was coming to conclusion, the power came on! But Wernher, instead of calling it a day, said: "May I now have the first slide, please?"'[6]

Charles 'Chuck' Hewitt similarly found out just how irrepressible von Braun could be. In the mid-1970s he and the rocket scientist travelled to Dallas for a reception to drum up support for a future benefit banquet for the newborn National Space Institute (NSI). Von Braun was the NSI's first president, and through the sponsorship of Uhl and Fairchild, Washington lawyer Hewitt had been hired as its executive director. For the reception, Hewitt and von Braun had gathered some of Dallas's social and corporate elite to interest them in sponsoring the big dinner. Plans included having Bob Hope as master of ceremonies and getting space programme enthusiast John Denver to sing, Hewitt recalled.

Before the reception he briefed von Braun: 'Well, Wernher, these are the major points: I will talk a little bit, but they really have come to hear you. You'll want to keep focused on this event we're planning, and the need for funding of the space programme, and all we're doing in space research, and this kind of stuff.'

'Okay, I got it,' von Braun assured him.

The reception got under way, and the two men entered the room. Hewitt said his brief piece and introduced von Braun. The scientist began with some general remarks. Abruptly, a lady raised her hand and asked him a question about . . . nuclear fission. That was nowhere on the agenda.

'Of course, Wernher proceeds to talk for 45 minutes on nuclear fission!' Hewitt marvelled. 'Now, this is a cocktail party. Not a person went for a drink. He had them all just mesmerised. About half the people in the room were basically socialites, people who normally would never be interested in nuclear fission. But when he spoke on it, it was just so fascinating.' In the end, 'it went over very well', recalled Hewitt. 'He asked me later, "Did I get a little carried away?" I said, "Yes, but it worked out okay." He was so focused – but on the question asked.'[7]

At Fairchild, von Braun worked on a number of projects. Early on, he visited Jordan, Iran and other countries to support Fairchild's then-current foreign sales effort for the A10 anti-tank combat aircraft. Uhl also assigned him the task of corporate strategic planning, which von Braun undertook. Speech-making and a wide range of other public relations duties also fell his way.

He returned to another old speciality: testifying on Capitol Hill. In a 1973 appearance before the Senate Committee on Aeronautics and Space Science, he promoted a business interest of Fairchild and a subject of strong personal interest: communication satellites (comsats) for education. He testified that he'd been drawn to Fairchild Industries because he 'wanted to be helpful in expanding the tremendous potential of communications satellites, in particular their use for audiovisual education'.[8]

Von Braun had long believed the potential uses of satellites were infinite. Forced during his NASA service in Washington to cease his drumbeating for interplanetary travel for the time being, he had expanded his thinking on satellite applications from Earth orbit. One practical example he cited was the need for railroad managers to keep track of their rolling stock. Easy, suggested von Braun: just put a relatively inexpensive transponder in each railcar, and then a detecting satellite could fill a computer with real-time information on the identity, location, speed and direction of every piece of equipment in the inventory. The space scientist also theorised that a network of satellites could work such wonders, if desired, as tracking the movements of large schools of fish by detecting from orbit the natural oil slicks in the oceans caused by the bigger fish eating the smaller fish in their path.[9]

But the greatest applications, in his view, lay in using comsats to bring universal education to poor, remote corners of the globe. His own visits to isolated villages in India and the Yucatán, for instance, in his pre-Fairchild travels, reinforced his conviction that with powerful satellites circling above, any mud-hut hamlet could gain access to classroom instruction by means of a simple television receiver. And so, wrote von Braun, these orbiters could be employed 'to help break the stranglehold of illiteracy, which still retards the progress and well-being of a substantial part of mankind more than any other factor'.[10]

The telecommunications satellite ATS6 was an unusually large, direct-broadcast comsat system with a big steerable antenna, high-powered transmitter and unlimited potential. Fairchild built three of them – the main unit, a back-up and a spare. The primary unit was launched perfectly by an Air Force Atlas–Agena D vehicle into distant geosynchronous, or stationary, orbit in 1974. It was right up von Braun's alley.

'Wernher loved the ATS6 satellite,' Uhl recalled. 'It showed how space could benefit mankind.' A series of educational, medical-related and other demonstration experiments using the supersatellite were set up at several places around the world, including India and Alaska. Von Braun visited every locale participating in the programme.[11]

India had the most extensive programme. 'It developed a set of little stations all through India, even in villages where they had no electricity,' Uhl said. 'They had a television set hooked up to a bicycle with a generator, and somebody would pedal the bicycle to operate the television. They made antennas out of chicken wire. It was a very successful programme.' To persuade India to buy into the ATS6 programme, von Braun had met with US Ambassador Patrick Moynihan and Prime Minister Indira Gandhi. He made the sale.[12]

For much of 1975–6, the giant comsat beamed lessons to eager Indian children and adults gathered around communal television sets in almost 2,500 isolated villages.

ATS6 and other Fairchild business projects kept von Braun hopping around the planet throughout his time with the company. He travelled to India often, to Alaska several times, to Brazil and Venezuela, to the Paris Air Show. He went to Spain, where he had lunch with Prince Juan Carlos and Princess Sophia at the palace. He went to Britain, for breakfast at last with Second World War nemesis Duncan Sandys, and a meeting with the Queen. Von Braun thoroughly enjoyed the travel, but as he told a former associate back at NASA's Marshall Center, engineer-manager Bob Schwinghamer, 'It is wearing me out.'[13]

Uhl experienced at least one characteristically manic von Braun jaunt. It was to Alaska in the mid-1970s, and it involved one of Fairchild's leftover ATS units. 'One day Wernher said, "Ed, we ought to go up to Alaska and check the [operational] satellite performance, try to sell that spare satellite we have, and we can do a little hunting and fishing." So off we went.' They left on a Thursday, reached an Alaskan hunting camp Friday, shot a bear Saturday, shot a moose and fished Sunday, flew up to Prudhoe Bay on Monday, spent the week visiting the villages and base stations involved in the educational and medical project, briefed the governor in Anchorage on Friday, and returned home that weekend. 'It was a typical von Braun trip,' remembered Uhl. 'I had purchased a Polaroid camera with a lot of film and taken that along. Every place we went, everybody wanted to have their picture taken with von Braun. He spent his time signing pictures. I spent my time snapping pictures.'[14]

Von Braun travelled in December 1972 to his familiar haunt, Cape Kennedy, for the night launching of Apollo 17, the last in the truncated series of manned lunar-landing flights. Its flight crew consisted of Navy men Gene Cernan and Ron Evans, and Harrison 'Dr Rock' Schmitt, who had earned a doctorate in geology. Von Braun had made sure his Fairchild secretary, Patricia Webb, could attend the historic event. Von Braun's visit included an interview at the launch site with Walter Cronkite. Always the optimist, von Braun expressed his hope that man would be 'back on the Moon within a decade'.[15] President Richard Nixon, the man who had led the dismantling of the high-priority space programme, called Apollo 17 'the end of the beginning'.

Back at the Marshall Center, as von Braun looked on in late 1972 from several hundred miles north, life was about to take a sharp turn for the worse for the original rocket team's German-born engineers and scientists still working for NASA. Most had managed to survive the deep cutbacks sustained by the Huntsville centre from the late 1960s into the early 1970s. Many, if not most, had figured their days were numbered when von Braun moved to Washington in March 1970. But under Eberhard Rees as head of the Marshall Center, a relative stability in the Germans' lives had been maintained.

. In January 1973, however, Rees unexpectedly resigned. He took advantage of a new NASA general offer of early-retirement incentives. Rees also felt he should step down so that, at 64, he could go out 'on top'. He had led several space advances – starting the High Energy Astronomical Observatory (HEAO) satellites and the Large Space Telescope (later renamed Hubble), among other projects – during his brief tenure succeeding von Braun.[16] William R. Lucas, a seasoned MSFC scientist-manager, had been groomed to take over after Rees.

'You see, Bill Lucas was the first [native-born] American to break what I called the "Kraut Line",' recalled Joseph M. Jones, retired Marshall Center chief of public affairs. 'Von Braun tapped him early to be a star. And he was in a position to succeed Eberhard Rees. Lucas was deputy to Rees then, and he would have been the natural successor. That was the plan.' But then NASA headquarters 'threw a monkey wrench' into the plan, as Jones put it. Washington sent in the 'heavy', Rocco Petrone, instead, 'to put the axe to the decks under these Germans', Jones observed.[17]

MSFC's new boss, a burly, American-football-playing West Pointer, was a retired Army colonel who had worked well with von Braun and the team in the early missile and space days, especially at Cape Canaveral. Petrone arrived in Huntsville in early 1973 amid rumours he was there to clean up the house. The word was that he had come in as a hatchet man charged with cutting the launch vehicle development centre down to size in the post-Apollo era. Petrone denied all such rumours to news reporters.

Formal orders from NASA-Washington soon arrived, however: cut some 1,000 of the centre's government personnel, cut more than 2,000 support contractor employees and cancel two of the three planned Skylab orbital space stations that were to use leftover Saturn hardware. Some saw the hand of Houston in the grim news, considering the clout that the Manned Spacecraft Center (later in the year renamed the Lyndon B. Johnson Space Center) carried at NASA headquarters.

Soon Petrone's name 'around Huntsville was kind of mud, because he would [be coming] down here with a big axe', the MSFC's James Shepherd recalled. 'But he wasn't devious. Rocco was a man of integrity, he was honest, and he wanted [the downsizing] done right. Some said [that meant] "Rocco's way". Maybe so, but he wanted it done right.'[18]

The remaining rocket men from Germany had trouble believing that. The heavy reductions struck especially hard at them, most of whom had spent twenty-eight years in US government service. All but a few were forced out, faced with the choice of retirement or demotion and reassignment. Von Braun chose not to oppose the purge publicly. According to Ernst Stuhlinger, he counselled his ageing comrades to take solace in their unique accomplishments and accept their fate with stoic dignity amid changing times and changing national priorities.[19] The downgrading of the Peenemünders was said to have been partly the result of their lack of civil service rating points given for past military service. Ex-Luftwaffe Sgt Walter Wiesman (né Wiesemann), the youngest and one of the most outgoing members of the German team, joked that he planned to ask if he could count his wartime service. He chose early retirement.

Von Braun, despite his philosophical advice to them, made no secret that he had expected the Germans to have been treated better at the hands of their old cohort, the Italian-American Petrone. The latter earned the nickname 'Il Duce' while at the Marshall Center. Petrone departed in March 1974 for NASA headquarters and a promotion. He was at the Huntsville centre little more than a year before turning over the leadership of the much-reduced force to Lucas. Petrone had wielded his axe skilfully. Almost none of the smattering of former Germans still around held major posts; two years later, only eight remained at any level. The consensus was that good-soldier Petrone, probably without malice, had carried out his orders to 'Americanise' Marshall while sharply reducing its overall payroll.[20]

'The system forced us out,' Walter Jacobi later observed. 'We grew too old, I suppose. Some may disagree with me, but there was just no spark without von Braun.' Karl Heimburg, who had been the MSFC's long-time Test Lab chief, noted another factor in the exodus. He left NASA 'because the center began to lose its manpower and I saw all those young people going out the door with families to feed. I felt guilty. All of us old men decided to get the hell out.'[21]

On the plus side, von Braun had been pleased in 1973 to see the Skylab space station programme, which he had helped conceive at the Marshall Center during the 1960s, materialise. A leftover Saturn V behemoth from Project Apollo launched the 70-ton, Marshall-managed Skylab into Earth orbit in May 1973. The cavernous prototype station, a modified upper stage of the moon rocket, was visited eleven days later by a crew of three, led by Naval officer 'Pete' Conrad, who stayed up for a month. Second and third flight-crew visits lasted two and three months, respectively, giving the United States a significant boost in spaceflight man-hours.

At about the same time, von Braun had a hand in the creation of the previously mentioned National Space Institute (NSI) as a private-sector organisation to tap and expand grassroots and high-profile support for the space programme and to lobby for it. The NSI was the brainchild of recently replaced NASA administrator Jim Fletcher. He had observed a decline in public interest in space, although many big names in influential or high-visibility positions, such as then-California Governor Ronald Reagan, the Revds Billy Graham and J. Fulton Sheen, Walter Cronkite, Hugh Downs, Johnny Carson, Charles Lindbergh and Henry Kissinger, remained outspoken supporters of the national effort.

Fairchild's Ed Uhl jumped aboard the NSI ship. He not only enlisted von Braun's help for the fledgling organisation but also lent Fairchild marketing executive Tom Turner to the project and soon put Washington attorney and space enthusiast Chuck Hewitt on the payroll to provide NSI staff executive leadership. The institute was founded in June 1974, with von Braun as president. About $500,000 was soon raised from the aerospace community to begin financing the NSI's activities of educating the American taxpayer and voter on the benefits of the space effort and of developing public support for it.[22]

Hugh Downs was already an ardent supporter of the space programme and one of the most visible figures in American television in the mid-1970s when von Braun approached him about taking a key leadership post with the new organisation. 'When he asked me to take over the presidency of the National Space Institute, I said I thought a scientist or engineer ought to fill that post,' Downs recalled. 'He said, "I'll become chairman for as long as I'm able, and you have got a promise from Fletcher to be vice president. So, since we need a communicator as president, the three of us will be like a sandwich – you in the middle and scientists above and below."' Downs, who would go on to log more hours on network commercial television than anyone else in the US before retiring in 2000, said yes to his friend Wernher.[23]

Other VIPs among early NSI backers included Bob Hope, Ethel Merman, Hugh O'Brien, James Van Allen, Alan Shepard, Senators Barry Goldwater and Hubert Humphrey, Congressman Olin E. 'Tiger' Teague, the Revd J. Fulton Sheen, Jacques Cousteau, James Wyeth, Isaac Asimov, Arthur C. Clarke and Ben Bova.

Comedian Hope was master of ceremonies at what proved to be the successful NSI fundraiser in Dallas that von Braun and Hewitt had helped engineer. Hope told attendees at this posh affair that he was happy to accept the invitation to join the organisation's first board of governors, 'because sooner or later I must go to another planet – the taxes are killing me here!'

As it developed, von Braun was too ill to attend the banquet held in late 1975. Not long after he joined Fairchild, Uhl had advised him of the company policy that all executives have periodic full medical examinations. In the early summer of 1973, while in Texas on other business, von Braun kept an appointment with his friend Dr James R. Maxfield Jr at his Maxfield Radiological Center in Houston. The physical would serve two purposes: carry out his employer's wishes and meet a requirement to keep his pilot's licence current.

Von Braun returned to Germantown with grim news for Uhl. The doctors had found an ominous shadow around his left kidney in the X-rays and had recommended immediate surgery. Von Braun worried that he'd been with the company such a short time and he didn't want to take an extended medical leave so soon. Uhl urged him not to think twice about it but to do precisely what the doctors had said, and take as long as he needed to recover.[24]

For the second time in recent years Maxfield and his medical colleagues admonished von Braun to have surgery without delay; and for the second time, the rocket scientist let it slide. After procrastinating all summer, however, he checked into Johns Hopkins University Hospital in Baltimore in September 1973. Surgeons removed his left kidney, which had a malignant tumour. The doctors said they believed they had removed all the cancerous tissue but ordered radiation treatments as a precaution. The patient recovered in due time and returned to work in good spirits.

Too Soon Dying

Chosen by the gods as rocketry's wunderkind, Wernher von Braun had at least nine lives. In Germany he survived launch-pad explosions, the impact of an incoming test V2 warhead that hurled him high in the air, hellish bombing raids, imprisonment by Himmler and the Gestapo, and a high-speed wartime car crash that killed his driver. He lived through a serious hepatitis attack on arrival in America, errant US test missiles and a near-miss in an aircraft, among other brushes with danger and death. John Bruce Medaris, the US Army general-turned-Anglican priest, observed as von Braun reached 60 that he must have 'an especially effective guardian angel' in view of 'the long succession of hazards to life and limb that you have managed to survive without apparent damage'.[1]

Yet now, barely into his 60s, the seemingly indestructible von Braun was facing an even more formidable hazard: cancer. When his doctor and friend Jim Maxfield discovered suspicious polyps in von Braun's colon in 1970 and recommended immediate surgery, von Braun had put it off.[2] 'Yes, he knew then that he had a problem, but he didn't yet realise it was already cancer,' von Braun's administrative secretary and confidante in Huntsville, Bonnie Holmes, later said. 'I think it was probably cancerous at the time he left [the MSFC], but he probably did not realise it. If he did know . . . he did not tell us.'[3]

She recalled that her long-time boss 'thought they would go away and not be a problem. He had always been in such good health. Except for his sinus problems and the operation he once had for that, he never had any health problems.' And yet, despite knowing that his beloved mother had died of colon cancer (in 1959) and that there can be a hereditary predisposition to it, Bonnie Holmes noted, 'he still took chances with his colon condition,[4] which is very unfortunate'.[5]

He had, however, apparently recovered from having a cancerous kidney removed in 1973. After surgery, radiation sessions, and a brief further convalescence, von Braun went back to work at Fairchild. There he ran into his friend Erik Bergaust, the veteran rocket and space writer, hunting companion and fellow employee. Von Braun immediately joked that his doctor had told him, 'Now that you have only one kidney left, you should drink more . . .' While he purposely let the sentence trail off, Bergaust supplied the missing word: 'water'.[6]

Bergaust asked von Braun what he would do when – and if – he ever retired. 'Read,' he had answered. There was never enough time, even for voracious reader von Braun.[7]

In what turned out to be his last return visit to Rocket City, USA, von Braun accepted an invitation from the City of Huntsville to attend the opening ceremonies on 14 March 1975 for the sprawling Von Braun Civic Center. With the dedication of the municipal complex of arts, entertainment, and conference facilities coming only nine days before von Braun's 63rd birthday, the *Huntsville Times* editorially suggested he accept it as an early birthday gift. Said the beaming von Braun: 'Can you imagine – getting a $15 million art and cultural centre for your birthday?!'

Unveiled during the Von Braun Civic Center's opening was a large oil portrait of von Braun with his head in the starry night heavens and his fundament and feet planted on the rocks of a mountain top. It hung at the entrance to the concert hall. During brief return business trips to Huntsville the year before, von Braun had sat several times for portrait painter C.E. 'Ed' Monroe Jr. The Huntsville-born artist and magazine illustrator had recently returned home after a successful career based in New York and Connecticut. During the first sitting, von Braun had told Monroe, a hunting buddy, that he hoped the painter was receiving a handsome commission for the work.

'Wernher,' replied Monroe, 'I'm doing this as a gift to the community – for posterity.'

'Ed,' von Braun had jibed, 'what has posterity ever done for you?'[8]

In July 1975 von Braun travelled to Cape Kennedy for the first public meeting of the NSI and the Saturn IB launching for the joint

United States–USSR Apollo–Soyuz manned mission in Earth orbit. Once again, old friend Walter Cronkite interviewed him on CBS News. Afterwards, dressed in a loud tropical shirt during a visit to the outdoor rocket exhibit at NASA's Kennedy Space Center, von Braun asked a friend, 'Will you take my picture in front of the Saturn V?' He paused, and then added, 'I may not be here again.'[9]

In addition to his tiring business-related travel for Fairchild, von Braun took his family on holiday to Canada's North Bay wilderness in August 1975. There, he experienced rectal bleeding. Typically, he shrugged it off. It reappeared with a vengeance a few weeks later, during a follow-up business visit with Ed Uhl to Alaska. Uhl did not let him ignore the ominous signs this time but 'sent him off to the hospital' for surgery as soon as they returned home, Uhl remembered.[10]

At Johns Hopkins University Hospital in Baltimore, surgeons discovered an advanced tumour in von Braun's colon. They removed the malignant growth and a section of the large intestine. Newspaper reports quoted unnamed hospital sources as saying that the surgery was related to 'a kidney ailment'. The private prognosis for the patient's long-term survival was not good.

Von Braun again returned to work. 'His spirit was always up, but his energy slowly decreased,' recalled Uhl. Between hospital stays and periods of rest at home, von Braun remained as active as possible at Fairchild as the months slid into 1976. That included speaking and writing about his favourite subject and making limited public appearances.[11]

It also meant keeping abreast of doings back at NASA's Marshall Center. 'You know, he was always interested in what was going on in the center, long after he left,' recalled William R. Lucas, who occupied von Braun's former post there as director for a record twelve years. 'During the final months that he was sick, he had Marshall send him new reports, which we would send in bundles. He would read them in bed.'[12]

In February 1976 in Washington, von Braun held what turned out to be his final news conference. He pressed once again for a more vigorous national space programme. He said space could make 'great contributions' to helping 'unemployment, balance of trade, increase in food production, protecting the environment, developing health care, energy, world peace'.

Early 1976 also saw a fast-failing von Braun participate in a Fairchild educational film targeted at young people. He advised them to do their best to prepare for a new age of space. He said that is where the answers lie for the future. 'It's your turn,' he emphasised. That became the film's title.

In the early months of 1976 von Braun also managed to come to the aid of UAH (the University of Alabama, Huntsville) when the university he had earlier nurtured sought to create a multi-million-dollar solar research institute on its campus. A Washington breakfast meeting with several pivotal US senators was arranged for 12 March. 'He was pretty sick, but he was there,' recalled long-time aerospace engineer-manager David Christensen, then overseeing the UAH solar research programme. 'He was moving very slow. This was one of his last hurrahs.' Von Braun found the strength to speak briefly for the proposed institute. Even in his weakened state he was impressive, remembered Christensen. 'He closed by saying, "Huntsville helped give you the moon, and I do not see why Huntsville cannot also give you the Sun."'[13]

The solar research institute project did not succeed, but not for lack of von Braun's support.

Christensen recalled another event in early 1976 that involved von Braun. This time the venue was Johns Hopkins University. He gave a lecture on satellites in education, and afterwards someone asked about astronauts going to Mars. 'I vividly remember him lighting up like a Christmas tree and giving this spiel about how we could have and should have followed through with the plan to go to Mars by 1982,' said Christensen. 'The thing that struck me is how, even though ill, he just turned on immediately when he started talking about Mars. He got very dynamic in explaining it all, as nobody else could, because Mars had really been his dream.'[14]

He gave his last videotaped interview in April, to US television's Hugh Downs at the Smithsonian's National Air and Space Museum. He cited an array of current space applications and benefits worldwide and said these were only the beginning.

Also that spring, von Braun attended the NASA headquarters retirement party for colleague William Pickering, the long-time head of the Jet Propulsion Laboratory in California. 'One of the strongest impressions I now have of von Braun is of him at that wingding they put on for me,' Pickering later recalled. 'He was very ill. His coming meant a lot to me. It was his last public appearance.'[15]

Von Braun and his friend and eventual co-biographer, Fred Ordway, continued their collaborative relationship all through 1976. Years later Ordway, having checked the journal he has kept daily since boyhood, recalled the day in January of that year that the two spent together. He had arranged for von Braun and several Fairchild colleagues to visit the new Energy Research and Development Administration in Washington, where Ordway worked. The afternoon visit ended with von Braun turning to him with a reminder of a National Space Institute reception the two needed to attend that evening and suggesting they afterwards share a quiet dinner out. 'So, please call Maria and tell her I'll be late,' von Braun had added.

'He was afraid she'd put her foot down, scold him, and tell him to get right home,' recalled Ordway, who had 'been through that routine before' with the couple. 'Anyway, she answered the phone and I said, "Hi, Maria. This is Fred Ordway —", but before I could say anything more, she interrupted, "Yes, I know, Wernher's going to be late. And he made you call me."'[16]

Maria's objections notwithstanding, the two men stayed in the city and had their quiet dinner. Von Braun's mind was still sharp, Ordway remembered. 'He was very keen on geography, and we tested each other's knowledge that evening', as they had on earlier occasions. Two months later, on 23 March 1976, Ordway, von Braun and a mutual friend from Germany celebrated the rocket pioneer's 64th birthday with lunch at a Washington restaurant. Ordway quoted from his journal: 'He was quite subdued that day, clearly suffering from his battle with cancer. It was sad to watch him walk hesitantly in and out of the restaurant with the aid of a cane. Since our January dinner, he had clearly gone downhill.'[17]

As the cancer spread, the now-silver-haired von Braun began spending more and more time in Alexandria Hospital, especially after May 1976. Maria was at his side, and their three children were frequent visitors. His oldest and closest associates, men such as Eberhard Rees and Ernst Stuhlinger, visited from Huntsville. Relatively newer friends, too, such as Neil Armstrong and Hugh Downs, came by.

'I visited him whenever I could,' Uhl recalled. 'Between visits I'd collect papers, reports, and documents, and we'd talk about all that was happening with space and the company. One day, he looked very weak but his spirit was good. He had his usual twinkle in his eye.

I said, "Wernher, how're you doing?" He said, "Well, Ed, I've had so many blood transfusions I can say truly that I'm a full-blooded American."' The two men enjoyed a quiet laugh over that. Then Uhl gave von Braun the reading material he'd brought, and they chatted for a while. The visitor soon noticed the patient was tiring.

'Wernher, I'd better go,' said Uhl. 'Your doctor has me on a short leash.'

Von Braun reached under his bed and pulled out a long object. 'Before I go,' he said, 'I want you to have my rifle.'

'Wernher,' replied a moved Uhl, accepting the gun, 'thank you very much. I'll clean it for you, I'll oil it for you, and I'll keep it for you – for our next trip.'[18]

There were no dry eyes between the two friends as Ed Uhl left the room.

Writing was something that von Braun could continue doing even as his physical condition and pain worsened throughout 1976. Sometimes he had to work almost in secret, though. He and Ordway were collaborating on a new book about solar-system discoveries. From his hospital bed von Braun insisted on reviewing drafts of chapters and adding material. He called Ordway from the hospital on 28 July. 'His voice sounded extremely weak', the historian wrote in his journal. Von Braun was somehow making progress on their chapter on Jupiter. 'We talked for at least twenty minutes,' remembered Ordway. 'He reminded me once again that if he suddenly hung up, it meant that either the nurse or Maria had arrived. He wasn't supposed to be working and didn't want to get scolded!'[19]

The collaborators spoke by phone about their book-in-progress until mid-January 1977, when von Braun was unable to work on it any further.[20]

'Almost a confessional', a colleague described von Braun's last major piece of solo writing. Sometime in 1976 he accepted the invitation of the Lutheran Church of America to present a major paper at its synod at the University of Pennsylvania that autumn. He worked on his eighty-two-page paper for several months in the hospital. He entitled it 'Responsible Scientific Investigation and Application'. When the time came for its presentation, he was too ill to appear. He had a surrogate step in and read it for him to the large assembly; other speakers there included Jonas Salk, Arthur M. Schlesinger Jr, Archibald Cox, James Baldwin and Norman Cousins.[21]

The highly personal essay, a blend of the philosophical and the practical, covered an array of 'the grim problems besetting humanity' and von Braun's thoughts on remedies. The subject areas ranged across science and technology, morality and taboos, the motivation for scientific study, a prioritisation of scientific research needs, environmental pollution, the relationship between religion and science, and human survival.

He addressed everything from the issue of scientific ethics and controls, in what later became a brave new world of genetic tinkering, to the infinite beneficial uses of Earth-orbiting data-gatherers. 'Such information from satellites,' he wrote, 'has been called as epoch-making as the first use of fire as a tool, or the first practical use of the wheel.' The surface had only been scratched in the use of satellites to help educate the illiterate billions on Earth, he stressed. He termed the human brain 'man's most precious resource' and deplored the fact that 'the vast majority of mankind uses it most sparingly. The reason is inadequate education.'

As for the true impetus for space travel and exploration, the rocket pioneer stated he was convinced that 'the answer lies rooted not in whimsy but in the nature of man. I guess it is all just in the basic makeup of man as God wanted him to be. I happen to be convinced that man's newly acquired capability to travel through outer space provides us with a way out of our evolutionary dead alley.'

His summing-up before the Lutheran gathering also reflected his tightening embrace of his own Christian faith as he faced death. He revisited what had become his familiar rationale that science and religion are fully compatible. 'In this reaching of the new millennium through faith in the words of Jesus Christ, science can be a valuable tool rather than an impediment. The universe revealed through scientific inquiry is the living witness that God has indeed been at work. Understanding the nature of the creation provides a substantive basis for the faith by which we attempt to know the nature of the Creator.'[22]

Decades of shared reflections by the rocket scientist such as those expressed in his deathbed essay led author Tom Wolfe, author of *The Right Stuff*, to declare repeatedly in later years, 'Wernher von Braun was the only philosopher NASA ever really had.' When Wolfe appeared on NBC News' top-rated *Today Show* in the autumn of 1998 to discuss his new novel, *A Man in Full*, and

John Glenn's recent return to space, he did so but then hastened to bring up von Braun:

He was a member of the German Wehrmacht, and he had a Teutonic accent – had everything but a dueling scar on his cheek – and so they couldn't bring him forward as 'NASA's philosopher'. But he was the philosopher, and he said, 'The importance of the space program is not surpassing the Soviets in space. The importance is to build a bridge to the stars, so that when the Sun dies, humanity will not die. The Sun is a star that's burning up, and when it finally burns up, there will be no Earth . . . no Mars . . . no Jupiter.' And he said, 'You have to find a way, because humans are the only thinking creatures that we know of in the entire universe, and we have to build a bridge to save this particular species.' I think that's a grand thought, and it should be the thought that everybody has in supporting the space program.[23]

On 1 January 1977, the hopelessly ill rocket pioneer resigned from Fairchild Industries. Later that month, outgoing US President Gerald Ford awarded von Braun the National Medal of Science. Having been bedridden in the hospital for months at that point, he was much too sick to receive it publicly, so Ford asked Uhl to make a private presentation. 'Several weeks went by before I could get the doctors to agree,' Uhl recalled. 'One day Maria called me. She met me outside his room. She prepared me for his appearance.'[24] When they entered the room, the visitor was glad she had.

'He looked like a skeleton, with only skin draped over his bones,' Uhl remembered. 'Here was this big, robust guy, whittled down to a pitiful thing.'[25] Uhl continued: 'I said, "Wernher, President Ford has awarded you the National Medal of Science, and as you know, that is the highest honor our country can give a scientist." He turned to Maria, tears in his eyes, and said, "Isn't this a great country? I came here with all that I owned in a cardboard box, somewhere between a former enemy and not yet a citizen, and we were given all the opportunities of citizenship. This country has treated me so well. And now the president is giving me this high honor."'[26]

Uhl prepared to leave, but von Braun asked him to stay. The visit continued for two hours. Uhl had known von Braun for more than thirty years, and he had grown to admire him enormously: 'He never looked at rockets really as a weapon, but as a means to reach and explore space.' This was the last time he saw his friend alive.[27]

Many of von Braun's friends and colleagues over the years had assumed the vigorous, jut-jawed space evangelist would live for ever. Arthur C. Clarke was one of them. In a 60th-birthday letter, Clarke noted that in 1958 he had written a short story set in 2001, 'Out of the Cradle, Endlessly Orbiting'. In it, a Russian spacecraft designer remarks that he hadn't seen von Braun 'since that symposium we arranged in Astrograd on his eightieth birthday, the last time he came down from the Moon'. Clarke added in the letter to his friend: 'That would be a very nice self-fulfilling prophecy. So – see you in Astrograd in '92, and [at] Clavius Base for your centennial [in 2012].'[28]

In the same vein, rocket-team member Walter Wiesman – who considered his leader 'the closest to the "Man for All Seasons" in this slightly imperfect world' – had closed his sixtieth birthday greetings to von Braun with: 'I have no doubt you will make 90 without too much effort. Just remember, please, it ain't necessarily clean living that will get you there. But what am I telling you? I heard that . . . from you years ago!'[29]

As von Braun faced the inevitable, he grew to cherish even more his family, close friends and the time remaining to him. Ever the loquacious one, he also grew reflective about his life and work, acknowledging his satisfaction with living to see many of his dreams come true. But there were doubts, too. Near the end he had sought reassurance from several old and dear colleagues that he, and they, had done the right thing in developing lethal missiles, because expediting the space age was always their one true goal, was it not? He was given their reassurances, said Stuhlinger.[30]

Von Braun had chosen not to write his memoirs or a full-blown autobiography. He was always too busy. Furthermore, as he had said in rejecting several close associates' urgings that he do so, he would inevitably leave out individuals who could rightly expect to be included, according to Stuhlinger and others. Perhaps he had other reasons as well. In any case, he would leave to others the telling of his journey through time . . . and space.

Von Braun spent the final weeks of his life in the hospital 'in seclusion and silence', with 'his last credo' being 'Thy will be done', his favourite passage from the Lord's Prayer.[31]

Years earlier, the irrepressibly curious von Braun had given the following as his favourite quotation by an American: 'If I should reach my last moments with perfect lucidity, I feel sure that I would enter eternity with my eyes wide open and a feeling of intense curiosity.'[32] It was not to be. In mid-June 1977 his two-year ordeal of ever-failing health ended. His body – and formidable brain – ravaged by a wicked disease that science had not yet conquered, this man of science and technology died quietly in the early hours of Thursday 16 June. The pied piper of space, a resident of planet Earth for sixty-five years and three months, had piped his last beguiling tune.

Time had always been so precious a commodity to Wernher von Braun that he strove not to waste a moment. One of his favourite quotations was from Thoreau: 'As if you could kill time without injuring eternity!' 'Time', von Braun once observed, 'puts a limit on us all. We can overcome all other dimensions, but in turn, we are overcome by time.'[33]

Before the day was over – and purposely before word of his death spread – he was buried in Ivy Hill Cemetery, in a churchyard in Alexandria. His family and a few friends were present. No news reporters, photographers or television cameras; no admiring fans – and no potential troublemakers. His simple headstone was engraved:

WERNHER VON BRAUN, 1912–1977
Psalms: 19:1

The Old Testament reference was to his long-time favourite passage of scripture, a soaring Psalm of David: 'The heavens declare the glory of God; and the firmament showeth His handiwork.'

Epilogue

'Moon Rocket King Is Dead' blared a London newspaper headline. 'Father of the American Space Age', another paper called him. With public announcement of von Braun's death withheld a full day, the news media broke the story on 17 June 1977. Press coverage that day and the following was widespread and in-depth. The *New York Times* ran a lengthy front-page obituary on 18 June by its space writer, John Noble Wilford. His lead opened: 'Wernher von Braun, the master rocket builder and pioneer of space travel . . .'. Inside, the story of his life and death ran for a full page. It noted, 'Dr. von Braun's name, perhaps more than any other, [is] synonymous with space travel.'

The Associated Press's obituary was written by one of von Braun's earliest and closest contacts in the news media, correspondent Howard Benedict. He opened the article with a restrained characterisation of von Braun as 'the German-born rocket pioneer whose talents helped the United States put men on the moon'. Benedict later noted it was far from the easiest dispatch he ever wrote. 'As I did, I remembered those evenings in the Silver Sands Motel [with von Braun as mentor], how they sparked an interest in a young reporter that kept him on the AP [space and missile] beat for more than three decades. And I recalled the dreams this great man had and the fervor with which he spoke of them – and how so many have gone unfulfilled.'[1]

President Jimmy Carter, US Naval Academy graduate and veteran submariner, issued a statement praising the departed rocketeer as 'a man of bold vision. To millions of Americans, Wernher von Braun's name was inextricably linked to our exploration of space and to the creative application of technology. Not just the people of our nation, but all the people of the world have profited from his work. We will continue to profit from his example.'

Press reaction in London to his death was, perhaps surprisingly, balanced. After all, little more than a generation before, his V2 missiles had killed British civilians, destroyed property and generally caused havoc in the country late in the Second World War. And yet, all nine of the major London newspapers recognised von Braun as a figure of history, one who led the way to the moon and beyond.

'Nazis' V2 Bomb Genius Is Dead', headlined the *Sun*. The tabloid opened its obituary with 'The man who invented the V2 rockets that struck terror into the hearts of Londoners . . .'. But the next paragraph described him as 'also the genius who designed the fabulous Saturn 5 rocket that put American Neil Armstrong on the Moon in 1969'. The *Daily Telegraph* pointed out that von Braun was 'once detested by the free world', yet it went on to run several upbeat quotations by him on his lifelong goals for space exploration. The paper praised him for 'eagerly' supporting the joint Apollo–Soyuz flight that united American astronauts and Soviet cosmonauts in orbit just two years before his death.

The *Daily Express* portrayed him in the most favourable light. 'Ironically,' it noted, 'von Braun died one day before the first manned [drop] test of the Space Shuttle *Enterprise* – the reusable rocket ship expected to be the workhorse of the American space programme for the next decade. . . . It was designed on von Braun's basic philosophy to expand man's role in space exploration.' And the *Daily Mail* editorialised: 'Legend says that von Braun – the only household name that the space age has thrown up – was the man who "gave us the Moon". Legend, for once, is quite right.'

'I will always remember his integrity, his love and concern for other people,' long-time secretary Bonnie Holmes told local reporters upon his death. 'He thought every human being should have his dignity. Of course, he'll be remembered for his achievements. But to his family and those of us who were close to him, he'll be remembered for his character.'

The day after he died, a memorial service was held at von Braun's former command post, NASA's George C. Marshall Space Flight Center, inside the gates of Redstone Arsenal. The service was beamed via closed-circuit television to its thousands of employees. It featured a eulogy by then-director William R. Lucas. 'Many of you have personal knowledge of how great a loss this is,' he said. 'Many of you, as did I, worked with him for twenty or twenty-five years, or more. You know what his contributions to space were, and in a larger sense, to humanity itself . . . and how immeasurable is the void he leaves.'

The Sunday following his death, von Braun's former church in Huntsville held a memorial service. Conducting it at the pre-Civil War Episcopal Church of the Nativity was the Revd George B. Wood. He had been a US Army airborne combat chaplain in the Normandy invasion and elsewhere in Europe during the Second World War. The rugged priest spoke of the native-German space leader – and one-time member of the Nazi Party – as 'a lover of freedom'. Wood added: 'If you read between the lines of his life's work and words, you sense a smoldering resentment and anger against oppression and tyranny.'[2]

Von Braun's former US military boss, Revd John Bruce Medaris, conducted a memorial mass for the departed rocket pioneer at the Episcopal Church of the Good Shepherd in Winter Park, Florida, not far from Cape Canaveral. In Huntsville, Father Medaris also took part in an Army tribute to von Braun in the Bicentennial Chapel at Redstone.

Several eulogists made the point that death had now robbed von Braun of the opportunity to climb aboard a rocket ship and thunder aloft as he had for so long dreamed of doing. Eugene Emme, history committee chairman for the American Institute of Aeronautics and Astronautics, addressed that point in a preface to their obituary. 'Cruel fate denied Wernher von Braun the chance to buy his ticket as a passenger bound for an excursion in space – his boyhood dream and manhood goal. Because of Wernher von Braun, however, almost everyone has been brought to the realization that we have been passengers on a spaceship all along – Spaceship Earth. Posterity will not forget him.'[3]

Summing up the tribute paid to von Braun by the American Institute of Aeronautics and Astronautics was Fred Durant, then the deputy director (for astronautics) of the Smithsonian's National Air and Space Museum. 'Most people dream of what they would like to see happen in the future. Few make dreams come true. Wernher von Braun did. He dreamed of man leaving the Earth and exploring the Moon and planets. . . . He not only had a dream, but he made his dream come true for all of us.'[4]

The memorial ceremonies themselves climaxed, six days after his death, with a national service held at the National Cathedral in Washington. The Very Revd Francis B. Sayre Jr, a great friend of von Braun's and a supporter of space exploration, conducted the service before more than 500 mourners. The cathedral's dean eloquently prayed:

We thank Thee, God of the Heavens,
For a man of new beginnings;
For one whose vaulting imagination was restive in Earth's tiny
    envelope and would not be prisoner there;
For a man of history though still a small boy inside;
For a friend and father whose eager life stood in the doorway of
    the Great Beyond . . . [5]

As the praises of von Braun sounded amid the soaring organ strains of Bach and Brahms, an object of fitting interest held a paramount position high above the proceedings from its permanent setting in a lofty window of the cathedral. Revd Sayre had previously secured it for placement there. It was a small rock from the moon.

Eulogies at the service were delivered by three of von Braun's colleagues: James Fletcher, former administrator of NASA; Ernst Stuhlinger, retired chief scientist at the Marshall Center; and Michael Collins, former astronaut and then-director of the Smithsonian's National Air and Space Museum. Collins flew the four-day Gemini 10 mission in Earth orbit in July 1966 and was the command module pilot on the nine-day Apollo 11 flight in July 1969 that first took humans to the surface of the moon. In his eulogy, Collins may have best expressed what Wernher von Braun's life had meant to so many:

I first met Wernher von Braun in 1964 and last saw him in 1976. During these twelve years, the manned space program progressed from its infancy, through adolescence, to its present state of maturity – in large part because of Wernher and his Saturn rocket program. Ten Saturn I, nine Saturn IB, and thirteen Saturn V rockets were launched, a total of thirty-two flights, all successful, all without loss of life, all on peaceful flights flown without weapons. Saturns sent nine astronauts up to Skylab, which was in itself a converted Saturn V upper stage, and kept three of them in space for eighty-four days. And finally, a last Saturn sent an American crew up to join a Russian spacecraft in Earth orbit.

Who was the man behind these incredibly successful statistics? Wernher von Braun was a study in contrasts. He was, at the same time, a visionary and a pragmatist, a technologist and a humanist. From his youth he had dreamed of flying to the far corners of the solar system, yet it was the next flight which seemed the most important one to him. He was a master of the intricacies of his machines, with their innumerable pipes, valves, pumps, tanks, and other vital innards, yet he realized that his rockets could only be as successful as the people who made them. And he assembled an extraordinarily talented team, people who worked well with each other, and who were totally devoted to Wernher.

In short, he was a leader, with the versatility that leaders of genius must possess. Because he worked with rockets, I would call him a rocket man, but that is a cold term, and he was anything but cold. He was a warm and friendly man, interested in everyone around him, no matter who they were. He had a marvelous knack for explaining his machines in simple, understandable, human language. And he never seemed too busy to share his ideas – and he was full of them – with others.

Wernher von Braun believed that the desire to explore is a fundamental part of mankind's nature.

Although he is no longer with us, I like to think that his spirit is around us, exploring, traveling through the far reaches of the universe, traveling paths of which he long dreamed, and which more than any other man, he has brought closer to reality.[6]

Letter on Goddard Patents

In early 1975 a West Point cadet wrote to von Braun asking about the persistent accusation that he and his team in Germany basically copied the patents and work of American rocket pioneer Robert Goddard and performed little original engineering in developing the V2. Von Braun, who was not in good health, chose to respond in detail to the young engineering student's apparently sincere question:

> Around 1930, when I was eighteen years old and a member of the German Society for Spaceflight . . . Dr. Robert Goddard was one of the great international names in the concept of flight through space, and I counted him among my boyhood heroes. I had read his booklet 'A Method of Reaching Extreme Altitudes' which described the multistage principle and presented some advanced ideas on how to improve the performance of solid fuel rockets. In subsequent years, when I developed liquid rockets for the German Ordnance Corps which ultimately led to the V2, I occasionally saw illustrations (e.g., a Goddard proposed aerial torpedo), or statements (e.g., Goddard says: 'Man can reach the moon') in aviation journals. However, at no time in Germany did I or any of my associates ever see a Goddard patent. I was not even aware of the fact that Goddard worked in the field so dear to my own heart, namely liquid-propellant rockets, let alone that even as early as 1926 he had successfully launched the world's first liquid-propellant rocket.[1]

Von Braun, who had arrived in America only weeks after Goddard's death, stated in his letter how, in 1950, the US Army had asked him to review and analyse 'a large stack of Goddard patents' related to a lawsuit filed against the government by the Goddard estate and the Guggenheim Foundation, an early sponsor of Goddard's work. The main contention was that the V2s brought to America and some of the new postwar US rockets infringed the reclusive American rocket innovator's patents, and so the government should pay royalties to his estate.[2] Von Braun further noted in his letter to the cadet:

> The government lawyers apparently took the position that at least the V2 design could not possibly have violated Goddard patents, because they had been highly classified and had, therefore, not been available to the German engineers. Moreover, even if any infringement existed, the United States would automatically have acquired all rights to the V2 technology by virtue of the captured V2 missile falling in the category of war booty under international law.

Von Braun wrote that he was asked to do 'a detailed written assessment' for use by the court, regarding whether any V2 engineering designs had infringed Goddard's patents. He stated that he could certify 'there were indeed infringements all over the place' – from the application concepts of jet vanes to turbo-pumps to guidance-and-control gyroscopes. Von Braun added in his letter:

> All the Goddard patents I saw were classified and had never been published, even as late as 1950. I was fully unaware of them while in Germany, and even in the United States I saw them for the first time only five years after my arrival, and upon receipt of a secret clearance.
>
> All the patents I reviewed in 1950 rendered impressive proof that Dr. Goddard had indeed a brilliant and imaginative mind. The patents covered not only design features actually (but unwittingly) used in the V2, but numerous alternate options.

The ailing von Braun concluded to the West Pointer: 'It might be of interest that (maybe in part as a result of my affirmative report) the lawsuit led to an amicable settlement, under which the US Government paid a generous sum to the Goddard estate and Guggenheim Foundation.'[3]

In March 1957 von Braun had been invited to address a banquet of the Worcester Engineering Society in the Massachusetts community where Goddard had begun his pioneering work. He asked to be taken – by Goddard's widow and test flight camera operator, Esther – to the site of the launching of history's first liquid-propelled rocket by the visionary physics teacher thirty-one years earlier, on 16 March 1926. Finding no historical marker there, von Braun began a campaign that evening to erect a proper one. His efforts, through the American Rocket Society, succeeded with the 1960 dedication of a memorial at the site.[4] Other honours in Goddard's memory followed, often with von Braun's involvement.

Letter on Moral Responsibility in Hitler's Germany

Wernher von Braun gave serious attention to the following letter that he received in January 1971 while working at NASA headquarters in Washington. It came from a stranger, a resident of Long Island, New York. The misgivings expressed, and questions raised, by this concerned American citizen undoubtedly reflected those that others harboured about the former enemy rocket scientist who had emigrated from Germany.

Dear Dr. von Braun:
I cannot understand why you did not use your influence and stature in World War II to try and save the Jewish people.

Surely a man of your greatness must have comprehended the enormity of the wrong being done.

I respect your ability as a scientist but cannot yet respect you as a human being until given an explanation of your silence during those times.

I am not a crank nor do I bear any malice towards you. But I would truly appreciate an explanation.
Sincerely,
Alan Fox.

On 22 January, von Braun wrote this extensive response, a copy of which was obtained from his Washington correspondence files.

Dear Mr. Fox:
In reply to your letter of January 11, I am giving you the following reply.

During the years immediately prior to the beginning of World War II it was obvious to anyone living in Europe that political persecution existed in several totalitarian countries. In the early years of Hitler's regime in Germany this persecution took many forms but the most obvious was the vilification of the Jewish people in the Nazi press. Most thinking Germans saw this for what it was, creating a necessary scapegoat for the desperate unemployment rate to rally the masses behind the Hitler government. However, most including myself did not believe even in our most violent nightmares that this overt antagonism would ultimately lead to anything like Auschwitz (of which I heard for the first time after the war). Until the outbreak of the war, Jews were welcomed as officers and enlisted men [NCOs] in the German Army, and social contacts were widely maintained with Jewish friends.

I confess to no deep psychological thinking on this matter during these times. I thought that when the political objective of the anti-Jewish campaign had been reached a new scapegoat would be found. Stalin's series of persecutions of the Kulaks, the Army officers (Tuchachevski, et al.), the Trotskiists, the Intelligentsia and the Russian Jews seemed to set a most likely pattern. During these years I was, of course, a young engineer with very little interest in politics, and rather engrossed in my studies on the potential of space flight and my rocket experiments. I felt very fortunate when I gained support for my work in the form of some money and facilities from the German Army. I did not have any more scruples in accepting this support than, say, the Wright Brothers may have had when they signed their first contract with the US War Department.

In 1939, when war was declared, our rocket work was directed to producing weapons. Most of my time before and during the war was spent at an experimental rocket station at Peenemünde, a remote spot on the Baltic Coast. Our days were spent in designing, building, and testing.

I have often been asked how could I produce weapons of war, and I have read many essays on the moral aspects of this general question, which I guess is as old as war itself and thus as old as mankind. From my own experience, I can only say this: when your country is at war, when friends are dying, when your family is in constant danger, when the bombs are bursting around you and you lose your own home, the concept of a just war becomes very vague and remote and you strive to inflict on the enemy as much or more than you and your relatives and friends have suffered.

There was another aspect, too: our knowledge of what was happening in Germany and the world was rather limited by the Nazi propaganda machine. In private discussions with friends, one would occasionally discuss things like the existence of concentration camps, in which all kinds of opponents of the Hitler regime, including Jews, were held. But I do not remember ever having heard of a single incident of an atrocity, let alone of deliberate mass killings of civilians. If you find this hard to believe, you have merely to ask yourself how long after the event it was that you first heard about the massacre at Mylai in Vietnam, and this in a country with a free press eager to unearth unpleasant facts, rather than in one with a rigidly controlled press determined to protect tightly held state secrets and to withhold anything from public purview that Hitler wanted the population not to know about.

You ask me why I did not lend my influence to save the Jewish people. First, as I just said, I truly was not aware that atrocities were being committed in Germany against anyone. I knew that many prominent Jewish, Catholic and Protestant leaders had been jailed for their opposition to the government. I also suspected from the fact that I had lost sight of my own Jewish friends, that many Jews had either fled the country or were held in concentration camps. But being jailed and being butchered are two different things.

Secondly, while I may have been of some importance to the German Army's rocket programs, I certainly did not wield any political influence over anyone outside of Peenemünde. I, myself, was arrested by Himmler and needed the influence of my commanding general to save myself.

As you know, the extent of the actual suffering and the criminal mass slaughter of the Jewish people became known to the world only many months after the hostilities ended, and it was only then that I learned of these things myself. I was deeply shocked and have ever since been ashamed of having been associated with a regime that was capable of such brutality. And, along with many millions of my former fellow countrymen who learned about these atrocities only after the war was over, I know that our generation must accept our share of the guilt for what happened.

Sincerely yours,
Wernher von Braun.

Notes

Chapter 1: That Accursed Blessing

1. Edward G. Uhl, talk given at Wernher von Braun Exploration Forum, Huntsville, Alabama, 21 September 1993.
2. Frederick I. Ordway III, interview with the author, Huntsville, 12 November 1998. All interviews are by the author unless specified otherwise. When a phone interview gives a location, the person being interviewed is at that location. Although the great bulk of the Huntsville interviews were in person, a few people preferred phone interviews.
3. *Careers in Astronautics and Rocketry*, co-authored by Wernher von Braun, Carsbie C. Adams, and Frederick I. Ordway III (New York: McGraw-Hill, 1962); von Braun earlier wrote the foreword to Adams's *Space Flight* (New York: McGraw-Hill, 1958), which was written 'with the collaboration of' Ordway, Heyward E. Canney Jr and Ronald C. Wakeford.
4. Carsbie C. Adams, trip journal and telephone interview, Spotsylvania, Virginia, 16 December 1998.
5. Stuhlinger recounted this story to me, which is also in a book he co-authored with Frederick I. Ordway III: *Wernher von Braun: Crusader for Space* (Melbourne, Fla.: Krieger Publishing, 1994), 278.
6. Frederick I. Ordway III, interview, Huntsville, 12 November 1998.
7. *Ibid.*
8. Francis L. 'Frank' Williams, telephone interview, Slidell, Louisiana, 11 November 1998.
9. Cover story on von Braun, 'The Seer of Space', *Life* (18 November 1957), 136.
10. This quotation and the following quoted letters are from the Wernher von Braun correspondence files, Wernher von Braun Library and Archives, US Space and Rocket Center, Huntsville, Alabama.
11. The matter of 'pressure' on von Braun to accept Nazi Party membership and an SS officer's commission is treated fully, with citations, in chapter 5; Linda Hunt, in *Secret Agenda: The United States Government, Nazi Scientists and Project Paperclip, 1945 to 1990* (New York: St Martin's Press, 1991), tries to connect him to the Holocaust; Christopher Simpson, in *Blowback: America's Recruitment of Nazis and Its Effects on the Cold War* (New York: Weidenfeld & Nicolson, 1988), 30, notes that famous photographs by US liberators of slave labourers at the Mittelwerk factory's Dora concentration camp were later

misconstrued to be of Jewish victims of the Holocaust; Chris Kraft, in *Flight: My Life in Mission Control* (New York: Dutton, 2001), 83, quotes Robert Gilruth on the subject of shifting allegiances.

12. Tom Lehrer song 'Wernher von Braun' from the live album *That Was the Year That Was*, released on 1 July 1965 by Warner Brothers.

13. Thomas L. Shaner, interview, Huntsville, 4 March 1999.

14. Frederick I. Ordway III, interview, Huntsville, 12 November 1998. Such episodes evidently didn't harm the Brucker–von Braun relationship. In 1972, when Wernher von Braun turned 60, Brucker's widow, Clara, wrote the scientist a warm letter. She recalled the 'countless times' her husband had spoken of seeing the Explorer launched as the thrill of his life. She also wrote: 'His great faith in your ability prompted him to battle the skeptics' in high places trying to stop early work on the Saturn boosters. 'We both loved and admired you for your great contribution to our Country's Space Program.'

15. Thomas L. Shaner, interview, Huntsville, 4 March 1999.

16. 'Movie to Hit Red Tape in Space Delay', *New York Herald Tribune*, 22 November 1959.

17. William R. Lucas, interview, Huntsville, 3 November 1998.

18. Israel M. Levitt, telephone interview, Philadelphia, 22 June 1999.

19. *Aviation Week & Space Technology*, top 100 aerospace figures, ranked overall and by category, 23 June 2003. Von Braun friend and associate Fred Ordway accepted the award in Paris for the von Braun family.

Chapter 2: To the Manor Born

1. Wernher downplayed his noble birth with the rocket team he came to lead in Germany, although he did include his title of baron (*Freiherr*) on his official letterhead. There, and later in the democratic United States, where he dropped the title, his team included several other barons. It was not until the post-Second World War years in the United States that a German count joined the Wernher von Braun team. Friedrich von Saurma, who had been an aeronautical engineer and test pilot on the Luftwaffe side at the Peenemünde R&D centre, stood one aristocratic notch higher. Friends said von Braun delighted in introducing Count von Saurma and quickly adding, 'He outranks me'.

2. Edward G. Uhl, talk at Wernher von Braun Exploration Forum, Huntsville, 21 September 1993.

3. Cover story on Wernher von Braun, 'Reach for the Stars', *Time*, 17 February 1958.

4. Magnus von Braun, *Von Ostpreussen bis Texas* [*From East Prussia to Texas*] (Stollhamm, Germany: Helmut Rauschenbusch Verlag, 1955); quoted by Bob Ward, 'Von Braun: Dead Set on Going Somewhere', *Huntsville Times*, 9 March 1975, CC-45.

5. *Ibid.*

6. *Time*, 'Reach for the Stars', 22.

7. Stuhlinger and Ordway, *Crusader for Space*, 10.

8. Bob Ward, *Wernher von Braun Anekdotisch* (Esslingen, Germany: Bechtle Verlag, 1972). Translated by Kurt Wagenseil.

9. Stuhlinger and Ordway, *Crusader for Space*, 10.

10. Decades later and an ocean away, the Hindemith connection would arise again. The Boston Pops Orchestra and its famed conductor, Arthur Fiedler, gave a concert in Huntsville in the 1960s. Arrangements were made for the orchestra to tour the Marshall Space Flight Center and for the maestro and Mrs Fiedler to visit von Braun. In preparation, the Fiedlers had read up on rocketry and space. 'At a reception that night Fiedler told me that the conversation never got around to space,' recalled Walter Weisman, a Wernher von Braun team member, in a paper presented at the 26 March 1992 meeting of the Alabama chapter of the American Institute of Aeronautics and Astronautics. Von Braun had mentioned his semester of piano studies with Hindemith to the Fiedlers. 'Well,' related Wiesman, 'Hindemith happened to be the maestro's favourite composer, and the half hour passed with space never coming up!'

11. Stuhlinger and Ordway, *Crusader for Space*, 117.

12. Wernher von Braun, 'Space Man – The Story of My Life', *American Weekly*, 20 July 1958 (part one of a three-part series), 8.

13. Ward, *Wernher von Braun Anekdotisch*.

14. Heather M. David, *Wernher von Braun* (New York: Putnam, 1967), 23.

15. Drew Pearson and Jack Anderson, 'Wernher von Braun: Columbus of Space', *True* magazine, February 1959, 22.

16. *Time*, 'Reach for the Stars', 22.

17. Von Braun, 'Space Man', part 1, 8.

18. Daniel Lang, 'A Romantic Urge', *New Yorker*, 21 April 1951, 75–93.

19. *Time*, 'Reach for the Stars', 22.

20. Ernst Stuhlinger, 'Wernher von Braun', eulogy written for the *Huntsville Times*, 17 June 1977.

21. It marked the start of a prolific, lifelong writing output by von Braun that included hundreds of magazine and newspaper articles, essays, technical papers and several books. One was an imaginative non-fiction account of a manned mission to his favourite destination in space, the planet Mars.

22. Ward, 'Dead Set on Going Somewhere', CC-45.

Chapter 3. Pioneering Rocketry

1. Helen B. Walters, *Wernher von Braun: Rocket Engineer* (New York: Macmillan, 1964), 20.

2. Ward, *Wernher von Braun Anekdotisch*.

3. Ernst Stuhlinger, 'How It All Began – Memories of an Old-timer', 20 July 1999, Wernher von Braun Library and Archives, US Space and Rocket Center.

4. Werner C. Krug, article in the *Birmingham (Ala.) News*, 18 November 1951.

5. Von Braun, 'Space Man', part 1, 26.

6. Ward, *Wernher von Braun Anekdotisch*.

7. Von Braun, 'Space Man', part 1, 9.

8. Wernher von Braun, 'Reminiscences of German Rocketry', *Journal of the British Interplanetary Society* 15, no. 3 (May–June 1956), 127; also Current Biography (1952), 607.

9. Constantine D.J. Generales Jr, 'Recollections of Early Biomedical Moon-Mice Investigations', Proceedings of the First and Second History Symposia of the International Academy of Astronautics, Belgrade, Yugoslavia, 26 September 1967, published in Frederick C. Durant III and George S. James, eds, *First Steps*

toward Space (Washington, DC: Smithsonian Institution Press, 1979; reprinted by Univelt, San Diego, 1985).

10. Von Braun, 'Reminiscences of German Rocketry', 128.

11. Von Braun and Generales met again, in the United States, at a 1950s scientific symposium on the subject of space medicine. The rocket scientist autographed a copy of the programme for his old friend, with the inscription: 'To the "World's First Space Doctor" (remember Zurich, 1931?).' And in 1960, von Braun further honoured his first friend from America by naming his son Peter Constantine von Braun.

12. von Braun, 'Reminiscences of German Rocketry', 128; also *Current Biography* (1952), 607.

13. Lang, 'A Romantic Urge'.

14. Walter Dornberger, V2 (New York: Viking, 1954), 27.

15. Von Braun, 'Reminiscences of German Rocketry', 131.

16. Lang, 'A Romantic Urge'.

17. Wernher von Braun, 'Why I Chose America', *American Magazine*, July 1952.

18. Erik Bergaust, *Wernher von Braun* (Washington, DC: National Space Institute, 1976), 550–1.

19. Patrick Moore, *Space: The Story of Man's Greatest Feat of Exploration* (Garden City, NY: Natural History Press [for the American Museum of Natural History], 1969), 75.

20. It would be a decade before the two would reconnect as colleagues and co-authors in America.

21. Von Braun, 'Reminiscences of German Rocketry', 134.

22. *Time*, 'Reach for the Stars', 23.

23. Ernst Stuhlinger, 'Wernher von Braun', paper given at the International Space Hall of Fame, Alamagordo, New Mexico, October 1976.

24. Arthur Rudolph was among the few rocket men who became Nazi Party members before Hitler came to power. He later contended he did so mainly because of his vehement opposition to the Communists and also in hope of advancing his own career. But then Rudolph 'left the party two or three years later because he was in strong disagreement with its ethnic [Jewish] policies', Ernst Stuhlinger told me in a memo dated 28 September 2001. Later, however, as Rudolph rose within the rocket team, he rejoined the party at the same time as most of the other top members of the team signed up with the Nazis.

25. William J. Cromie, 'Wernher von Braun – Dean of Rocketry', *Shreveport (La.) Journal*, p. 10-D, and other newspapers, 27 October 1967.

26. Arthur Rudolph, letter, 23 March 1972, Wernher von Braun's 60th birthday album: 'X + 60 and Counting'.

27. Otto Kraehe, letter from Paris to Christel Ludewig McCanless, Huntsville, July 1999.

28. Lang, 'A Romantic Urge'.

29. *Ibid.*

Chapter 4: Peenemünde Priority

1. Walters, *Werner von Braun*, 41.

2. Ernst Klauss, letter, March 1972, Wernher von Braun's 60th birthday album: 'X + 60 and Counting'.

3. Rudolf Hermann, quoted in Stuhlinger and Ordway, *Crusader for Space*, 34–5.
4. Letter (German-born writer's identity unclear), March 1972, included in 'X + 60 and Counting'.
5. Dornberger, V2, 53.
6. Helmut Horn, letter, 25 February 1972, 'X +60 and Counting'.
7. Rudolf Hermann, 23 March 1972, 'X + 60 and Counting'.
8. Dorette Kersten (Schlidt), interview, Huntsville, 21 September 1998.
9. Stuhlinger and Ordway, *Crusader for Space*, 11.
10. Ward, *Wernher von Braun Anekdotisch*.
11. Von Braun, 'Reminiscences of German Rocketry'.
12. Dennis Piszkiewicz, *Wernher von Braun: The Man Who Sold the Moon* (Westport, Conn.: Praeger, 1998), 1932.
13. Ward, *Wernher von Braun Anekdotisch*.
14. *Ibid.*
15. Michael J. Neufeld, *The Rocket and the Reich: Peenemünde and the Coming of the Ballistic Missile Era* (New York: Free Press/Simon and Schuster, 1995), 81–2.
16. Hermann Oberth, letter, 25 February 1972, 'X + 60 and Counting'.
17. Richard Lehnert, letter, 12 February 1972, 'X + 60 and Counting'.
18. Klaus H. Scheufelen, letter, 25 February 1972, 'X + 60 and Counting'.
19. Fritz K. Mueller, letter, 23 March 1972, 'X + 60 and Counting'.
20. Bernhard R. Tessmann, letter, 23 March 1972, 'X + 60 and Counting'.
21. *Ibid.*
22. Dorette Kersten (Schlidt) interview, Huntsville, 21 September 1998.
23. *Ibid.*
24. *Ibid.*
25. *Ibid.* Rumours persisted for years of a wartime marriage. They stemmed from the fact that in April 1943 von Braun had filed a request with the appropriate SS office for permission to marry a Berlin woman. Michael J. Neufeld wrote of the filing in 'Wernher von Braun, the SS, and Concentration Camp Labor: Questions of Moral, Political, and Criminal Responsibility', *German Studies Review* 25, no. 1 (25 January 2002), 59. Evidently this was the same woman von Braun was visiting on his weekend flights. She was a physical therapist or exercise instructor, Dorette Kersten (Schlidt) told me in an 18 December 2003, interview. But von Braun's mother eventually 'put an end' to the romance. 'The Baroness had better things in mind for him.' In an interview held on 30 November 2003, his close associate Stuhlinger told me that, to his knowledge, von Braun did not marry the Berlin woman or anyone else during the war. Neufeld likewise concluded in his previously cited paper that 'the proposed marriage to a Berlin woman never took place'.
26. William R. Lucas, interview, Huntsville, 3 November 1998.
27. Frederick I. Ordway III and Mitchell R. Sharpe, *The Rocket Team: From the V2 to the Saturn Moon Rocket – The Inside Story of How a Small Group of Engineers Changed World History* (New York: Crowell, 1979), 40–1, citing an 18 June 1971 interview with Hoelzer.
28. Lee B. James, interview, Huntsville, 28 August 1998.
29. George Barrett, 'Visit with a Prophet of the Space Age', *New York Times Sunday Magazine* (20 October 1957), 87.
30. Von Braun, 'Reminiscences of German Rocketry', 140.
31. Cromie, 'Wernher von Braun – Dean of Rocketry'.

32. Arthur C. Clarke noted in his prophetic 1951 book, *The Exploration of Space*, that it was a matter of record that von Braun had remarked in the early 1940s, 'Oh, yes, we shall get to the moon – but of course I daren't tell Hitler yet.'

33. Ernst Stuhlinger, prepared remarks at American Institute of Aeronautics and Astronautics panel discussion, Huntsville, 14 December 1998.

34. Walter W. Jacobi, letter, March 1972, 'X + 60 and Counting'.

35. Von Braun, 'Reminiscences of German Rocketry', 144.

Chapter 5: Encounters with Hitler

1. Stuhlinger and Ordway, *Crusader for Space*, 28.

2. Von Braun, 'Reminiscences of German Rocketry', 130.

3. Von Braun, 'Why I Chose America', 111.

4. Barrett, 'Visit with a Prophet of the Space Age', 88; also Pearson and Anderson, 'Wernher von Braun', 24–5.

5. Lang, 'A Romantic Urge'.

6. Bergaust, *Wernher von Braun*, 55–6.

7. *Ibid.*, 60.

8. Ward, *Wernher von Braun Anekdotisch*.

9. Stuhlinger and Ordway, *Crusader for Space*, 28, 30.

10. Hans Hueter, 'Submarine Launched Ballistic Missiles', historical report for US Army missile agency, 21 April 1960.

11. Lang, 'A Romantic Urge'.

12. *Ibid.*

13. After the war, however, Gen Dwight D. Eisenhower, the Supreme Allied Commander in Europe, was to state that if the A4/V2 had been deployed earlier it would have disrupted and probably prevented the Normandy invasion.

14. Joseph C. Moquin, interview, Huntsville, 10 October 1990.

15. Pearson and Anderson, 'Wernher von Braun', 25; also Ward, *Wernher von Braun Anekdotisch*, 27; as for Himmler's death, he killed himself after capture by biting into a phial of poison on 23 May 1945.

16. Michael J. Neufeld, guest lecture, University of Alabama in Huntsville, 17 November 1998.

17. Neufeld, *The Rocket and the Reich*, 178.

18. Gerhard H.R. Reisig, interview, Huntsville, 8 December 1998.

19. Charles Hewitt, telephone interview, Alexandria, Virginia, 9 February 1999.

20. Israel M. Levitt, telephone interview, Philadelphia, 22 June 1999.

21. Dorette Kersten (Schlidt), interview, Huntsville, 21 September 1998.

22. Neufeld, 'Wernher von Braun, the SS, and Concentration Camp Labor'.

23. *Ibid.*

24. Frederick I. Ordway III, telephone interview, Huntsville, 11 November 1998.

25. Michael J. Neufeld, guest lecture, University of Alabama in Huntsville, 17 November 1998.

26. *Ibid.*

27. Charles Hewitt, telephone interview, Alexandria, Virginia, 9 February 1999.

28. *Ibid.*

29. Lang, 'A Romantic Urge'.

30. Albert Speer, *Inside the Third Reich – Memoirs* (New York: Macmillan, 1970), cited in Stuhlinger and Ordway, *Crusader for Space*, 30.

31. Dieter K. Huzel, *Peenemünde to Canaveral* (Englewood Cliffs, NJ: Prentice-Hall, 1962).
32. Neufeld, *The Rocket and the Reich*, 47.
33. *Ibid.*, 184–5; also, Ordway and Sharpe, *The Rocket Team*, 38–9.
34. Michael J. Neufeld, guest lecture, University of Alabama in Huntsville, 17 November 1998.
35. Dorette Kersten (Schlidt), interview, Huntsville, 21 September 1998.
36. *Ibid.*
37. Hermann Oberth, preface to Ward, *Wernher von Braun Anekdotisch*.
38. Dorette Kersten (Schlidt), interview, Huntsville, 21 September 1998.
39. Stuhlinger and Ordway, *Crusader for Space*, 333.
40. Von Braun, 'Reminiscences of German Rocketry', 142.
41. *Ibid.*
42. *Ibid.*
43. Wernher von Braun speech, National Military-Industrial Conference, Chicago, 17 February 1958, reprinted in *Chemical & Engineering News* (3 March 1958), 54.
44. Gerhard H.R. Reisig, letter to editor (unpublished), *Huntsville Times*, 4 December 1998.
45. Von Braun, 'Space Man', part 1, 28.
46. Von Braun, 'Reminiscences of German Rocketry'.
47. *Life*, 'Seer of Space', 139.
48. Von Braun, 'Reminiscences of Germany Rocketry'.
49. Von Braun, 'Why I Chose America', 112 (italics in original).
50. Michael J. Neufeld, guest lecture, University of Alabama in Huntsville, 17 November 1998.
51. Von Braun, 'Reminiscences of German Rocketry', 143.
52. *Time*, 'Reach for the Stars', 24.
53. Pearson and Anderson, 'Wernher von Braun', 26.

Chapter 6: Comes Now the V2

1. Von Braun, 'Reminiscences of German Rocketry', 143.
2. David L. Christensen, interview, Huntsville, 7 September 2003.
3. Authors differ as to the exact dates (within 6–8 September) as well as the geography of the launch sites. The most detailed source on this point is Dennis Piszkiewicz, *The Nazi Rocketeers: Dreams of Space and Crimes of War* (Westport, Conn.: Praeger, 1995), 173–4. No author has specified from where in 'western Germany' the mobile V2 battery fired on those dates.
4. Stuhlinger and Ordway, *Crusader for Space*, 333.
5. Manchester *Guardian*, quoting Wernher von Braun's earlier comments on the occasion of his death, June 1977.
6. Wernher von Braun, letter to L.J. Carter of the British Interplanetary Society, 29 September 1949, in the correspondence files, Wernher von Braun Library and Archives, US Space and Rocket Center, Huntsville. There is no evidence that any of the von Braun boys ever lived in England, however.
7. Von Braun, 'Reminiscences of German Rocketry', 144.
8. Lang, 'A Romantic Urge'.
9. Bergaust, quoting unnamed von Braun associate in *Wernher von Braun*, 95.

10. A.V. 'Val' Cleaver, letter, 16 February 1972, 'X + 60 and Counting'.
11. Walter Dornberger, epilogue to *Astronautical Engineering and Science: From Peenemünde to Planetary Space – Honoring the Fiftieth Birthday of Wernher von Braun*, edited by Ernst Stuhlinger et al. (New York: McGraw-Hill, 1963).
12. In the United States von Braun received a number of patents for his technical inventions. They ranged from a means of lengthening a rocket's propellant tanks (and thus increasing its thrust and range) within the same outside dimensions to a manoeuvrable 'bottle suit' system for astronauts' extra-vehicular activities. As one of several examples of von Braun's personally interceding with technical fixes of engineering problems, NASA engineer-manager Bob Schwinghamer recalled the time he briefed von Braun on a vacuum pump planned for a space-flight hardware subsystem. 'You're taking a vacuum pump into the vacuum of space?!' von Braun exploded. He suggested a different kind of pump, which Schwinghamer's group went on to use successfully.
13. Bergaust, Wernher von Braun letter to enquiring West Point cadet reprinted in *Wernher von Braun*.
14. Most of the reclusive Goddard's voluminous patents and papers were kept under wraps, for personal and government security reasons, until after the war and the American physicist's death in 1945, before he and von Braun could meet. In 1950, in connection with a claim of patent infringement filed with the US government by Goddard's widow, estate and the Guggenheim Foundation, von Braun was asked by officials to assess the matter. See Appendix I for details.
15. Georg von Tiesenhausen, telephone interviews, Huntsville, 21 and 24 June 2004; and Hans Hueter, 'Submarine Launched Ballistic Missiles', an historical report to the US Army missile agency, 20 April 1960. See also Ordway and Sharpe, *The Rocket Team*, 255; and Neufeld, *The Rocket and the Reich*, 55. Interviewee von Tiesenhausen told me that a postwar friend commented to him that 'it was just as well' that the Prüfstand XII project was cut short and no V2s were fired from sea at New York, 'considering your future and that you came to the United States!'
16. Thomas Franklin [Hugh McInnish], *An American in Exile: The Story of Arthur Rudolph* (Huntsville, Ala.: Christopher Kaylor Publishers, 1987).
17. Neufeld, *The Rocket and the Reich*, 213, 226–7.
18. *Ibid.*, 207.
19. Neufeld, 'Wernher von Braun, the SS, and Concentration Camp Labor'.
20. Stuhlinger and Ordway, *Crusader for Space*, 44. Information on Gerhard H.R. Reisig from December 2003 interview by Ernst Stuhlinger.
21. *Ibid.*
22. Neufeld, *The Rocket and the Reich*, 202.
23. Gerhard H.R. Reisig, letter to editor (unpublished), *Huntsville Times*, 4 December 1998.
24. Neufeld, *The Rocket and the Reich*, 227–8.
25. Stuhlinger and Ordway, *Crusader for Space*, 44.
26. Ernst Stuhlinger, written response to queries by author, 13 March 2000.
27. Von Braun letter to Albin Sawatski, director of Mittelwerk, 15 August 1944. Text of letter included in Ernst Stuhlinger, 'Wernher von Braun and

Concentration Camp Labor: An Exchange', *German Studies Review*, 26 January, 2003, 121–4.

28. Neufeld, 'Wernher von Braun, the SS, and Concentration Camp Labor', 69.
29. André Sellier, a Mittelbau-Dora survivor, member of Sadron's work unit, and author of a 1998 history of the camp (Sellier, *Histoire du camp de Dora*, 117–18), cited in Neufeld's 'Wernher von Braun, the SS, and Concentration Camp Labor', 68.
30. Charles Sadron, 'A la Usine de Dora', in *Témoignages Strasbourgeois: De l'Université aux Camps de Concentration* (Paris, 1947), 198–9, cited in Neufeld's 'Wernher von Braun, the SS, and Concentration Camp Labor', 68.
31. Ernst Stuhlinger, interview, Huntsville, 16 August 2003.
32. Neufeld, 'Wernher von Braun, the SS, and Concentration Camp Labor', 69.
33. V.H. Bilet and J.D. McPhilimy, 'Production and Disposition of German A-4 (V2) Rockets', staff study no. A-55-2167-N1, DDC no. ATI-18315, March 1948, Headquarters Air Material Command, Wright Field, Dayton, Ohio, as cited in Stuhlinger and Ordway, *Crusader for Space*, 42.
34. Stuhlinger and Ordway, *Crusader for Space*, 40, 42.
35. Ernst Stuhlinger, memorial essay on Wernher von Braun on what would have been his 90th birthday, for the *Huntsville Times*, 23 March 2002.
36. Huzel, *Peenemünde to Canaveral*, 117.
37. Wernher von Braun, letter to *Paris Match* editors, 26 April 1966. The contents of the letter were disclosed, in large part, by Stuhlinger and Ordway in *Crusader for Space*, 51.
38. Stuhlinger and Ordway, *Crusader for Space*, 53.
39. Michael J. Neufeld, guest lecture, University of Alabama in Huntsville (UAH), 17 November 1998.
40. Michael J. Neufeld, guest lecture, Athens (Ala.) State University, 11 October 2000.
41. Gerhard H.R. Reisig, letter to editor (unpublished), *Huntsville Times*, 3 December 1998, in response to Neufeld's UAH lecture and resultant news coverage.
42. Ernst Stuhlinger et al. (unnamed), 'Memorandum', 10 October 1998, in advance of Neufeld's UAH lecture.
43. Ordway and Sharpe, *The Rocket Team*, 251, 406.
44. *Ibid.*, 79.
45. *Ibid.*, 245.
46. Neufeld, *The Rocket and the Reich*, 264.
47. Hans Dollinger, *The Decline and Fall of Nazi Germany and Imperial Japan* (New York: Bonanza Books/Crown, 1967), 275.
48. *Financial Times*, quoting in June 1977, on Wernher von Braun's death.
49. *Ibid.*
50. 'Film on von Braun Has a Stormy Bow', *New York Times*, 20 August 1960; see also 'Von Braun Gets Boos, Cheers', *Huntsville Times*, 21 August 1960.
51. Von Braun, 'Reminiscences of German Rocketry', 144–5.
52. Bob Ward, *A Funny Thing Happened on the Way to the Moon* (New York: Fawcett, 1969), 50; the A9/10 'Amerika rocket' is referred to also by von Braun, 'Reminiscences of German Rocketry', 144–5, and by Ernst Stuhlinger, 'The "America Rocket": It Was Just a Name', *Huntsville Times*, 26 March 2000.
53. Ordway and Sharpe, *The Rocket Team*, 254.

54. James L. Daniels Jr, 'A Biography of Wernher von Braun', Appendix 1 in Stuhlinger et al., *Aeronautical Engineering and Science*, 369.
55. Walters, *Wernher von Braun*, 79–80.
56. Edward D. Mohlere, interview, Huntsville, 20 September 1998.
57. Ordway and Sharpe, *The Rocket Team*, 118–19.
58. Von Braun, 'Why I Chose America', 112.
59. Lang, 'A Romantic Urge'.

Chapter 7: Bound for America

1. Wernher and Maria have been referred to in print variously as second and first cousins. Von Braun wrote on at least one occasion that they were second cousins. But the couple shared the same von Quistorp grandparents, making them first cousins.
2. Wernher von Braun, 'Space Man – The Story of My Life', *American Weekly*, 27 July 1958 (part two of a three-part series), 11–12.
3. Hannes Luehrsen, letter, March 1972, 'X + 60 and Counting'.
4. Ordway and Sharpe, *The Rocket Team*, 256–63.
5. Willy Ley, *Rockets, Missiles, and Space Travel*, rev. edn (New York: Viking, 1957), 244.
6. Von Braun, 'Why I Chose America', 112.
7. Lang, 'A Romantic Urge'.
8. Bergaust, *Wernher von Braun*, 92.
9. Winston S. Churchill, *Triumph and Tragedy* (Boston: Houghton Mifflin, 1953).
10. *Time*, 'Reach for the Stars', 24.
11. Ordway and Sharpe, *The Rocket Team*, 2.
12. Lang, 'A Romantic Urge'.
13. Bill O'Hallaren, letter to the editor, *New Yorker* (26 May 1951), 106.
14. Walters, *Wernher von Braun*, 91.
15. Lang, 'A Romantic Urge.'
16. John G. Zierdt, interview, Huntsville, 21 October 1998.
17. Johann J. Klein, letter, March 1972, 'X + 60 and Counting'.
18. Sometimes 'Ludi'.
19. Stuhlinger and Ordway, *Crusader for Space*, 63.
20. *Ibid.*
21. John Barbour and the Associated Press, *Footprints on the Moon* (New York: Associated Press, 1969), 17.
22. Von Braun mini-biography, in Shirley Thomas, *Men of Space: Profiles of the Leaders in Space Research, Development, and Exploration*, vol. I (Philadelphia: Chilton, 1960), 142.
23. Ward, *Wernher von Braun Anekdotisch*.
24. Stuhlinger and Ordway, *Crusader for Space*, 65.
25. *Ibid.*, 245.
26. Wilhelm Angele, letter, 23 March 1972, 'X + 60 and Counting'.
27. Stuhlinger and Ordway, *Crusader for Space*, 66.
28. *Ibid.*
29. The *Huntsville Times* and other press accounts of the naturalisation ceremonies on 14 April 1955 for Wernher von Braun and some forty other German rocketeers and their families.

30. Thomas, *Men of Space*, 142.
31. Stuhlinger and Ordway, *Crusader for Space*, 227.
32. Long after what Toftoy's widow, Hazel, called those 'tough years' in immediate postwar Germany had passed, she reminded von Braun of his appearance on the 1950s episode of the popular Ralph Edwards's *This Is Your Life!* television show that honoured the then-major-general. Even then, she wrote to von Braun in a congratulatory letter of 20 February 1972, 'You emphasized his [Toftoy's] humanitarian efforts.'
33. 'If people only knew the screening von Braun and the V2 team members went through,' retired US Army Maj-Gen John G. Zierdt told me in a 1999 interview. 'Believe me, they were checked out thoroughly' before being invited individually to America. Character checks went so far as to include interviews with early teachers about possible playground bullying tendencies and youthful Fascist leanings, Zierdt recalled. The general, whose career included Second World War service in Europe, said he had access to the Peenemünders' security records. Zierdt said he had no knowledge that the dossiers of some were sanitised to make them more palatable for import by America. Assigned to Redstone Arsenal for ten of the eleven years during the period from 1956 to 1967, he worked with von Braun and the team at various Army missile agencies there early in that period, and later was commanding general.
34. Stuhlinger and Ordway, *Crusader for Space*, 66.
35. All told, some 140 German missile experts eventually came to the United States, a few as late as the mid-1950s. It took that long for some to slip out through the Soviet bloc's Iron Curtain and reach the West.
36. Ward, *Wernher von Braun Anekdotisch*.
37. Mike Marshall, 'Launching a Dream', *Huntsville Times*, 24 September 1995.
38. Howard Benedict, Associated Press, 'The Rocketeers of Peenemünde', in the *Buffalo (NY) News* (p. F-6) and other newspapers, 20 January 1985.
39. Among the more notable German missile men who ended up in Russian hands was Helmut Gröttrup, who was eventually repatriated to Germany. He was one of the two Peenemünde rocketeers that Himmler had the Gestapo arrest with von Braun in March 1944.
40. Lang, 'A Romantic Urge'.
41. Thomas, *Men of Space*, 135.
42. Lang, 'A Romantic Urge.'
43. *Ibid.*
44. Stuhlinger and Ordway, *Crusader for Space*, 260.
45. Dornberger, after his release by the British, emigrated to the United States. He worked first as a consultant to the Air Force and then joined Bell Aircraft as a vice president. He retired from the company in 1960 and moved to Mexico, where he maintained contact with von Braun and other V2 associates. Walter Dornberger died in 1980, during a visit to Germany.
46. Erich W. Neubert, letter, 23 March 1972, 'X + 60 and Counting'. Thirteen years would pass before Neubert laid eyes again on his homeland, when he visited family and old friends.
47. Thomas, *Men of Space*, 143.
48. Bergaust, *Wernher von Braun*, 124.
49. Walters, *Wernher von Braun*, 96.
50. Von Braun, 'Why I Chose America', 114.

Chapter 8: A Fort Called Bliss

1. Von Braun, 'Why I Chose America', 112.
2. Stuhlinger and Ordway, *Crusader for Space*, 227.
3. Dorette Kersten (Schlidt), interview, Huntsville, 21 September 1998.
4. Stuhlinger and Ordway, *Crusader for Space*, 76.
5. Lt-Col William E. Winterstein, retired, letter, 23 March 1972, 'X + 60 and Counting'.
6. Ernst Stuhlinger, 'German Rocketeers Find a New Home', talk at Burritt Museum, Huntsville, 21 September 1995.
7. Stuhlinger and Ordway, *Crusader for Space*, 73.
8. Pam Rogers, 'A Member of the Old Team Looks Back', *Redstone Rocket*, Redstone Arsenal, Huntsville (5 February 1986), 9.
9. Konrad Dannenberg, interview, Huntsville, 10 September 1998.
10. Lang, 'A Romantic Urge'.
11. Stuhlinger, 'German Rocketeers Find a New Home'.
12. Stuhlinger, telephone interview, Huntsville, 28 February 2001. See also Bergaust, *Wernher von Braun*, 130–31.
13. Johann J. Klein, letter, 29 February 1972, 'X + 60 and Counting'. The motto comes from an old European cavalry saying that jump-masters would use to exhort raw trainees having trouble getting their mounts to leap ditches in open terrain. Adopted by von Braun in urging young rocketeers to tackle tough tasks with gusto, the full maxim goes: 'Throw your heart over the ditch, and your horse will follow!'
14. Werner K. Gengelbach, letter, 25 February 1972, 'X + 60 and Counting'.
15. Wolfgang Steurer, letter, March 1972, 'X + 60 and Counting'.
16. *Ibid.*
17. Arthur C. Clarke, letter to editor of *Wireless World*, February 1945, cited by Stuhlinger and Ordway in *Crusader for Space*.
18. *Time*, 'Reach for the Stars', 24.
19. Bob Ward, quoting Wernher von Braun in 'United States Almost Lost Its "Mr. Space" 25 Years Ago, von Braun Letter Reveals,' *Huntsville Times*, 30 June 1972.
20. Cromie, 'Wernher von Braun – Dean of Rocketry', 10-D.
21. In November 2002 Magnus von Braun, who was in ill health in a nursing home in Phoenix, Arizona, sent word through his daughter in response to my queries that he had no recollection of the platinum episode. It may have occurred, it may not have; he simply could not remember, he said. He died in June 2003.
22. Conversations with Ernst Stuhlinger, Konrad Dannenberg, et al., in the autumn of 2002.
23. Stuhlinger and Ordway, *Crusader for Space*, 77.
24. Lang, 'A Romantic Urge'.
25. Stuhlinger and Ordway, *Crusader for Space*, 82–3.
26. *El Paso Times*, 1 July 1947.
27. Von Braun, 'Space Man – The Story of My Life', *American Weekly*, 3 August 1958 (part 3 of a three-part series), 14.
28. Dorette Kersten (Schlidt), interview, Huntsville, 21 September 1998.
29. Pearson and Anderson, *Wernher von Braun*, 25.

30. Von Braun, 'Space Man', part 3, 14.
31. Ward, *Wernher von Braun Anekdotisch*.
32. Thomas, *Men of Space*, 144.
33. Werner Dahm, interview, Huntsville, 6 December 1998.
34. Thomas, *Men of Space*, 145.
35. Bergaust, *Wernher von Braun*, 151.
36. William C. Fortune, letter, 22 February 1972, 'X + 60 and Counting'.
37. Ward, *A Funny Thing Happened on the Way to the Moon*, 49–50; see also Daniel Lang, 'A Reporter at Large: What's Up There', *New Yorker*, (31 July 1948), 44–5.
38. Wernher von Braun letter to L.J. Carter of the British Interplanetary Society, 29 September 1949, correspondence files, Wernher von Braun Library and Archives, US Space and Rocket Center, Huntsville.
39. Hubertus Strughold, letter, 23 March 1972, 'X + 60 and Counting'.
40. Mike Marshall, 'Launching a Dream', *Huntsville Times*, 24 September 1995.
41. *Life*, 'Seer of Space'.
42. Interviews with Wernher von Braun rocket-team old-timers, Huntsville, late 1990s.
43. Wernher von Braun letter to then-Maj James Hamill, 12 January 1948, correspondence files, Wernher von Braun Library and Archives, US Space and Rocket Center, Huntsville.
44. Wernher von Braun, interview, Huntsville, mid-June 1972.
45. James P. Hamill, telephone interview, Washington, DC area, mid-June 1972.
46. Wernher von Braun, interview, Huntsville, mid-June 1972. Von Braun requested that I kill a proposed newspaper article about his newly discovered old letter. 'But this is a piece of history,' I protested. 'But history that never happened,' von Braun countered. The article, with mitigating comments by him and Jim Hamill, appeared in the *Huntsville Times* on 30 June 1972.
47. Hannes Luehrsen, interview, Huntsville, March 1965, for Bob Ward's, 'Signal for Germans' Move Here Came 15 Years Ago', *Huntsville Times* (1 April 1965), 2.
48. Wernher von Braun speech, Georgia state convention of the American Legion, Savannah, 25 July 1959.
49. Dorette Kersten (Schlidt), interview, Huntsville, 21 September 1998.
50. Mike Wright, NASA–MSFC historian, 'Huntsville and the Space Program', *Alabama Heritage* magazine (parts 1 and 2, spring and summer 1998), part 1, 42.

Chapter 9: New Home Alabama

1. David G. Harris, interview, Huntsville, January 1967.
2. Weldon Payne, *Web to the Stars: A History of the University of Tennessee Space Institute* (Dubuque, Iowa: Kendall/Hunt Publishing, 1992), 10.
3. Patrick W. Richardson, interview, Huntsville, 14 August 1998.
4. Bergaust, *Wernher von Braun*, 186.
5. Von Braun, 'Why I Chose America', 111.
6. Cromie, 'Wernher von Braun – Dean of Rocketry', D-11.
7. Paul O'Neil, 'The Splendid Anachronism of Huntsville', *Fortune*, June 1962, 238.
8. Hannes Luehrsen, interview for Ward's, 'Signal for Germans' Move Here', 2.

9. Werner Dahm, interview, Huntsville, 6 December 1998.
10. Jimmy Walker, letter, 14 February 1972, 'X + 60 and Counting'.
11. Louis Salmon, letter, 28 February 1972, 'X + 60 and Counting'.
12. M. Beirne Spragins, letter, 18 February 1972, 'X + 60 and Counting'.
13. Wernher von Braun correspondence files, Wernher von Braun Library and Archives, US Space and Rocket Center.
14. Author's recollection of Mr O'Neal's account, quoted in Bergaust, *Wernher von Braun*, 170.
15. M. Beirne Spragins, letter, 18 February 1972, 'X + 60 and Counting'.
16. In his spare lifestyle on modest Army civilian pay in the early 1950s, associates noted, von Braun drove a used Chevrolet and owned a cheap gramophone for playing his cherished classical records. He was a hunter who owned no shotgun. There was no television set in the von Braun household, by his decree.
17. Hans F. Gruene, letter, March 1972, 'X + 60 and Counting'.
18. Frances Gates Moore, interview, Huntsville, 24 May 1999.
19. Wernher von Braun letter to the British Interplanetary Society's A.V. 'Val' Cleaver, London, 30 July 1951, Wernher von Braun Library and Archives, US Space and Rocket Center.
20. Ward, *Wernher von Braun Anekdotisch*.
21. Patrick W. Richardson, interview, Huntsville, 14 August 1998.
22. *Ibid.*, and letters from Patrick Richardson, 28 February 1972, and Martha H. Richardson Simms [Rambo], 25 February 1972, 'X + 60 and Counting'.
23. Edward D. Mohlere, interview, Huntsville, 20 September 1998.
24. Barrett, 'Visit with a Prophet of the Space Age', 87.
25. Wernher von Braun, 'Teamwork: Key to Success in Guided Missiles', *Missiles and Rockets*, October 1956, 41.
26. James K. Hoey, letter, 23 March 1972, 'X + 60 and Counting'.
27. George C. Bucher, letter, 28 February 1972, 'X + 60 and Counting'.
28. Donald Bowden, telephone interview, Huntsville, 1 June 2000.
29. Larkin Davis, letter, 23 March 1972, 'X + 60 and Counting'.
30. Bonnie Holmes, letter, 23 March 1972, 'X + 60 and Counting'.
31. *Ibid.*
32. Stuhlinger and Ordway, *Crusader for Space*, 256.
33. Dale James, 'Behind the Scenes with von Braun', *Huntsville Times*, 8 September 1994.
34. Bob Ward, *The Light Stuff* (Huntsville, Ala.: Jester Books, 1982), 106.
35. Sam K. Hoffman, letter, March 1972, 'X + 60 and Counting'.
36. Arlie R. Trahern Jr, letter, 23 March 1972, 'X + 60 and Counting'.
37. John C. Goodrum Sr, interview, Huntsville, 28 January 2003.
38. James M. Gavin, letter, 14 February 1972, 'X + 60 and Counting'.
39. William R. Lucas, interview, Huntsville, 3 November 1998.
40. William R. Lucas, letter, March 1972, 'X + 60 and Counting'.
41. Wernher von Braun's father died in 1972 at the age of 94; his mother had died earlier, in 1959, of colon cancer. Their eldest son, career German diplomat Sigismund, lived to be 87, dying in 1998.
42. Tom Carney, 'FBI Files of Wernher von Braun', *Old Huntsville* magazine, 2001, 105.
43. Stuhlinger and Ordway, *Crusader for Space*, 227.

Chapter 10: Early Media Trail

1. Rick Chappell, remarks at Wernher von Braun Exploration Forum, Huntsville, 21 September 1993.
2. Fred L. Whipple, letter, 2 March 1972, 'X + 60 and Counting'.
3. Ron Miller and Frederick C. Durant III, *The Art of Chesley Bonestell* (London: Paper Tiger/Collins and Brown, 2001).
4. Chesley Bonestell, letter, 16 February 1972, 'X + 60 and Counting'.
5. Cornelius Ryan, letter, 23 February 1972, 'X + 60 and Counting'. Cornelius Ryan gained further renown as author of a trilogy of best-selling wartime histories, beginning with *The Longest Day*, which was set on D-Day.
6. *Ibid.*
7. A.V. Cleaver, letter, 6 April 1952, Wernher von Braun correspondence files, Wernher von Braun Library and Archives, US Space and Rocket Center.
8. Milton W. Rosen, letter, 16 February 1972, 'X + 60 and Counting'.
9. *Ibid.*
10. *Ibid.*
11. For full accounts of the *Collier's* series, see Fred L. Whipple's article, 'Recollections of Pre-Sputnik Days', in *Blueprint for Space* (Washington, DC: Smithsonian Institution Press, 1992), edited by Frederick I. Ordway III, and the chapter following Whipple's, 'The *Collier's* and Disney Series', by Randy Liebermann.
12. Jonathan Norton Leonard, *Flight into Space: The Facts, Fancies, and Philosophy* (New York: Random House, 1953), 79–80.
13. *Ibid.*, 80.
14. *Ibid.*, 101.
15. President Ronald Reagan came close to matching it, however, with his 1980s Strategic Defense Initiative (SDI) concept, disparaged as the 'Star Wars' plan by many on the political left and in the news media. Never taken much beyond research and early developmental phases, the SDI challenge was still credited as part of the US military build-up that helped force the break-up of the Soviet Union.
16. Adolf K. Thiel, [1972] letter undated, 'X + 60 and Counting'.
17. *Ibid.*
18. Von Braun's super-salesmanship was acknowledged by author Dennis Piszkiewicz in the title of his acerbic 1998 biography, *Wernher von Braun: The Man Who Sold the Moon*. Piszkiewicz's jaundiced slant on the whole von Braun saga had earlier surfaced in his 1995 book, *The Nazi Rocketeers*.
19. Adolf K. Thiel, telephone interview, Palos Verdes Estates, California, 1 October 1998.
20. Werner Dahm, interview, Huntsville, 6 December 1998.
21. William R. Bosche, letter, 10 February 1972, 'X + 60 and Counting'.
22. Wernher von Braun, letter to L.R. Shepherd of the BIS, Chilton, 3 March 1951, Wernher von Braun correspondence files, Wernher von Braun Library and Archives, US Space and Rocket Center.
23. Wernher von Braun, letter to Kenneth Gatland of the BIS and IAF, London, 24 April 1951, Wernher von Braun correspondence files, Wernher von Braun Library and Archives, US Space and Rocket Center.

24. Wernher von Braun, letter to Val Cleaver of the BIS and IAF, London, 12 May 1951, Wernher von Braun correspondence files, Wernher von Braun Library and Archives, US Space and Rocket Center.
25. Frederick C. Durant III, telephone interview, Chevy Chase, Maryland, 2 November 1998.
26. Wernher von Braun, letter to L.J. Carter of the BIS, 29 September 1949, Wernher von Braun correspondence files, Wernher von Braun Library and Archives, US Space and Rocket Center.
27. Compendium of Wernher von Braun quotations, Wernher von Braun Library and Archives, US Space and Rocket Center.

Chapter 11: Towards the Cosmos

1. David G. Harris, chief of Redstone Public Information Office, interview, Huntsville, January 1967.
2. Charles A. Lundquist, interview, Huntsville, 19 November 1998.
3. Stuhlinger and Ordway, *Crusader for Space*, 229.
4. *Ibid.*
5. Ward, *Wernher von Braun Anekdotisch*, 51.
6. Edward D. Mohlere, interview, Huntsville, 20 September 1998.
7. Wernher von Braun interview, 'Space Travel: Sooner Than You Think', *US News & World Report*, 9 September 1955, 68; reprinted by magazine as 'Space Travel: When It Is Coming . . . What It Will Be Like', 18 October 1957, 41.
8. Thomas, *Men of Space*, 150.
9. *Ibid.*
10. Ernst Stuhlinger, letter to Michael J. Neufeld of Smithsonian's National Air and Space Museum, 4 December 1998.
11. Gerhard B. Heller, letter, 23 March 1972, 'X + 60 and Counting'.
12. *Time*, 'Reach for the Stars', 24.
13. Ernst Stuhlinger, confidential government memorandum, October 1957 (see note 23 below).
14. James M. Gavin, letter, 14 February 1972, 'X + 60 and Counting'.
15. *Ibid.*
16. *Time*, 'Reach for the Stars', 24.
17. Ernst Stuhlinger, letter, 25 February 1972, 'X + 60 and Counting'.
18. R.P. Hazzard, letter, 23 March 1972, 'X + 60 and Counting'.
19. The 'C' in Jupiter-C stood for 'Composite', referring to the solid-rocket upper stages added to the elongated, liquid-fuelled Redstone missile; the 'Jupiter' designation was a bureaucratic name game to use funds from the budget of the real Jupiter IRBM.
20. *Time*, 'Reach for the Stars', 24.
21. Stuhlinger and Ordway, *Crusader for Space*, 128.
22. Mike Gray, *Angle of Attack: Harrison Storms and the Race to the Moon* (New York: W.W. Norton, 1992), 15–16.
23. Wernher von Braun's chief scientist, Ernst Stuhlinger, and top Soviet space scientist Leonid Sedov met a few days after the Sputnik shocker. This was on October 7–8 in Barcelona, at the eighth Congress of the International Astronautical Federation. As Stuhlinger wrote in a five-page 'For Official Use Only' memo dated 29 October 1957, Sedov asked him why the

United States was pinning its satellite hopes on the untried, sharply weight-limited Vanguard concept when it possessed proven, powerful boosters, and why did Dr von Braun go along with this? Sedov was incredulous when told von Braun had nothing to do with the decision and was not being consulted on Project Vanguard, Stuhlinger wrote. (Copy in author's files.)

Chapter 12: Nobody's Perfect

1. Foster A. Haley, interview, Huntsville, 29 March 1999.
2. Patrick W. Richardson, interview, Huntsville, 14 August 1998.
3. Evelyn Spearman, 'Lacey's Spring Native Remembers Her Boss', *Huntsville News*, 9 April 1970.
4. Thomas L. Shaner, interview, Huntsville, 4 March 1999.
5. *Ibid.*
6. David Hinkle, interview, Huntsville, 2 December 1998.
7. Stuhlinger and Ordway, *Crusader for Space*, 223; Ward, *Wernher von Braun Anekdotisch*.
8. Arthur C. Clarke, letter, 21 February 1972, 'X + 60 and Counting'.
9. Patrick W. Richardson, interview, Huntsville, 14 August 1998.
10. Ward, *Wernher von Braun Anekdotisch*.
11. Jonathan Eberhart, 'The House That Space Built', *Science News* 93, no. 13 (April 1968), 363.
12. Edward D. Mohlere, interview, Huntsville, 20 September 1998.
13. J.N. Foster, letter, 23 March 1972, 'X + 60 and Counting'.
14. William R. Lucas, interview, Huntsville, 3 November 1998.
15. *Ibid.*
16. Thomas L. Shaner, interview, Huntsville, 4 March 1999.
17. Stuhlinger and Ordway, *Crusader for Space*, 253. He excluded from that generalisation the launch of the first manned lunar landing mission, at 9.32 a.m. on 16 July 1969, from Cape Canaveral.
18. Thomas L. Shaner, interview, Huntsville, 4 March 1999.
19. Stuhlinger and Ordway, *Crusader for Space*, 278.
20. Thomas L. Shaner, interview, Huntsville, 4 March 1999.
21. Stuhlinger and Ordway, *Crusader for Space*, 276.
22. Thomas L. Shaner, interview, Huntsville, 4 March 1999.
23. J.N. Foster, interview, Huntsville, 2 September 1998.
24. Lee B. James, letter, 9 February 1972, 'X + 60 and Counting'.
25. Ward, *Wernher von Braun Anekdotisch*.
26. Iris von Braun, 'Wernher von Braun: As His Daughter Sees Him', *Huntsville Times*, 26 May 1963.
27. Bonnie Holmes, telephone interview, Eva, Alabama, 26 October 1998.
28. Thomas L. Shaner, interview, Huntsville, 4 March 1999.
29. James T. Shepherd, interview, Huntsville, 8 October 1998.
30. Stuhlinger and Ordway, *Crusader for Space*, 246.
31. Wernher von Braun paper, 'Responsible Scientific Investigation and Application', presented in absentia at the Lutheran Church of America synod, Philadelphia, 29 October 1976.
32. James L. Daniels Jr, interview, Huntsville, 19 August 1998.

33. Thomas L. Shaner, interview, Huntsville, 4 March 1999.
34. Leland F. Belew, interview, Huntsville, 28 October 1998.
35. Correspondence files, Wernher von Braun Library and Archives, US Space and Rocket Center.
36. Rudolf H. Schlidt, interview, Huntsville, 18 January 2001.
37. *Ibid.*
38. Ernst Stuhlinger, telephone interview, Huntsville, 28 February 2001.
39. Robert Lindstrom, interview, Huntsville, 6 October 1998.
40. Thomas L. Shaner, interview, Huntsville, 4 March 1999.
41. Peter Cobun, 'Catalyst, Inspirer, Promoter – von Braun Forged the Way', *Huntsville Times*, 17 June 1977.
42. William R. Lucas, interview, Huntsville, 3 November 1998. Lucas became a favourite of von Braun's, however. The space centre director earmarked him as an eventual successor.
43. James L. Kingsbury, telephone interview, Huntsville, 24 September 2001.
44. Thomas L. Shaner, interview, Huntsville, 4 March 1999.
45. *Ibid.*
46. Harry Atkins, interview, Huntsville, 27 May 1999.
47. Spearman, 'Lacey's Spring Native Remembers Her Boss'.
48. I. M. Levitt, telephone interview, Philadelphia, 22 June 1999.
49. John G. Zierdt, interview, Huntsville, 21 October 1998.
50. David Newby, telephone interview, Guntersville, Alabama, 22 December 1998.
51. Frederick I. Ordway III, telephone interview, Arlington, Virginia, 13 November 1998.
52. Wernher von Braun associate, anonymity requested, 1999 interview.
53. Robert Lindstrom, interview, Huntsville, 6 October 1998.

Chapter 13: New Age of Space

1. W.L. Halsey Jr, interview, Huntsville, 19 June 2004.
2. Gordon L. Harris, *$elling Uncle Sam* (Hicksville, NY: Exposition Press, 1976), 81.
3. Ernst Stuhlinger, 'Sputnik 1957: Memories of an Old-timer', talk given in Washington, DC, 4 October 1997.
4. John B. Medaris, with Arthur Gordon, *Countdown for Decision* (New York: Putnam, 1960), 155.
5. *Ibid.*
6. *Time*, 'Reach for the Stars', 24.
7. *Ibid.*
8. I.M. Levitt, Associated Press news story, 7 October 1957, *Huntsville Times* and elsewhere.
9. John Barbour and the Associated Press, *Footprints on the Moon* (New York: Associated Press, 1969), 20–21.
10. News report of Senator Lyndon Johnson's statement to the press, week of 5 October 1957. See Brian Trumbone, 'Sputnik, 1957', www.Stocksand News.com, 2000.
11. Truman F. Cook, letter, 25 February 1972, 'X + 60 and Counting'.
12. *Life*, 'Seer of Space', 136.

13. Wernher von Braun article for the Associated Press, 'Manned Flight to Moon in Five Years Possible – von Braun', in the *Birmingham (Ala.) News* and elsewhere, 10 November 1957.

14. *Life*, 'Seer of Space', 136.

15. Stuhlinger and Ordway, *Crusader for Space*, 132, 133.

16. Medaris with Gordon, *Countdown for Decision*, 168.

17. *Time*, 'Reach for the Stars', 25.

18. Ernst Stuhlinger, correspondence with author, from Huntsville, 28 March 1999; see also Stuhlinger and Ordway, *Crusader for Space.*

19. James A. Van Allen, correspondence with author, from Iowa City, Iowa, 16 March 1999; see also Van Allen, *Origins of Magnetospheric Physics* (Washington, DC: Smithsonian Institution Press, 1983).

20. John G. Zierdt, letter, 28 February 1972, 'X + 60 and Counting', retold in interview, Huntsville, 21 October 1998.

21. As events rapidly unfolded, Joachim Kuettner did not have a hand in sending the first man into space. The Russians beat him there with a young pilot named Yuri Gagarin. But he did help the von Braun team rocket the first American into space. Navy Cdr (later Adm) Alan Shepard always maintained he would have been the first man in space if von Braun had not added one more test flight with yet another chimpanzee.

22. Joachim Kuettner, 'Und Familie', letter, March 1972, 'X + 60 and Counting'.

23. *Time*, 'Reach for the Stars', 25.

24. Wernher von Braun, 'My Most Exciting Moment', *Popular Mechanics* (September 1960), 86.

25. *Ibid.*, 87.

26. Explorer 1 stayed aloft for thirteen years, far beyond expectations, and discovered the Van Allen radiation belts encircling the Earth.

27. *Time*, 'Reach for the Stars', 25.

28. Stuhlinger and Ordway, *Crusader for Space*, 140.

29. Wernher von Braun speech before National Military-Industrial Conference, Chicago, 17 February 1958; reprinted in *Chemical & Engineering News* (3 March 1958), 56.

30. *Time*, 'Reach for the Stars', 25.

31. William H. Pickering, telephone interview, Altadena, California, 16 September 1998.

32. In a 1972 letter wishing von Braun a happy 60th birthday, Bill Pickering alluded to the borrowed presidential white tie and asked: 'By the way, did you ever return it?' He did indeed, long-time von Braun secretary Bonnie Holmes confirmed to me decades later.

33. William H. Pickering, telephone interview, Altadena, California, 16 September 1998.

34. Correspondence files, Wernher von Braun Library and Archives, US Space and Rocket Center.

35. Hans Hueter, 'Submarine Launched Ballistic Missiles', historical report to US Army missile agency, 21 April 1960.

36. *Ibid.*

37. Von Braun, 'Space Man', part 1, 7.

38. Stuhlinger and Ordway, *Crusader for Space*, 165.

39. Hermann Oberth, letter, 25 February 1972, 'X + 60 and Counting'.

40. Donald R. Bowden, telephone interview, Huntsville, 1 June 2000.
41. *Ibid.*
42. Rankin A. Clinton Jr, interview, Huntsville, 16 January 2001.
43. During the dramatic 'thirteen days' of the 1961 Cuban missile crisis, Clinton and the then-head of his unit at Redstone, Carl Duckett, essentially set up camp in Washington and regularly briefed President John Kennedy, Attorney-General Robert Kennedy and senior security officials. The Army unit's aerial photo analysis of the secret Soviet missile sites in Cuba figured prominently in the tense episode. Duckett was later recruited to be a deputy director of the Central Intelligence Agency (CIA) and to create similar foreign missile intelligence capabilities within that agency. The Huntsville 'spook' unit evolved into the expanded Department of Defense Missile and Space Intelligence Center, still based at Redstone, with von Braun's foreign-intelligence coach – 'Randy' Clinton – as its director.
44. Rankin A. Clinton Jr, interview, Huntsville, 16 January 2001.
45. T. Keith Glennan, letter, 10 February 1972, 'X + 60 and Counting'.
46. Charles A. Lundquist, interview, Huntsville, 19 November 1998.
47. Kraft, *Flight*, 83, 127; Charles A. Lundquist, interview, Huntsville, 19 November 1998; Frank Williams, telephone interview, Birmingham, Alabama, 31 January 2004; Maxime Faget, telephone interview, Apollo, Texas, 12 July 2003.
48. Ernst Stuhlinger, telephone interview, Huntsville, 22 April 2003.
49. James T. Shepherd, telephone interview, Huntsville, 6 November 2001.
50. Josef Boehm, letter, February 1972, 'X + 60 and Counting'.
51. Howard Benedict, correspondence with author, from Titusville, Florida, 15 December 1998.
52. *Ibid.*
53. *Ibid.*
54. Vachel Stapler, interview, Huntsville, 29 November 2000.
55. Stuhlinger and Ordway, *Crusader for Space*, 231, 241–2.

Chapter 14: Challenge of the Moon

1. Robert Schwinghamer, telephone interview, Huntsville, 29 September 1998.
2. Andrew J. Dunar and Stephen P. Waring, *Power to Explore: A History of Marshall Space Flight Center, 1960–1990* (Washington, DC: NASA History Series, 1999), 1.
3. Harris, *$elling Uncle Sam*, 199.
4. *Ibid.*
5. Robert Schwinghamer, telephone interview, Huntsville, 29 September 1998.
6. Bill Easterling, '"Rocket City" Celebrates: Missile Gap Cut, von Braun Elated', *Huntsville Times*, 6 May 1961.
7. President Kennedy memo to Vice President Lyndon Johnson, 20 April 1961, NASA Historical Archives.
8. Wernher von Braun letter to Vice President Lyndon Johnson, 29 April 1961, NASA Historical Archives; also in Wernher von Braun Library and Archives, US Space and Rocket Center.
9. Cromie, 'Wernher von Braun – Dean of Rocketry', 11-D.
10. The Wernher von Braun lab chiefs: Kurt H. Debus, launch (until he became director of what evolved into the NASA–Kennedy Space Center); Ernst D. Geissler,

aeroballistics; Walter Haeussermann, guidance and control; Karl L. Heimburg, field test; Hans H. Hueter, systems support equipment; Helmut Hoelzer, computation; Hans H. Maus, fabrication and assembly; W.A. 'Willy' Mrazek, structures and mechanics; Erich W. Neubert, checkout and reliability; and Ernst Stuhlinger, research projects. Other Peenemünde veterans in high posts at the MSFC included Arthur Rudolph, Werner Kuers and Hermann Weidner.

11. Cromie, 'Wernher von Braun – Dean of Rocketry', 11-D.
12. Hugh Downs, letter to author from ABC News 20/20 offices, New York, 11 December 1998.
13. Lang, 'A Romantic Urge'.
14. Walter Haeussermann, paper presented to the American Institute of Aeronautics and Astronautics, Huntsville chapter, 1989 meeting; text shared with author 17 November 1998.
15. *Ibid.*
16. Kraft, *Flight*, 83.
17. Stuhlinger and Ordway, *Crusader for Space*, 250.
18. Robert Schwinghamer, telephone interview, Huntsville, 29 September 1998.
19. Original Glenn postcard and translation in Wernher von Braun Library and Archives, US Space and Rocket Center.
20. Bonnie Holmes, telephone interview, Eva, Alabama, 26 October 1998.
21. Cobun, 'Catalyst, Inspirer, Promoter'.
22. Dale James, 'Behind the Scenes with von Braun', *Huntsville Times*, 8 September 1994.
23. Copies of both the Wernher von Braun and Jacqueline Kennedy letters are in the Wernher von Braun Library and Archives, US Space and Rocket Center.
24. J.N. Foster, interview, Huntsville, 2 September 1998.
25. Lee B. James, interview, Huntsville, 28 August 1998.
26. Robert Schwinghamer, telephone interview, Huntsville, 29 September 1998.
27. Lee B. James, interview, Huntsville, 28 August 1998.
28. *Ibid.*
29. *Ibid.*
30. Wernher von Braun in an unpublished interview by then-NASA contract writer Robert Sherrod at NASA headquarters, 25 August 1970; unedited file copy of interview transcript provided to author in 1999.
31. Hugh Downs, letter to author from ABC News 20/20 offices, New York, 11 December 1998.
32. Cromie, 'Wernher von Braun – Dean of Rocketry', 11-D.
33. Mike Wright, 'Huntsville and the Space Program', parts 1 and 2, *Alabama Heritage* magazine (spring and summer 1998), part 1, p. 27.
34. Cromie, 'Wernher von Braun – Dean of Rocketry', 10-D.
35. *Ibid.*
36. Gene Bylinski, 'Dr. von Braun's All-Purpose Space Machine', *Fortune* (May 1967), 142–9, 214, 218–19.
37. Peter Cobun, 'A Footnote Is Enough', *Huntsville Times*, 8 August 1976.
38. Bernhard R. Tessman, letter, 23 March 1972, 'X + 60 and Counting'.
39. Leland F. Belew, interview, Huntsville, 28 October 1998.
40. Cobun, 'Catalyst, Inspirer, Promoter'.
41. David Newby, letter, 23 March 1972, 'X + 60 and Counting'.

42. Ruth von Saurma, letter, 28 February 1972, 'X + 60 and Counting'.
43. William R. Lucas, interview, Huntsville, 3 November 1998.
44. Wernher von Braun, 'Why Space Exploration Is Vital to Man's Future', *Space World* (September 1969): 31.
45. J.T. Shepherd, MSFC memorandum to Wernher von Braun, 'Subject: *Huntsville Times*', 15 December 1965. A copy of the memo was leaked years later to the author.
46. Ward, The *Light Stuff*, 143.
47. Paul Haney, telephone interview, St Petersburg, Florida, 2 April 1981.

Chapter 15: En Route to Victory

1. *Life*, 'Seers of Space', 138.
2. James S. Farrior, letter, 23 March 1972, 'X + 60 and Counting'.
3. Ernst Stuhlinger, letter, 25 February 1972, 'X + 60 and Counting'. This trip was in 1956.
4. Walter Cronkite, letter from CBS News offices, New York, 14 February 1972, 'X + 60 and Counting'.
5. Walter Cronkite, telephone interview, New York, 19 August 1999.
6. A. Emile Joffrion, interview, Huntsville, 15 December 1998.
7. *Ibid.*
8. Harry M. Rhett Jr, interview, Huntsville, December 1996.
9. Edward G. Uhl, talk given at the annual Wernher von Braun Exploration Forum, Huntsville, 21 September 1993.
10. Leland F. Belew, interview, Huntsville, 28 October 1998.
11. David H. Newby, correspondence with author, Guntersville, Alabama, 14 September 1998.
12. J.N. Foster, interview, Huntsville, 2 September 1998.
13. Edward T. Grubbs, interview, Decatur, Alabama, 21 October 1998.
14. George A. Fehler, interview, Decatur, Alabama, 21 October 1998.
15. Marshall's Gulfstream, the first of four purchased by NASA in the early 1960s, with von Braun's endorsement, was fitted out with a dozen passenger seats plus work tables, a bar and a TV set or two. One seat was designated as von Braun's. Through the years, the plane carried all seven of the original Mercury astronauts, the Apollo 11 flight crew, most of the other moon-walking astronauts and a lengthy roster of other big names in space. Four decades later, 'G1' was still going strong at MSFC, having logged well over 25,000 flight hours. It continued to be called 'Dr von Braun's plane'.
16. George A. Fehler, interview, Decatur, Alabama, 21 October 1998.
17. Edward T. Grubbs, interview (jointly with George A. Fehler), Decatur, Alabama, 21 October 1998.
18. Avid aviator and aerospace pioneer Wernher von Braun was enshrined posthumously in the early 1980s in the National Aviation Hall of Fame, in Dayton, Ohio, the home of a couple of other pioneers, name of Wright. . . .
19. Donald I. Graham Jr, letter, 23 March 1972, 'X + 60 and Counting'.
20. John C. Goodrum Sr, interview, Huntsville, 16 December 1998.
21. Edward T. Grubbs, interview (jointly with George A. Fehler), Decatur, Alabama, 21 October 1998.
22. *Ibid.*

23. Carsbie C. Adams, telephone interview, Spotsylvania, Virginia, 16 December 1998; and Adams's privately published (1999) expanded trip journal, 'Wernher von Braun in Troposphere Orbit with C.C. Adams'.

24. Their plight so moved von Braun that he later sought more substantial, organised aid for Tibetan refugees. According to Bonnie Holmes, he worked through the famed travel adventurer of the day, Lowell Thomas, who was attracted to the cause.

25. Carsbie C. Adams, telephone interview, Spotsylvania, Virginia, 16 December 1998; and Adams's journal, 'Wernher von Braun in Troposphere Orbit with C.C. Adams'.

26. Thomas L. Shaner, interview, Huntsville, 4 March 1999.

27. Alexander von Quistorp headed a major Berlin banking house. Alive until 1974, he was well aware of his famous son-in-law's ascendancy in the world of banking.

28. Patrick W. Richardson, interview, Huntsville, 14 August 1998.

29. Sarah Sanders Preston, telephone interview, Huntsville, 5 March 1999.

30. Edward D. Mohlere, interview, Huntsville, 20 September 1998.

31. Olin E. Teague, letter to Wernher von Braun from Washington, DC, 8 March 1971.

32. Lee B. James, interview, Huntsville, 28 August 1998.

33. Harry Atkins, interview, Huntsville, 27 May 1999.

34. Homer Hickam Jr was another coal-mining West Virginia youth influenced by von Braun towards a career with NASA. Inspired by Sputnik I to learn model rocketry, the teenager in Coalwood, West Virginia, received encouragement in correspondence with the space scientist. After his thirty years as a NASA engineer, lastly at the Marshall Center after von Braun had departed for Washington, Hickam captured his experiences as a young rocketeer in the best-selling 1998 book, *The Rocket Boys*. The book was soon made into the hit movie *October Sky*.

35. William H. Pickering, telephone interview, Altadena, California, 16 September 1998.

36. Senator Stennis remained a supporter that NASA could always count on. The sprawling Mississippi Test Facility was later named after him. A nine-storey observation tower there, for space officials and VIP guests to view static-rocket test firings, was named in 1998 in memory of von Braun.

37. James T. Shepherd, interview, Huntsville, 5 October 1998.

38. James B. Odom, interview, Huntsville, 23 September 1998.

39. J.N. Foster, interview, Huntsville, 2 September 1998.

40. Wernher von Braun, 'Rundown on Jupiter-C', *Astronautics* magazine (October 1958), 32.

41. Lee B. James, interview, Huntsville, 28 August 1998.

42. James T. Shepherd, interview, Huntsville, 8 October 1998.

43. Charles Lundquist, interview, Huntsville, 19 November 1998.

44. James T. Shepherd, interview, Huntsville, 5 October 1998.

45. *Ibid.*

46. James Splawn, interview, Huntsville, 15 October 2002.

47. As engineers' and astronauts' use of the tank increased over the years, NASA's Houston centre decided the facility wasn't such a bad idea after all. It had let the Marshall Center sneak one over on it, but no more, both Shepherd and

Splawn said a Houston official vowed in a meeting at MSFC. The Texas centre won approval to build its own water tank – bigger and better than Huntsville's. Later, in the continuing arch-rivalry between the two centres, Marshall was forced to shut down and mothball its tank – after von Braun was gone.

48. Edward D. Mohlere, interview, Huntsville, 20 September 1998.
49. James T. Shepherd, interview, Huntsville, 5 October 1998.
50. Gray, *Angle of Attack*, 254–5.
51. Harrison B. 'Stormy' Storms Jr died on 11 July 1992.

Chapter 16: Lunar Triumph

1. Ward, *Wernher von Braun Anekdotisch*.
2. Martin Caidin, *Rendezvous in Space* (New York: Dutton, New York, 1962), 311.
3. James T. Shepherd, letter, 23 March 1972, 'X + 60 and Counting'; also J.T. Shepherd, interview, Huntsville, 8 October 1998; and Walton G. Clarke, interview, Huntsville, 4 September 1998.
4. George E. Mueller was the NASA chief of Manned Space Flight during Project Apollo.
5. Eugene G. Cowart, interview, Huntsville, 17 August 2003.
6. Patrick W. Richardson, interview, Huntsville, 14 August 1998.
7. Thomas Franklin [Hugh McInnish], *An American in Exile: The Story of Arthur Rudolph* (Huntsville: Christopher Kaylor, 1987), 143.
8. Ibid., 146–9; William E. Winterstein Sr, *Gestapo USA: When Justice Was Blindfolded* (San Francisco: Robert D. Reed, 2002), 51, 140.
9. Franklin, *An American in Exile*, 156–8; Winterstein, *Gestapo USA*, 115–29.
10. Wernher von Braun, review for the Associated Press of the first ten years of the space age, in the *New York Times* (p. 80) and elsewhere, 24 September 1967.
11. Wernher von Braun interview, 'Has US Settled for No. 2 in Space?' *US News & World Report* (14 October 1968), 74.
12. Ibid., 75.
13. Pat R. Odom, interview and correspondence, Huntsville, 14 January 2003.
14. Although the Galileo project first won congressional approval in 1977, the spacecraft was not launched until 1989. Galileo arrived at Jupiter on 7 December 1995, circled the planet and proceeded to send back loads of new information, including dramatic photographs of the surface of the giant planet and its moons. Wernher von Braun did not live to see Galileo's success.
15. Ruth G. von Saurma, letter, 28 February 1972, 'X + 60 and Counting'.
16. J.N. Foster, interview, Huntsville, 2 September 1998.
17. William Greenburg, 'Von Braun Certain of Flight's Success', *Nashville Tennessean*, 11 July 1969, and elsewhere.
18. Ibid.
19. Harold Kennedy, 'Von Braun Looks Beyond the Moon as Flight Nears', *Birmingham (Ala.) News*, 11 July 1969, 4, and elsewhere.
20. Ibid.
21. Moscow's intended Moon-rocket, known as Type G or G Class, clustered several engines from advanced ICBMs and multiple combustion chambers to generate its designed millions of pounds of thrust. In summer 1969, after booster roll-

out for a static test at the Soviets' Baikonur Cosmodrome in Kazakhstan, an upper-stage leak caused a fire that destroyed the vehicle and devastated the launch facility, according to sources, including Roger E. Bilstein, *Stages to Saturn: A Technological History of the Apollo/Saturn Launch Vehicles* (Washington, DC: NASA, 1980), 388. In midsummer 1970, the Soviets tried again. Reports stated that the super-rocket was launched but broke apart before reaching Earth orbit. In November 1972 disaster struck yet again, with failure of the first stage. Moscow gave up on Type G after that. In the mid-1970s, however, the Soviets began development of the heavy-lift Energia booster for their Buran space shuttle. Buran made a successful unmanned flight in Earth orbit and a safe return; it never flew with cosmonauts aboard, and the programme was cancelled.

22. Anthony R. Curtis, editor, *Space Almanac* (2nd edn) (Houston: Gulf Publishing Company, 1992), 19–20.
23. Kennedy, 'Von Braun Looks Beyond the Moon'.
24. Lee B. James, interview, Huntsville, 28 August 1998.
25. Bergaust, *Wernher von Braun*, 420.
26. *Ibid.*
27. 'Prayer is the most important work of man,' von Braun was quoted as saying in a 1977 feature obituary by George W. Cornell, long-time religion writer for the Associated Press.
28. Stuhlinger and Ordway, *Crusader for Space*, 269.
29. Wernher von Braun speech to a joint session of the Alabama legislature, Montgomery, 29 July 1969; he made similar remarks before the Apollo 11 lunar landing.
30. William C. Fortune, letter, 22 February 1972, 'X + 60 and Counting'.
31. George A. Fehler and Edward T. Grubbs, joint interview, Decatur, Alabama, 21 October 1998.
32. Wright, 'Huntsville and the Space Program', part 2, 31.
33. Wernher von Braun correspondence files, Wernher von Braun Library and Archives, US Space and Rocket Center.
34. Wernher von Braun speech to a joint session of the Alabama legislature, Montgomery, 29 July 1969; he made similar remarks before the Apollo 11 lunar landing.
35. Bernard Weinraub, 'Huntsville's Joy Has a German Flavor', *New York Times* (25 July 1969), 29.
36. Howard Benedict, correspondence with the author, from Titusville, Florida, 15 December 1998.
37. Including Hollywood giants James Stewart, Ray Milland, Rosalind Russell, Irene Dunne and Fred MacMurray; a former actor, Ronald Reagan, then governor of California; former cowboy actors Gene Autry, Randolph Scott and Andy Devine; comedians Bob Hope, Jack Benny and Red Skelton; singers Tony Martin, Pat Boone and Connie Francis; other entertainers Lionel Hampton, Art Linkletter, Les Brown and Cesar Romero; Apollo 11 television commentators Walter Cronkite of CBS, Hugh Downs of NBC and Jules Bergman of ABC; union leaders George Meany and Walter Reuther; the Revd Billy Graham, Gens Omar Bradley and James Doolittle, aviation hero Charles Lindbergh, industrialist James S. McDonnell of McDonnell-Douglas Corporation, presidential advisers Henry Kissinger and Daniel Moynihan,

assorted presidential family members, space officials, major newspaper
publishers, ambassadors, and the like, according to the *New York Times*
(14 August 1969), 22.

38. Lee B. and Kathleen James, joint interview, Huntsville, 28 August 1998.
39. Bonnie Holmes, telephone interview, Eva, Alabama, 26 October 1998.
40. Bergaust, *Wernher von Braun.*
41. Cromie, 'Wernher von Braun – Dean of Rocketry', 10-D.
42. Stuhlinger and Ordway, *Crusader for Space*, 197.
43. Walter Cronkite, telephone interview, New York, 19 August 1999.
44. Walter Cronkite, quoted in Ralph Petroff, 'Apollo Program Laid High-Tech
    Groundwork', *USA Today* (22 June 1999), 3-E.
45. Walter Cronkite, telephone interview, New York, 19 August 1999.
46. Stuhlinger and Ordway, *Crusader for Space*, 234.

Chapter 17: Rocket City Legacy

1. William Joseph Stubno Jr, educational data in master's thesis, 'The Impact of
   the von Braun Board of Directors on the American Space Program', University
   of Alabama in Huntsville (UAH), 1980.
2. John L. McDaniel, letter, 23 March 1972, 'X + 60 and Counting'.
3. Wernher von Braun would continue to be memorialised in bricks and mortar
   in Huntsville. In 1975, the city named its new cultural and conference centre
   after him. NASA in 1994 designated the Marshall Center's three-building
   headquarters grouping the Wernher von Braun Office Complex. In May 2000,
   during a year-long community 'celebration' honouring von Braun, his old
   rocket team and their fifty years of achievements in Alabama, UAH named its
   refurbished Research Institute the Wernher von Braun Research Hall. And, in
   January 2004, a $39-million building complex bearing the name of Wernher
   von Braun opened at Redstone to house the Army Space and Missile Defense
   Command; a $23-million addition was announced to follow.
4. Foster A. Haley, interview, Huntsville, 30 June 1997.
5. *Ibid.*
6. Vernon Pizer, 'Alabama's Adopted Spacemen', *American Legion* magazine
   (January 1960), 38.
7. Harry L. Pennington, interview, Huntsville, 27 October 1998.
8. The Smithsonian's National Air and Space Museum holds title to the Saturn V
   on display in Huntsville.
9. Edward O. Buckbee, interview, Huntsville, 4 August 1998.
10. US Space Camp opened as a five-day camp of fun and learning for boys and
    girls, climaxing with a simulated space mission – just as von Braun had
    suggested. Camp experiences for teenagers, adults and aviation enthusiasts
    were added later. By 2003, Space Camp had attracted 375,000 attendees
    (mostly children, including the offspring of politicians and celebrities) from
    every state and many foreign lands. It was the location for much of the
    filming of the 1986 feature motion picture *Space Camp*.
11. Foster A. Haley, interview, Huntsville, 30 June 1997.
12. Wernher von Braun speech, Rotary International District Governors
    Conference, Huntsville, 12 April 1965.
13. Fred S. Schultz, letter, 28 February 1972, 'X + 60 and Counting'.

14. Wernher von Braun speech to Alabama Chamber of Commerce board of directors' quarterly meeting, Huntsville, 26 May 1965; Wernher von Braun Library and Archives, US Space & Rocket Center.
15. Cobun, 'Catalyst, Inspirer, Promoter'.
16. James T. Shepherd, interview, Huntsville, 8 October 1998.
17. Rankin A. Clinton Jr, interview, Huntsville, 16 January 2001, and correspondence with author, 14 February 2001.

Chapter 18: After Apollo, What?

1. James T. Shepherd, telephone interview, Huntsville, 6 November 2001.
2. Ernst Stuhlinger, telephone interview, Huntsville, 22 April 2003.
3. Maxime Faget, telephone interview, Apollo, Texas, 12 July 2003.
4. Eugene G. Cowart, interview, Huntsville, 17 August 2003.
5. Leland F. Belew, interview, Huntsville, 28 October 1998.
6. Stuhlinger and Ordway, *Crusader for Space*, 290.
7. *Ibid.*, 291.
8. Frank Williams, telephone interview, Slidell, Louisiana, 11 November 1998.
9. Konrad Dannenberg, interview, Huntsville, 10 September 1998.
10. Werner Dahm, interview, Huntsville, 6 December 1998.
11. J.N. Foster, interview, Huntsville, 2 September 1998.
12. Lee B. James, interview, Huntsville, 28 August 1998.
13. Dorette Kersten Schlidt, interview, Huntsville, Alabama, 21 September 1998.
14. Thomas L. Shaner, interview, Huntsville, 4 March 1999.
15. Stuhlinger and Ordway, *Crusader for Space*, 291.
16. Patrick W. Richardson, interview, Huntsville, 14 August 1998.
17. Jack Hartsfield, 'Von Braun's Feelings Are Mixed', *Huntsville Times*, 3 February 1970.
18. Thomas L. Shaner, interview, Huntsville, 4 March 1999.
19. Dale James, 'Behind the Scenes with von Braun: Bonnie Holmes Was His Trusted Secretary', *Huntsville Times*, 8 September 1994.
20. Robert Schwinghamer, telephone interview, Huntsville, 29 September 1998.
21. Thomas L. Shaner, interview, Huntsville, 4 March 1999.
22. A *Huntsville Times* reporter was present at von Braun's first talk to MSFC employees after word broke that he was leaving. Hartsfield, 'Von Braun's Feelings Are Mixed', *Huntsville Times*.
23. Bonnie Holmes, telephone interview, Eva, Alabama, 16 December 2000.
24. *Huntsville News*, coverage of VBCC opening, 25 February 1970.
25. Author's notes from the scene and Huntsville newspaper accounts, 25 February 1970. Other celebrations of space feats did follow at the Courthouse Square, as von Braun hoped. One was the first Space Shuttle flight of a home-grown astronaut, Jan Davis. Another came after astronaut-hero John Glenn's return to space in November 1998 aboard the shuttle as a fit 77-year-old. Early the following month he and his crewmates visited the Marshall Center to brief and thank employees. Then they gave a public report at the renamed Von Braun Center, enjoyed a downtown parade past cheering crowds and finally a lively programme of events on the old square. Glenn was the centre of admiring attention, despite his best efforts to just blend in.

26. Robert Schwinghamer, telephone interview, Huntsville, 29 February 1998.
27. John D. Hilchey, letter, 28 February 1972, 'X + 60 and Counting'.

Chapter 19: DC and the Gods

1. Peter Petroff, interview, Huntsville, 4 June 1999.
2. *Ibid.*
3. Beginning in the 1950s, German car-maker Daimler-Benz presented von Braun with a new, top-of-the-line Mercedes model every few years. The deal had major promotional value to the company. Before transferring to Washington, the scientist made a gift of an older-model Mercedes to the mechanic, Wolfgang Fricke, who had tended his cars for years in Huntsville. In 1975 von Braun travelled to Germany to accept an invitation to join the Daimler-Benz board of directors.
4. Julia E. Kertes, correspondence with author, from Louisville, Ohio, 30 December 1998.
5. *Ibid.*
6. Julia E. Kertes, follow-up correspondence with author, from Louisville, Ohio, 13 February 2001.
7. Thomas L. Shaner, interview, Huntsville, 4 March 1999.
8. *Ibid.*
9. George E. and Pamela Huxford Philyaw, joint interview, Huntsville, 12 April 2000.
10. James L. Daniels Jr, interview, Huntsville, 19 August 1998; and letter to von Braun, 24 February 1972, 'X + 60 and Counting'.
11. Various interviews with von Braun family friends, Huntsville, 1998–9.
12. Ernst Stuhlinger, editorial article, 'Wernher von Braun: "I Am an Evolutionary, Not a Revolutionary"', *Huntsville Times*, 17 June 1977.
13. Betty Beale society column in the *Washington Star*, March–April 1970; see also Ward, *Wernher von Braun Anekdotisch*, 85.
14. Frank Williams, telephone interview, Slidell, Louisiana, 11 November 1998.
15. James L. Daniels Jr, interview, Huntsville, 19 August 1998.
16. Julia E. Kertes, telephone interview, Louisville, Ohio, 10 December 1998.
17. Stuhlinger and Ordway, *Crusader for Space*, 294–5.
18. James L. Daniels Jr, interview, Huntsville, 19 August 1998.
19. *Ibid.*
20. Julia E. Kertes, correspondence with author, from Louisville, Ohio, 30 December 1998.
21. *Ibid.*
22. *Ibid.*
23. *Ibid.*
24. *Ibid.*
25. James L. Daniels Jr, interview, Huntsville, 19 August 1998.
26. James L. Daniels Jr, interview, Huntsville, 16 December 1998; and letter, 24 February 1972, 'X + 60 and Counting'.
27. Julia E. Kertes, correspondence with author, from Louisville, Ohio, 30 December 1998.

28. *Ibid.*
29. Julia E. Kertes, correspondence with author, from Louisville, Ohio, 15 December 2000.
30. Julia E. Kertes, correspondence with author, from Louisville, Ohio, 30 December 1998.
31. *Ibid.*
32. J.N. Foster, interview, Huntsville, 2 September 1998.
33. Julia E. Kertes, correspondence with author, from Louisville, Ohio, 30 December 1998.
34. Julia E. Kertes, telephone interview, Louisville, Ohio, 20 January 2003.
35. James L. Daniels Jr, interview, Huntsville, 19 August 1998.
36. J.N. Foster, interview, Huntsville, 2 September 1998.
37. Thomas L. Shaner, interview, Huntsville, 4 March 1999.
38. Early in his planning stint at NASA headquarters, von Braun had floated a dramatic idea to ignite Richard Nixon's imagination (assuming he had one). The concept presupposed that the 1970s would see a US space shuttle flying, a space station in orbit and plans laid for a manned Mars mission in the 1980s. It was Wernher von Braun's suggestion, according to Washington associates, that the President, in the last year of his presumed second term, would fly aboard the shuttle to the space station in July 1976 in a spectacular highlight of America's bicentennial celebration. But Nixon didn't bite. By 1976 it was all moot: there was no shuttle, no station and no Nixon – he had departed the White House in mid-term disgrace.
39. Thomas L. Shaner, interview, Huntsville, 4 March 1999; J.N. Foster, interview, Huntsville, 2 September 1998.
40. Thomas L. Shaner, interview, Huntsville, 4 March 1999.
41. Julia E. Kertes, telephone interview, Louisville, Ohio, 20 January 2003.

Chapter 20: Perigee in Washington

1. Peter Cobun, cited in 'A Footnote Is Enough', *Huntsville Times*, 8 August 1976.
2. Lee B. James, interview, Huntsville, 28 August 1998.
3. Frank Williams, telephone interview, Birmingham, Alabama, 21 November 2000.
4. Stuhlinger and Ordway, *Crusader for Space*, 301–3.
5. J.N. Foster, interview, Huntsville, 2 September 1998.
6. Lee B. James, interview, Huntsville, 28 August 1998.
7. J.N. Foster, interview, Huntsville, 29 April 2003.
8. Thomas L. Shaner, interview, Huntsville, 4 March 1999.
9. Ward, *The Light Stuff*, 70.
10. Hermann Weidner, letter, 23 March 1972, 'X + 60 and Counting'.
11. Edward D. Mohlere, interview, Huntsville, 20 September 1998.
12. Joseph J. Trento, interview with Thomas Paine, 18 July 1986, in *Prescription for Disaster: From the Glory of Apollo to the Betrayal of the Shuttle* (New York: Crown, 1987), 89–90.
13. *Ibid.*
14. John C. Goodrum Sr, interview, Huntsville, 18 December 1998.
15. Stuhlinger and Ordway, *Crusader for Space*, 303–4.
16. Ernst Stuhlinger, telephone interview, Huntsville, 21 September 2001.

17. Robert Sherrod, unpublished interview with Wernher von Braun, 25 August 1970, at NASA headquarters; copy of unedited transcript from Washington files shared with author in 1999.
18. Kathleen (Mrs Lee B.) James, telephone interview, Huntsville, 18 October 2000.
19. Frank Williams, telephone interview, Slidell, Louisiana, 11 November 1998.
20. James L. Daniels Jr, interview, Huntsville, 19 August 1998.
21. Frank Williams, telephone interview, Slidell, Louisiana, 11 November 1998.
22. *Ibid.*
23. Walter R. Dornberger, letter from Chapala, Jalisco, Mexico, 26 February 1972, 'X + 60 and Counting'.
24. Frank Williams, first, second and third telephone interviews, Slidell, Louisiana, 13 September, 7 November and 11 November 1998; and copy of the ship's log of *Josephine III.*
25. Frank Williams, first, second and third telephone interviews, Slidell, Louisiana, 13 September, 7 November and 11 November 1998.
26. *Ibid.*
27. Julia E. Kertes, telephone interview, Louisville, Ohio, 6 January 1999.
28. The album, 'X + 60 and Counting', a photocopy of which was given to each letter writer, was presented to von Braun in a leather-bound volume of the original letters. The album project harked back to a volume of space-related essays and other papers done in honour of his 50th birthday and published a year later by McGraw-Hill.
29. Frank Williams, telephone interview, Slidell, Louisiana, 11 November 1998.
30. J.N. Foster, interview, Huntsville, 2 September 1998.
31. Philip W. Smith, Newhouse News Service, 'Von Braun Will Retire, Take Job with Fairchild', published in the *Huntsville Times* and elsewhere, 26 May 1972.
32. Bob Ward, editorial page article, 'Von Braun's Departure – Why?' *Huntsville Times*, 28 May 1972.
33. *Ibid.*
34. Julia E. Kertes, correspondence with author, from Louisville, Ohio, 6 and 30 December 1998.
35. Julia E. Kertes, telephone interview, Louisville, Ohio, 6 January 1999.
36. Stuhlinger and Ordway, *Crusader for Space*, 307.
37. George Michael (né George Wilhelm) Low died in 1984, Fletcher in 1991 and Jim Webb and Tom Paine in 1992. In a touch of irony, Paine had been chosen in 1985 by the Reagan White House to chair a National Commission on Space and produce a report on the future of space exploration. The Paine Commission's 1986 report called for 'a pioneering mission for 21st-Century America' of supporting a broad-based effort on the 'space frontier' from Earth orbit to the moon and Mars. Incoming President George H.W. Bush announced that the aggressive plan deserved a high national priority. It went nowhere after Congress did not fund the increased space budgets needed to accomplish it. It came some fifteen years after von Braun and Paine had put forward their similarly ambitious twenty-year space plan for the nation.
38. Adolf K. Thiel, telephone interview, Palos Verdes Estates, California, 1 October 1998.

Chapter 21: On the Private Side

1. Edward G. Uhl, talk given at the annual Wernher von Braun Exploration Forum, Huntsville, 21 September 1993.
2. *Ibid.*
3. Edward G. Uhl, telephone interview, Trappe, Maryland, 14 December 1998.
4. *Ibid.*, and talk at Wernher von Braun Exploration Forum, Huntsville, 21 September 1993.
5. Edward G. Uhl, talk at Wernher von Braun Exploration Forum, Huntsville, 21 September 1993.
6. Thomas Turner, correspondence with author, from San Antonio, 11 December 1998.
7. Charles C. Hewitt, telephone interview, Alexandria, Virginia, 3 February 1999.
8. Stuhlinger and Ordway, *Crusader for Space*, 311.
9. Charles Mitchelmore, Chicago Daily News Service, news article on the twenty-third Congress of the International Astronautical Federation (IAF) in Vienna, *Huntsville Times* and elsewhere, 15 October 1972.
10. Wernher von Braun, preface to L.B. Taylor, *For All Mankind* (New York: Dutton, 1975).
11. Edward G. Uhl, talk at Wernher von Braun Exploration Forum, Huntsville, 21 September 1993; and telephone interview, Trappe, Maryland, 14 December 1998.
12. *Ibid.*
13. Robert Schwinghamer, telephone interview, Huntsville, 29 September 1998.
14. Edward G. Uhl, talk at Wernher von Braun Exploration Forum, Huntsville, 21 September 1993; and telephone interview, Trappe, Maryland, 14 December 1998.
15. Stuhlinger and Ordway, *Crusader for Space*, 310; also, Bergaust, *Wernher von Braun*, 440.
16. Ernst Stuhlinger, telephone interview, Huntsville, 4 January 2004.
17. Joseph M. Jones, interview, Huntsville, 6 October 1998.
18. James T. Shepherd, interview, Huntsville, 8 October 1998.
19. Ernst Stuhlinger, interview, Huntsville, 29 September, 2001; Stuhlinger and Ordway, *Crusader for Space*, 316–17.
20. A sole member of the original rocket team remained at the Marshall Center as late as 2005. German-born Werner Dahm, in his 80s, determinedly stayed put as chief aerodynamicist.
21. Cobun, 'A Footnote Is Enough'.
22. Actually, the NSI began life as the National Space Association. In April 1975, less than a year after its founding, the name was changed to the National Space Institute. In 1987 the NSI merged with the L-5 Society, and they became the National Space Society.
23. Hugh Downs, correspondence with author, from ABC News's 20/20 offices in New York, 11 December 1998.
24. Stuhlinger and Ordway, *Crusader for Space*, 265, 325.

Chapter 22: Too Soon Dying

1. John Bruce Medaris, letter, 23 March 1972, 'X + 60 and Counting'.
2. Bonnie Holmes, telephone interviews, Eva, Alabama, 16 December 2000 and 3 January 2001.
3. *Ibid.*
4. Von Braun's counterpart in the Soviet Union, Sergei Korolyov, had earlier contracted colon cancer. Independently the two space leaders had expressed a desire to meet. They never did. Korolyov, the chief designer of the Soviets' space programme, died in January 1966 from complications arising from colon surgery.
5. Bonnie Holmes, telephone interviews, Eva, Alabama, 16 December 2000 and 3 January 2001.
6. Bergaust, *Wernher von Braun*, 469–70.
7. *Ibid.*, 362.
8. C.E. Monroe Jr, telephone interview, Guntersville, Alabama, 7 February 1999.
9. Erik Bergaust, National Space Institute twelve-page memorial publication in tribute to Wernher von Braun, 1977.
10. Edward G. Uhl, talk at Wernher von Braun Exploration Forum, Huntsville, 21 September 1993; and telephone interview, Trappe, Maryland, 14 December 1998.
11. *Ibid.*
12. William R. Lucas, interview, Huntsville, 3 November 1998.
13. David L. Christensen, interview, Huntsville, 21 May 1999.
14. *Ibid.*
15. William H. Pickering, telephone interview, Altadena, California, 16 September 1998.
16. Frederick I. Ordway III, memorandum to author, from Arlington, Virginia, 29 June 1999.
17. *Ibid.*
18. Edward G. Uhl, talk at Wernher von Braun Exploration Forum, Huntsville, September 1993.
19. Frederick I. Ordway III, memorandum to author, from Arlington, Virginia, 29 June 1999.
20. The book was published after Wernher von Braun's death as *New Worlds: Discoveries from Our Solar System* (Garden City, NY: Anchor Press/Doubleday, 1979). In autumn 1976, the cancer-ravaged scientist had also managed to review most of the manuscript for the history *The Rocket Team*, by Frederick I. Ordway and Mitchell R. Sharpe, and to write a foreword for it. That book was also published in 1979.
21. Wernher von Braun paper presented at the Lutheran Church of America synod, Philadelphia, 29 October 1976; reprinted in *The Nature of a Humane Society*, edited by H. Ober Hess (Philadelphia: Fortress Press, 1977).
22. *Ibid.*
23. Transcript, NBC's *Today* programme, New York, 11 November 1998.
24. Edward G. Uhl, talk at Wernher von Braun Exploration Forum, Huntsville, 21 September 1993.
25. Edward G. Uhl, telephone interview, Trappe, Maryland, 14 December 1998.

26. *Ibid.*, and talk at Wernher von Braun Exploration Forum, Huntsville, 21 September 1993.
27. *Ibid.*
28. Arthur C. Clarke, letter, 21 February 1972, 'X + 60 and Counting'.
29. Walter Wiesman, letter, 1 March 1972, 'X + 60 and Counting'.
30. Ernst Stuhlinger, interview, Huntsville, 29 September 2001.
31. Stuhlinger and Ordway, *Crusader for Space*, 329.
32. This quotation from Leon Bloy was cited in a letter dated 13 February 1963 by von Braun's Marshall Center chief of information, Bart J. Slattery Jr, in response to a citizen's enquiry.
33. Stuhlinger and Ordway, *Crusader for Space*.

Epilogue

1. Howard Benedict, correspondence with the author, from Titusville, Florida, 15 December 1998.
2. Peter Cobun, 'Man of Space, Earth, Spirit – von Braun Is Bid Farewell', *Huntsville Times*, 20 June 1977.
3. Eugene Emme, preface to the American Institute of Aeronautics and Astronautics' (AIAA's) Wernher von Braun obituary, June 1977.
4. Frederick C. Durant III, AIAA's Wernher von Braun obituary, June 1977.
5. Francis B. Sayre Jr, Wernher von Braun memorial service programme, Washington National Cathedral, 22 June 1977.
6. Printed here with permission from Michael Collins; excerpts appeared in Stuhlinger and Ordway, *Crusader for Space*, 329–30.

Appendix I: Letter on Goddard Patents

1. Wernher von Braun correspondence files, Wernher von Braun Library and Archives, US Space and Rocket Center.
2. In March 1999, *Time* magazine included Goddard in its special issue on the twentieth-century's '100 Greatest Minds'. Von Braun and his team received parenthetical mention. The worn anecdote was repeated about a captured German missile man who, when asked about the origin of the V2, supposedly replied, 'Why don't you ask your own Dr Goddard? He knows better than any of us.' The anonymous quotation is apocryphal, prominent team member Ernst Stuhlinger told me in 2002. He said he had no idea who might have made the oft-repeated remark, and doubted it was ever made.
3. Wernher von Braun correspondence files, Wernher von Braun Library and Archives, US Space and Rocket Center.
4. Erik Bergaust, *Reaching for the Stars: A Biography of the Great Pioneer in Space Exploration, Wernher von Braun* (Garden City, NY: Doubleday, 1960), 24–5.

Bibliography

Books by Wernher von Braun

Across the Space Frontier. With Willy Ley, Fred L. Whipple, Joseph Kaplan *et al.* Cornelius Ryan, (ed.). New York: Viking, 1952.

Careers in Astronautics and Rocketry. With Carsbie C. Adams and Frederick I. Ordway III. New York: McGraw-Hill, 1962.

Conquest of the Moon. With Fred L. Whipple and Willy Ley. Cornelius Ryan, (ed.). New York: Viking, 1953.

Exploration of Mars. With Willy Ley. New York: Viking, 1956.

First Men to the Moon. New York: Holt, Rinehart & Winston, 1960.

History of Rocketry and Space Travel. With Frederick I. Ordway III. New York: Crowell, 1966. (Rev. edn 1969, 1975, and, also with Dave Dooling, 1985.)

Man on the Moon. London: Sidgwick & Jackson, 1953.

The Mars Project. Urbana: University of Illinois Press, 1953.

Moon. With Silvio A. Bedini and Fred L. Whipple. New York: Harry N. Abrams, 1974.

The Rocket's Red Glare. With Frederick I. Ordway III. New York: Doubleday, 1976.

Space Frontier. New York: Holt, Rinehart & Winston, 1963.

Books about Wernher von Braun

Andrews, John Williams. *A.D. Twenty-One Hundred: A Narrative of Space*. Boston: Branden Press, 1969.

Barbour, John, and the Associated Press. *Footprints on the Moon*. New York: Associated Press, 1969.

Bergaust, Erik. *Reaching for the Stars: A Biography of the Great Pioneer in Space Exploration, Wernher von Braun*. Garden City, NY: Doubleday, 1960.

——. *Rocket City USA: The Story of Huntsville, Alabama, Home of the Big Boosters*. New York: Macmillan, 1963.

——. *Wernher von Braun*. Washington, DC: National Space Institute, 1976.

Bilstein, Roger E. *Stages to Saturn: A Technological History of the Apollo/Saturn Launch Vehicles*. Washington, DC: National Aeronautics and Space Administration, 1980.

Carpenter, M. Scott, L. Gordon Cooper Jr, John H. Glenn Jr, Virgil I. Grissom, Walter M. Schirra Jr, Alan B. Shepard Jr, and Donald K. Slayton. *We Seven: By the Astronauts Themselves.* New York: Simon & Schuster, 1962.

Collins, Michael. *Carrying the Fire: An Astronaut's Journeys.* New York: Farrar, Straus & Giroux, 1974.

Cunz, Dieter. *They Came from Germany: The Story of Famous German-Americans.* New York: Dodd, Mead, 1966.

David, Heather M. *Wernher von Braun.* New York: Putnam, 1967.

De Maeseneer, Guido. *Peenemünde: The Extraordinary Story of Hitler's Secret Weapons V1 and V2.* Vancouver, Canada: AJ Publishing, 2001.

Dickson, Paul. *Sputnik: The Shock of the Century.* New York: Walker & Company, 2001.

Dornberger, Walter. *V2.* New York: Viking, 1954.

Dunar, Andrew J., and Stephen P. Waring. *Power to Explore: A History of Marshall Space Flight Center, 1960–1990.* Washington, DC: National Aeronautics and Space Administration, 1999.

Durant, Frederick C., III, and George S. James (eds). *First Steps toward Space.* Washington, DC: Smithsonian Institution Press, 1979; reprinted by Univelt, San Diego, 1985.

Fallaci, Oriana. *If the Sun Dies.* New York: Atheneum, 1966.

Freeman, Marsha. *How We Got to the Moon: The Story of the German Space Pioneers.* Washington, DC: Twenty-first Century Associates, 1993.

Goodrum, John C. *Wernher von Braun: Space Pioneer.* Huntsville, Ala.: Strode Publishers, 1969.

Gray, Mike. *Angle of Attack: Harrison Storms and the Race to the Moon.* New York: W.W. Norton, 1992.

Huzel, Dieter K. *Peenemünde to Canaveral.* Englewood Cliffs, NJ: Prentice-Hall, 1962.

Klee, Ernst, and Otto Merk. *The Birth of the Missile: The Secrets of Peenemünde.* New York: Dutton, 1965.

Kraft, Chris. *Flight: My Life in Mission Control.* New York: Dutton, 2001.

Lasby, Clarence G. *Project Paperclip: German Scientists and the Cold War.* New York: Atheneum, 1971.

Leonard, Jonathan Norton. *Flight into Space: The Facts, Fancies, and Philosophy.* New York: Random House, 1953.

Ley, Willy. *Rockets, Missiles, and Space Travel.* New York: Viking, 1957.

——, *Rockets, Missiles, and Men in Space.* New York: Viking, 1968.

Mailer, Norman. *Of a Fire on the Moon.* Boston: Little, Brown, 1970.

McGovern, James. *Crossbow and Overcast.* New York: William Morrow, 1964.

Medaris, John B., with Arthur Gordon. *Countdown for Decision.* New York: Putnam, 1960.

Neufeld, Michael J. *The Rocket and the Reich: Peenemünde and the Coming of the Ballistic Missile Era.* New York: Free Press/Simon & Schuster, 1995.

Piszkiewicz, Dennis. *Wernher von Braun: The Man Who Sold the Moon.* Westport, Conn.: Praeger, 1998.

Ordway, Frederick I., III, and Mitchell R. Sharpe. *The Rocket Team: From the V2 to the Saturn Moon Rocket – The Inside Story of How a Small Group of Engineers Changed World History.* New York: Crowell, 1979. (Rev. edn, Burlington, Ontario: Apogee, 2003).

Stuhlinger, Ernst, and Frederick I. Ordway III. *Wernher von Braun: Crusader for Space*. 2 vols. Melbourne, Fla.: Krieger Publishing, 1994.

Stuhlinger, Ernst, Frederick I. Ordway III, Jerry C. McCall, and George C. Bucher, (eds). *Astronautical Engineering and Science: From Peenemünde to Planetary Space – Honoring the Fiftieth Birthday of Wernher von Braun*. New York: McGraw-Hill, 1963.

Thomas, Shirley. *Men of Space: Profiles of the Leaders in Space Research, Development, and Exploration*, Vol. 1. Philadelphia: Chilton, 1960.

von Braun, Magnus. *Von Ostpreussen bis Texas [From East Prussia to Texas]*. Stollhamm, Germany: Helmut Rauschenbusch Verlag, 1955.

Walters, Helen B. *Wernher von Braun: Rocket Engineer*. New York: Macmillan, 1964.

Ward, Bob. *Wernher von Braun Anekdotisch*. Esslingen, Germany: Bechtle Verlag, 1972.

——. *The Light Stuff*. Huntsville, Ala.: Jester Books, 1982.

Wilford, John Noble. *We Reach the Moon: The New York Times Story of Man's Greatest Adventure*. New York: Bantam, 1969.

Williams, Beryl, and Samuel Epstein. *The Rocket Pioneers on the Road to Space*. New York: Julian Messner, 1958.

Wolfe, Tom. *The Right Stuff*. New York: Farrar, Straus & Giroux, 1979.

Index